Lecture Notes in Mathematics

Found[illegible]

Edit[illegible]

F[illegible]

1. Lecture Notes aim to report new developments in all areas of mathematics - quickly, informally and at a high level. Monograph manuscripts should be reasonably self-contained and rounded off. Thus they may, and often will, present not only results of the author but also related work by other people. They may be based on specialised lecture courses. Furthermore, the manuscripts should provide sufficient motivation, examples and applications. This clearly distinguishes Lecture Notes from journal articles or technical reports which normally are very concise. Articles intended for a journal but too long to be accepted by most journals, usually do not have this "lecture notes" character. For similar reasons it is unusual for doctoral theses to be accepted for the Lecture Notes series.

2. Manuscripts should be submitted (preferably in duplicate) either to one of the series editors or to Springer-Verlag, Heidelberg. In general, manuscripts will be sent out to 2 external referees for evaluation. If a decision cannot yet be reached on the basis of the first 2 reports, further referees may be contacted: the author will be informed of this. A final decision to publish can be made only on the basis of the complete manuscript, however a refereeing process leading to a preliminary decision can be based on a pre-final or incomplete manuscript. The strict minimum amount of material that will be considered should include a detailed outline describing the planned contents of each chapter, a bibliography and several sample chapters.

Authors should be aware that incomplete or insufficiently close to final manuscripts almost always result in longer refereeing times and nevertheless unclear referees' recommendations, making further refereeing of a final draft necessary.

Authors should also be aware that parallel submission of their manuscript to another publisher while under consideration for LNM will in general lead to immediate rejection.

3. Manuscripts should in general be submitted in English.

Final manuscripts should contain at least 100 pages of mathematical text and should include

- a table of contents;
- an informative introduction, with adequate motivation and perhaps some historical remarks: it should be accessible to a reader not intimately familiar with the topic treated;
- a subject index: as a rule this is genuinely helpful for the reader.

Continued on inside back-cover

Lecture Notes in Mathematics

Edited by J.-M. Morel, F. Takens and B. Teissier

Editorial Policy
for the publication of monographs

1. Lecture Notes aim to report new developments in all areas of mathematics – quickly, informally and at a high level. Monograph manuscripts should be reasonably self-contained and rounded off. Thus they may, and often will, present not only results of the author but also related work by other people. They may be based on specialized lecture courses. Furthermore, the manuscripts should provide sufficient motivation, examples and applications. This clearly distinguishes Lecture Notes from journal articles or technical reports which normally are very concise. Articles intended for a journal but too long to be accepted by most journals, usually do not have this "lecture notes" character. For similar reasons it is unusual for doctoral theses to be accepted for the Lecture Notes series.

2. Manuscripts should be submitted (preferably in duplicate) either to one of the series editors or to Springer-Verlag, Heidelberg. In general, manuscripts will be sent out to 2 external referees for evaluation. If a decision cannot yet be reached on the basis of the first 2 reports, further referees may be contacted: the author will be informed of this. A final decision to publish can be made only on the basis of the complete manuscript, however a refereeing process leading to a preliminary decision can be based on a pre-final or incomplete manuscript. The strict minimum amount of material that will be considered should include a detailed outline describing the planned contents of each chapter, a bibliography and several sample chapters.
Authors should be aware that incomplete or insufficiently close to final manuscripts almost always result in longer refereeing times and nevertheless unclear referees' recommendations, making further refereeing of a final draft necessary.
Authors should also be aware that parallel submission of their manuscript to another publisher while under consideration for LNM will in general lead to immediate rejection.

3. Manuscripts should in general be submitted in English.
Final manuscripts should contain at least 100 pages of mathematical text and should include
- a table of contents;
- an informative introduction, with adequate motivation and perhaps some historical remarks: it should be accessible to a reader not intimately familiar with the topic treated;
- a subject index: as a rule this is genuinely helpful for the reader.

Continued on inside back-cover

Lecture Notes in Mathematics 1768

Springer
Berlin
Heidelberg
New York
Barcelona
Hong Kong
London
Milan
Paris
Tokyo

Markus J. Pflaum

Analytic and Geometric Study of Stratified Spaces

Springer

Author

Markus J. Pflaum
Department of Mathematics
Humboldt University
Rudower Chaussee 25
10099 Berlin, Germany

E-mail: pflaum@mathematik.hu-berlin.de

Cataloging-in-Publication Data applied for

Mathematics Subject Classification (2000):
58Axx, 32S60, 35S35, 16E40, 14B05, 13D03

ISSN 0075-8434
ISBN 3-540-42626-4 Springer-Verlag Berlin Heidelberg New York

Springer-Verlag Berlin Heidelberg New York
a member of BertelsmannSpringer Science+Business Media GmbH

http://www.springer.de

Printed in Germany

Typesetting: Camera-ready TEX output by the author

SPIN: 10852611 41/3142-543210/du - Printed on acid-free paper

For Stephanie and Konstantin

Contents

Introduction

Expressed in a more intuitive terminology, stratified spaces are collections of differentiable manifolds which are glued together in an appropriate way. This characteristic feature becomes apparent in the original name "complexes of manifolds" [188] or "manifold collections" [189, Sec. 11] by HASSLER WHITNEY for a predecessor of the modern notion of a stratified space. WHITNEY's article [188] from the year 1947 can be regarded as the birth date of an abstract theory of stratifications. Nevertheless, mathematicians have considered already before 1947 topics, which nowadays are treated within the theory of stratified spaces, like for example at the end of the nineteenth century, when algebraic geometers began to study singularities or when interest in simplicial complexes and triangulations of algebraic varieties began. The reason for the introduction of the so-called "complexes of manifolds" in [188] was the observation that the boundary of a noncompact bounded submanifold of some Euclidean space can often be decomposed in locally finitely many lower dimensional manifolds. This point of view has been taken up again in modern geometric analysis and for instance forms the basis of the concept of a manifold-with-boundary or a manifold-with-corners (see for example MELROSE [126, 127] or 1.1.19). RENE THOM noticed in his work [166] of 1955 that his iterated sets of singularities of a smooth mapping $f : \mathbb{R}^n \to \mathbb{R}^m$ are manifolds for certain generic functions and comprise a "manifold collection" in the sense of WHITNEY [189]. WHITNEY proved 1957 in [190] that every real algebraic variety can be decomposed in finitely many manifolds, and ŁOJASIEWIC [113] could show an analogous result for real analytic sets. So already at the end of the 50s some of the most important applications of stratification theory had been discovered.

The term "stratification" originates from THOM; he coined it 1962 in his work [167]. After THOM's famous article *Ensembles et morphismes stratifiés* [169] had appeared 1969, the new expression was generally accepted.

After the first years of stratification theory the main interest was to examine the topological properties of stratified spaces. In this time, more precisely in the 60s and 70s, mathematicians studied various criteria on a stratified space which should entail properties of the space which were needed for the intended application. As a conditio sine qua non it was required, as in the definition introduced in Section 1.1, that a decomposition or stratification of a topological space in locally finitely many manifolds should satisfy the "condition of frontier". This condition says that for every two manifolds R, S of the decomposition such that R has nonempty intersection with the topological closure of S the manifold R lies entirely in the closure of S. The most significant conditions on a stratified space besides the "condition of frontier"

M.J. Pflaum: LNM 1768, pp. 1 - 9, 2001

are topological local triviality introduced by THOM in [167] as well as the famous conditions (A) and (B) which were introduced by WHITNEY in [192].

Local triviality says essentially that the considered stratified space looks in the neighborhood of each one of its points like a trivial fiber bundle over the stratum of the point. The Whitney conditions on the other hand make a statement about the behavior of the tangent spaces of a stratum near the boundary. The reader will find detailed explanations on this as well as on other conditions in Section 1.4.

One of the main problems which the founders of stratification theory considered in the early years was the question under which conditions a stratified space is locally trivial. WHITNEY showed by a counter example that his condition (A) does not imply local triviality, but conjectured that this would hold for condition (B). THOM [169] succeeded to prove this conjecture. The essential idea which led THOM to the goal was the invention of so-called controlled vector fields which have a continuous flow though they themselves are in general not continuous. Moreover, THOM could show with the method of controlled vector fields his famous isotopy lemmas the first of which entails topological local triviality.

The American mathematician JOHN N. MATHER, who at that time mostly worked on questions of stability of differentiable maps, simplified the constructions of [169] and created in the so far unpublished note [122] an axiomatic setup for THOM's ideas which nowadays is known under the name "control theory". In the work [122] which was intended as the first chapter of a not yet completed book on the topological stability of differentiable maps MATHER describes in detail on over seventy pages, how one proves the two isotopy lemmas of THOM and topological local triviality. Moreover, he supplies all the necessary tools in particular very general existence and uniqueness results for tubular neighborhoods.

At this point let us enumerate several of the "topological" high points of stratification theory:

- For every semianalytic set there exists a semianalytic triangulation. (ŁOJASIEWIC [114], 1964).
- Every semianalytic set possesses a Whitney stratification by strongly analytic manifolds (ŁOJASIEWIC [115], 1965).
- Real and complex algebraic varieties possess a stratification satisfying Whitney's condition (A) and (B) (WHITNEY [191], 1965).
- Subanalytic sets can be Whitney stratified (HIRONAKA [86], 1973 and HARDT [77, 78], 1975).
- Subanalytic sets and proper light subanalytic mappings possess a triangulation (HIRONAKA [87], 1975 and HARDT [79], 1977).
- Whitney stratified sets can be triangulated (GORESKY [62], 1978 and JOHNSON [96], 1983).
- GORESKY–MCPHERSON [63, 64] invent 1980 intersection homology for pseudomanifolds and prove a duality theorem à la POINCARÉ for intersection homology

groups with complementary perversity; such a duality result does in general not hold for singular homology of stratified spaces.

- Every proper topologically stable smooth mapping between manifolds can be triangulated (VERONA [180], 1984).

- Based on the work of MOSTOWSKI [130], PARUSIŃSKI [141] proved that every subanalytic set possesses a Lipschitz stratification.

Concerning nontopological aspects of stratified spaces, more intensive studies on the geometry and analysis on stratified spaces began only at the end of the 70's, beginning of the 80's. One of the initiators hereby was CHEEGER [38, 39, 41], who mainly considered questions of spectral theory of spaces with conic singularities and of Riemannian pseudomanifolds. One of the essential objects in the spectral theory of operators on regular as well as singular spaces is the deRham complex which lies densely embedded in the Hilbert space $L^2(\Omega^\bullet \mathsf{M})$ of L^2-forms with respect to an a priorily chosen Riemannian metric. According to CHEEGER [39], the most important question now is, under which prerequisites a L^2-version of the theorem of STOKES holds in the singular case as well. The L^2-STOKES' theorem implies that the Dirac operator of the deRham complex possesses a natural selfadjoint extension, a property which BRÜNING–LESCH [28] called "uniqueness of ideal boundary condition" and which has been considered in the cited work even for more general Hilbert complexes. Up to now one knows only for cones and horns (CHEEGER [39]) and complex conformal cones (BRÜNING–LESCH [29]) that the L^2-STOKES' theorem is true. In some more recent work corresponding results have been proved for semianalytic surfaces with isolated singularities (GRIESER [67]) and for complex varieties X with isolated singularities (PARDON–STERN [139], GRIESER–LESCH [68]), where in the last case one has to make the restriction $k \neq \dim_{\mathbb{C}} \mathsf{X}, \dim_{\mathbb{C}} \mathsf{X} - 1$ on the order k of the considered L^2-forms. The results obtained so far all indicate that the L^2-STOKES' theorem holds for all compact complex algebraic varieties, a conjecture which is generally shared. The L^2-STOKES' theorem is closely connected to L^2-cohomology, hence to intersection homology. In the first paragraph of the fifth chapter, which treats deRham cohomology on stratified spaces, some remarks on L^2-cohomology have been included.

Asymptotic expansions, in particular for the heat kernel, play an important role not only for the spectral theory of compact manifolds but also for the one of stratified spaces. Studies on the asymptotics for singular spaces have been initiated by CHEEGER [40] in the year 1983. In the fundamental work of BRÜNING–SEELEY [31] the so-called "singular asymptotics lemma" was introduced which emerged as an essential tool for asymptotic methods under the presence of singularities. Using the singular asymptotics lemma BRÜNING–SEELEY [32, 34] were able to study asymptotics for spaces with cone like singularities. Also in connection with asymptotic expansions is the work BRÜNING–LESCH [30] on the spectral theory of algebraic curves. Concerning orbit spaces of Riemannian G-manifolds M (with M and G compact) BRÜNING–HEINTZE [26, 27] have given the asymptotic expansion of the operator on G\M induced by the Laplacian on M.

A promising path for a better understanding of analytic properties of singular spaces lies in the ansatz to set up a theory of pseudodifferential operators on singular spaces. Hereby one aims at constructing parametrices for appropriate differential operators on stratified spaces with the help of a suitable symbol calculus. The most important contributions in this direction come from MELROSE and SCHULZE, who independently developed appropriate pseudodifferential calculi. The ansatz of SCHULZE is explained in detail in the two books [148, 149]. In a more recent work [150] SCHULZE has constructed operator algebras for manifolds with singularities with the help of an appropriately adapted functional structure. The so-called b-calculus, which was developed by MELROSE for manifolds-with-boundary, is explained in [126] and generalized in [127] to manifolds-with-corners. A significant success of this approach is among other results the signature formula of HASSELL–MAZZEO–MELROSE [81] for manifolds-with-corners of codimension 2.

An undisputed goal of geometric analysis is to search for an index formula for stratified spaces. A first important result in this direction is the index formula for orbifolds by KAWASAKI [99, 100]. The 1998 article by LOTT [117] on signatures of S^1-quotients, which naturally are stratified spaces, can be rated as a further indication that such index formulas seem to be promising.

The noncommutative geometry by CONNES [45] but also the algebraic index theorems of NEST–TSYGAN [133] have shown the importance of Hochschild and cyclic homology for geometry and analysis. Therefore mathematicians began to study Hochschild and cyclic homology of function algebras over singular spaces. Let us mention in chronological order contributions in this direction by BRYLINSKI [35], WASSERMANN [182], BLOCK–GETZLER [17], TELEMAN [165] and BRASSELET–LEGRAND [23]. BRYLINSKI and BLOCK–GETZLER have studied cyclic homology in the equivariant case. WASSERMAN calculated by application of various techniques the Hochschild and cyclic homology for function algebras on orbifolds. TELEMAN introduced a new method for the computation of Hochschild homology with the help of jets. Using this method he could prove the topological version of the theorem of HOCHSCHILD–KOSTANT–ROSENBERG [88] which already had been shown by CONNES [45] for the case of compact manifolds (see also PFLAUM [143] for the noncompact case). Moreover, TELEMAN showed a topological HOCHSCHILD–KOSTANT–ROSENBERG theorem for piecewise differentiable functions over a simplicial complex. BRASSELET–LEGRAND finally calculated the Hochschild homology and periodic cyclic homology of function algebras on controlled spaces, where the considered functions have poles of prescribed order along the strata. The main result of their work is that intersection homology can be interpreted as periodic cyclic homology of these function algebras.

After this tour through the history of stratification theory let me explain the matter of concern of this work. As already has been mentioned in the historical notes, the topological properties of stratified spaces are well studied. Concerning the geometry and analysis of stratified spaces progress has not flourished as far as for the topological aspects. Looking at the present status of research one notices that there exist several deep results on the spectral geometry, the index theory etc. of stratified spaces with a very particular structure like for example a conic one, but that geometric-analytic considerations for more general classes of stratified spaces seem to be afflicted with insurmountable difficulties.

Under the global headline "smooth structure" this work is intended as a systematic contribution to the geometric analysis of stratified spaces. It is to be shown by concrete examples that "smooth structures" form a useful concept which allow deeper insight to the geometric-analytic structure of a stratified space. Moreover, this work is a monograph explaining the theory of stratified spaces from the basics on and thus should fill a gap in the existing literature on stratified spaces. A view on the content now explains at best which particular topics will be treated in this work.

The first chapter has a twofold purpose, namely first to describe the basic notions decompositions and stratifications together with the most important stratification conditions and secondly to introduce a meaningful functional structure on stratified spaces which is appropriate for analytic and geometric considerations.

Let me explain at this point, why I regard it necessary to give a detailed exposition of the basic notions of the theory of stratified spaces. One can find in the mathematical literature quite a variation on the notion of a stratified space which sometimes might lead to a lack of clarity. For example, in case G is a (compact) Lie group, it is not immediate to tell precisely, what the strata of the canonical stratification of a G-manifold M respectively of its orbit space $\mathsf{G}\backslash\mathsf{M}$ by orbit types are. Namely, the subset $\mathsf{M}_{(\mathsf{H})} \subset \mathsf{M}$ of orbit type (H) (see Chapter 4) associated to a closed subgroup $\mathsf{H} \subset \mathsf{G}$ is in general not a manifold, rather only its connected components comprise manifolds. Hence the decomposition of M into the $\mathsf{M}_{(\mathsf{H})}$ – and likewise the one of $\mathsf{G}\backslash\mathsf{M}$ into the sets $\mathsf{G}\backslash\mathsf{M}_{(\mathsf{H})}$ – is not a stratification. If one now tries to obtain a stratification of M by decomposition into the connected components of the $\mathsf{M}_{(\mathsf{H})}$, then it is not immediately clear, why the "condition of frontier" should be satisfied. Moreover, the decomposition of M into the connected components of the $\mathsf{M}_{(\mathsf{H})}$ is in general too fine. Thus one has to pursue another path to get a stratification of M or $\mathsf{G}\backslash\mathsf{M}$. The solution of this problem lies in the idea to regard a stratification of a topological space X not as a global object in form of a decomposition, but as a local object in shape of a mapping which assigns to every point x a set germ such that this induces locally decompositions. Transferred to the case of a G-manifold M, one obtains a stratification in this sense, if one associates to every point $x \in \mathsf{M}$ resp. every orbit $\mathsf{G}x$ the germ $\mathcal{S}_x$ of the set $\mathsf{M}_{(\mathsf{G}_x)}$ resp. $\mathsf{G}\backslash\,\mathsf{M}_{(\mathsf{G}_x)}$; here G_x is the isotropy group of x. The reader will find details for this procedure in Section 4.3.

The idea to regard stratifications as assignments of particular set germs goes back to MATHER [123] and comprises the probably clearest definition of a stratified space. Therefore I have used it in this monograph (see Def. 1.2.2). The definition via set germs might first seem very abstract, but it has the advantage that from such a definition it follows unequivocally what to understand by a stratum, namely a piece of the coarsest decomposition inducing the set germ map $\mathcal{S}$ (see Proposition 1.2.7).

We do not want to withhold from the reader that some authors require, differently to this work, local triviality in the definition of stratified space. The reason is that for some topological considerations of stratified spaces as for example for the intersection homology of GORESKY–MACPHERSON [63, 64, 65] local triviality is indispensable. But in particular cases local triviality might be difficult to prove and is not necessary for many other considerations. Therefore we did not include local triviality in the definition of a stratified space, in particular as this is not the case in MATHER [123, 122] either.

Section 1.3 in the first chapter forms the backbone of this monograph; it is dedicated to smooth structures which are indispensable for all further considerations in this work. To be able to study a stratified space X under the view points of analysis and geometry, it is necessary to have an algebra of appropriate functions on X, which in general is not intrinsically given by the stratified space. Expressed in the language of algebraic geometry, one additionally needs a structure sheaf for X. In the mathematical literature there appear various function algebras on stratified spaces like for example the algebra of continuous or on every stratum smooth functions or the algebra of controlled functions. But these function algebras contain in general too many functions and are often not even locally finitely (topologically) generated. On the other hand one knows for several classes of stratified spaces already canonical candidates for a meaningful functional structure, like for example the algebraic and analytic varieties with the restricted smooth functions of the ambient $\mathbb{C}^n$ or orbit spaces of the form $G\backslash M$ with the sheaf induced by the G-invariant smooth functions on M. The concept of a smooth structure introduced in 1.3 naturally includes all the mentioned examples of stratified spaces with a functional structure.

But what is a smooth structure in "down to earth terms"? One can regard a smooth structure as an equivalence class of local embeddings of the stratified space in an appropriate $\mathbb{R}^n$, which is in an abstract way in accordance with WHITNEY's original conception of a stratified space in his article *Complexes of manifolds.* The local embeddings are the so-called singular charts; their exact definition is borrowed from the definition of charts on differentiable manifolds. From a smooth structure one can derive canonical sheaves, first the one of smooth functions, but also the sheaf of Whitney functions which has its significance in the extension theory of smooth functions. In Section 1.7, where we will present the first results on the extension theory of smooth functions, we introduce the notion of Whitney functions flat on a closed subset Z and the notion of Whitney functions tempered relative Z. If one multiplies a Whitney function flat on Z with a Whitney function tempered relative Z, then one obtains again a function flat on Z. This feature is very useful for "gluing" together smooth functions and will be used often in this work. Finally, the reader will find in 1.6 several explanations on the notion of a rectifiable curve for spaces with a smooth structure, though the significance of this will become only clear in the section on Riemannian metrics.

In the second chapter it will be shown that the functional structures introduced in the first chapter are suitable to do differential geometry on stratified spaces. Of course one has to expect cutbacks in comparison with the geometry of manifolds, but it is amazing how many geometric structures can be transferred to stratified spaces with a smooth structure. As first examples we have to name stratified tangent bundles and vector fields (see Sections 2.1 and 2.2). Hereby, WHITNEY's condition (A) gets a new interpretation: it guarantees that the stratified tangent bundle of a stratified space with a smooth structure exists and carries a smooth structure itself. Besides the stratified tangent bundle there exists also a stratified cotangent bundle, but unlike in the manifold case the stratified tangent and cotangent bundle of a stratified space with smooth structure are in general not isomorphic. Metric aspects dominate the forth section of the second chapter. The notion of a Riemannian metric will be carried over to stratified spaces and it is shown, how one can use such a Riemannian metric under certain regularity assumptions for measuring the length of curves and for the definition of a geodesic distance. Naturally, the question then arises, whether one can construct on every regular stratified space like in the manifold case a Riemannian metric such that the geodesic distance becomes a complete metric. This questions gets a positive answer in Theorem 2.4.17. According to GROTHENDIECK [72] one can assign to every commutative algebra in a canonical way the (noncommutative) algebra of differential operators. In Section 2.5 we will apply the definition of GROTHENDIECK to the sheaf of smooth functions; one thus obtains the differential operators on stratified spaces. The end of the second chapter is dedicated to themes from mathematical physics. The concept of a Poisson stratified space will be explained and several examples will be introduced. We close the chapter with some remarks on the quantization of singular symplectic spaces.

The topic of the third chapter is MATHER's control theory. The ideas of MATHER will be supplemented by an important addition, namely by curvature moderate tubes and control data. The notion "curvature moderate" introduced in this work is new in the mathematical literature. It regulates the behavior of the tangent spaces of a stratum near the boundary in higher order than the Whitney conditions do. In other words the property that a stratum is curvature moderate means nothing else than that the stratum curves near the boundary in a controlled way and that the same holds for the higher derivatives of the curvature. As a consequence the projection and tubular function of a curvature moderate tubular neighborhood of the stratum S are tempered relative ∂S This fact will turn out to be very useful for the extension theory of smooth functions over a stratum (see Theorem 3.8.3), and forms an essential ingredient in the proof of the Poincaré lemma for Whitney forms 5.4.4. Furthermore, we will derive in the third chapter consequences of MATHER's control theory like the first isotopy lemma of THOM. Partially wecwill supplement these results by curvature moderate versions. Chapter 3 ends with an introduction to a new class of stratified spaces, the so-called cone spaces. It is proved that cone spaces can be Whitney stratified.

An important class of stratified spaces is given by orbit spaces. They will be treated in Chapter 4. The results of this chapter are mostly well-known, though scattered in the literature. Particular attention is laid on the detailed construction of the stratification of a G-manifold M and of its orbit space $G\backslash M$ by orbit types. Afterwards it is shown with the help of the theorem of SCHWARZ that the G-invariant functions on M induce a smooth structure for the orbit space.

Chapter 5 is dedicated to an analytic-topological topic, namely the deRham theory on stratified spaces with a smooth structure. From the sheaf of smooth functions one obtains a sheaf of Kähler differentials, hence of differential forms and the deRham complex. For some spaces like for example cone spaces the cohomology of the deRham complex coincides with the cohomology of the underlying topological space, but this is not the case in general and often the deRham complex does not provide the right cohomology. The situation becomes completely different when considering the exterior complex of Kähler differentials of Whitney functions or in other words the complex of Whitney forms. Its cohomology coincides for a large class of curvature moderate stratified spaces with the singular cohomology, thus one obtains a deRham theorem for Whitney forms.

The last chapter of this work deals with topological Hochschild homology. In the first two sections we supply the necessary theoretical foundations for a topological version of Hochschild homology. In the mathematical literature on the Hochschild homology of function algebras one often transfers results of "formal" homological algebra to a topological version without reasoning. That this is justified indeed, with some caution, will be shown in 6.1 and 6.2. Afterwards we will apply in Section 6.3 the obtained methods to define continuous Hochschild homology and to derive some of its basic properties. In the last section of Chapter 6 we compute the Hochschild homology for smooth functions on a manifold and on certain cone spaces.

Summarizing, the main and new results of this work are the following:

- The notion of a smooth structure for a stratified space is introduced. Examples of such spaces with a smooth structure are given by algebraic and analytic varieties, by orbit spaces of proper Lie group actions and by singular Marsden–Weinstein reduced spaces.

- It is proved that every regular (A)-stratified space X (even with possibly infinite dimension or infinitely many strata) can be equipped with a Riemannian metric μ in such a way that X is a length space with the metric given by the geodesic distance defined by μ (Theorem 2.4.17).

- The notion of a curvature moderate tubular neighborhood and of curvature moderate control data is introduced. It is shown that the "curvature moderate" versions of the existence and uniqueness results for tubular neighborhoods hold true (Section 3.5). For strongly curvature moderate stratified spaces the existence of curvature moderate control data in each differentiability order $m \in \mathbb{N}$ is proved (Theorem 3.6.9).

- It is shown that for orbit spaces $G\backslash M$ which do not contain any strata of codimension 1 the deRham cohomology coincides with the basic and the singular cohomology (Theorem 5.3.5). Moreover, a theorem going back to KOSZUL, which says that for compact G the basic and the singular cohomology of $G\backslash M$ are canonically isomorphic, will be generalized to proper G-actions (Corollary 5.3.3).

- For an (A)-stratified space with curvature moderate control structure and tempered resolutions of its strata in every order $m \in \mathbb{N}$ it is shown that the Poincaré lemma holds for Whitney forms and that the cohomology of Whitney forms coincides with the singular cohomology (5.4.4 and Corollary).

- For certain cone spaces the Hochschild homology is calculated and it is shown that on such spaces a theorem à la HOCHSCHILD–KOSTANT–ROSENBERG is true (Theorem 6.4.7).

This work became considerable longer than originally planned, but nevertheless many aspects in connection with the topic of this book could not or only marginally be considered. First we have to name here semialgebraic, semianalytic and subanalytic sets, which have many points in common with this work and often served as examples, but could not be treated in detail, as this would have further enlarged the size of this Habilitation thesis. By the same reason we did not include intersection homology, but it is planned for future projects to clarify the connection between the Whitney–deRham cohomology introduced here and intersection homology.

At this point let us make some remarks about where the theory of smooth structures on stratified spaces could tend to. It is necessary to further clarify the theory of differential operators on stratified spaces with smooth structure, in particular it would be desirable to come to a microlocalization of such differential operators or in other words to a theory of stratified pseudodifferential operators. The present work already gives some hints which should be pursued. For example, the Poincaré lemma for Whitney–deRham forms has shown that for the study of partial differential equations on stratified spaces the theory of jets and Whitney functions is not only helpful but even indispensable. This point of view should be extended further and applied to other partial differential equations on stratified spaces. The future will show, in how far the ideas of this work will contribute to current questions like index theory on stratified spaces, Hodge theory, L^2-cohomology etc. I am convinced that a systematic geometric language like in this work will lead to the right path to the solution of the present problems in the geometric analysis of stratified spaces.

After this overview I would now like to invite the reader to a concrete tour through the theory of stratified spaces and close this introduction in the hope that with the present work some little progress has been made on the way to the geometry of tame stratified spaces in the sense of GROTHENDIECKS *Esquisse d'un Programme* [73].

Acknowledgments I would like to thank Jochen Brüning for his continuous support and all the constructive discussions. Particular thank goes to Matthias Lesch for explaining many topics in geometric analysis during his time in Berlin as well as for his advice, whether of a mathematical or of a nonmathematical kind. I thank Daniel Grieser for discussions on smaller or larger problems in connection with this work and Martin Bordemann for stimulating discussions on aspects of singular symplectic spaces and their quantization.

Essential ideas for this work have been obtained during a research stay supported by the DFG at Berkeley in October 1997 and during a four week long research stay at the IHES, Bures-sur-Yvette, in fall 1998. I thank both institutions and the DFG for their support.

Finally I would like to thank my wife Stephanie for her loving help during this Habilitation project and my son Konstantin for his encouraging smile in the last weeks.

Notation

For the convenience of the reader we will explain in the following the most important symbols used in this work.

Symbol	Meaning	Reference
$\mathfrak{a}, \mathfrak{b}$	matrices or linear mappings, elements of an algebra	
$\mathfrak{a}.v$	evaluation of the map $\mathfrak{a}$ at a vector v	
A, Z	locally closed subset of a topological (often stratified) space	
$\mathcal{A}, \mathcal{B}$	real or complex algebras	
$B, B_\varepsilon(x)$	ball resp. ball with radius ε around a point x of a metric space	
$B^n_\varepsilon(x)$	ball with radius ε around a point $x \in \mathbb{R}^n$	
$\mathbb{C}$	set of complex numbers	
$\mathcal{C}^m_X$, $\mathcal{C}^\infty_X$	sheaf of m-times differentiable resp. smooth functions on X	p. 26
$\mathcal{C}^m_{X,x}$, $\mathcal{C}^m_x$	stalk of the sheaf of $\mathcal{C}^m$-functions on X with footpoint $x \in X$	p. 28
$D_v f(x)$, $Df(x).v$	derivative of a differentiable function f at the footpoint x in direction of v	
$D^k f(x).(v_1, \cdots, v_k)$	higher derivatives of f at the footpoint x in direction of $v_1, \cdots, v_k$	
$\mathcal{D}$, $\mathcal{D}^k$	sheaf of differential operators of order k on an (A)-stratified space resp. presheaf of differential operators	p. 81
$e_1, \cdots, e_n$	canonical basis of $\mathbb{R}^n$	
E, F, N	vector bundles	
$\mathcal{E}^m_{X,\mathcal{U}}$, $\mathcal{E}^\infty_{X,\mathcal{U}}$	sheaf of Whitney functions of class $\mathcal{C}^m$ resp. $\mathcal{C}^\infty$ on X, defined with respect to an atlas $\mathcal{U}$	p. 43

Symbol	Meaning	Reference
$\mathsf{g}, \mathsf{h}, \mathsf{k}$	elements of a Lie group	
$\mathfrak{g}, \mathfrak{h}$	Lie algebras	
$\mathsf{G}, \mathsf{H}, \mathsf{K}$	Lie groups	
i, j, k, l	indices, most often natural numbers	
$i_\nu\alpha$	insertion of a vector ν in the differential form α	p. 70, p. 205
I, J	index sets	
$\mathcal{I}, \mathcal{J}$	ideals of an algebra $\mathcal{A}$	
$\mathcal{J}^{m,c}(Z; A)$	ideal of the jets of class $\mathcal{C}^m$ on A which are flat on Z of order c	p. 53
k	cone chart	p. 148
K	compact set	
l	link chart	p. 148
m, k	elements of the set $\mathbb{N} \cup \{\infty\} \cup \{\omega\}$, express most often order of differentiability	
$\mathfrak{m}, \mathfrak{n}$	elements of an $\mathcal{A}$-module	
$m, n, \tilde{n}, N$	natural numbers, most often dimensions	
M, N	differentiable manifolds	
$\mathcal{M}, \mathcal{N}$	modules of an algebra $\mathcal{A}$	
$\mathcal{M}^{m,c}(Z; A)$	algebra of relatively Z of order c tempered jets on $A \setminus Z$ of class $\mathcal{C}^m$	p. 57
$\mathbb{N}$	set of natural numbers including 0	
$\mathbb{N}^{>0}$	set of positive integers	
p, q	polynomials	
P	principal bundle	
P, Q	projection operators	
R, S	pieces or strata of a decomposed space	p. 15, p. 25
$\mathbb{R}$	set of real numbers	
$\mathbb{R}^{>0}, \mathbb{R}^{\geq 0}, \mathbb{R}^{<0}, \mathbb{R}^{\leq 0}$	set of positive, nonnegative, negative resp. nonpositive real numbers	
$\mathbb{RP}^n$	n-dimensional real projective space	
S_n	permutation group in n variables	
S^n	n-sphere	
Sym^k	functor of the k-times symmetric tensor product	

Symbol	Meaning	Reference
TM, TX	tangent bundle of the manifold M resp. the stratified space X with smooth structure	p. 63
T	tubular neighborhood	p. 92
T^n_m	standard tubular neighborhood of $\mathbb{R}^m$ in $\mathbb{R}^n$ for $m \leq n$	p. 93
$\mathsf{U}, \mathsf{V}, \mathsf{W}, \mathsf{O}$	open sets in a topological space	
v, w	vectors, elements of a vector bundle	
v_x	elements of a vector bundle with footpoint x	
V, W	vector fields on a manifold or an (A)-stratified space	p. 65
V_x	evaluation of the vector field V at the footpoint x	
$\mathcal{V}, \mathcal{W}$	vector spaces	
$\mathsf{x}, \mathsf{y}, \mathsf{z}$	points of a stratified space	
x, y	singular charts of a stratified space, differentiable charts on a manifold	p. 26
$\mathsf{x}^1, \cdots, \mathsf{x}^n$	components of a singular chart with values in $\mathbb{R}^n$	
X, Y	decomposed or stratified spaces	p. 15, p. 23
$\mathfrak{X}^\infty(\mathsf{M}), \mathfrak{X}^\infty(\mathsf{X})$	space of smooth vector fields on a manifold resp. an (A)-stratified space	p. 65
$\mathbb{Z}$	set of integers	
α, β	multiindices in $\mathbb{N}^n$	
α, β, ω	differential forms on a manifold or a stratified space with smooth structure	p. 68
δ_μ	geodesic distance corresponding to the Riemannian metric μ	
$\iota_i : \mathbb{R} \to \mathbb{R}^n$	injection of the i-th coordinate	
ι^n_m	canonical injection of $\mathbb{R}^m$ in $\mathbb{R}^n$ via the first m coordinates for $m \leq n$	
Λ	Poisson bivector on a stratified space	p. 83
Λ^k	functor of the k-times antisymmetric tensor product	

Symbol	Meaning	Reference
μ, η, θ	Riemannian metric on a manifold or a stratified space, also scalar product on a vector bundle	p. 71
$\pi_i : \mathbb{R}^n \to \mathbb{R}$	projection onto the i-th coordinate	
π^n_m	canonical projection of $\mathbb{R}^n$ onto $\mathbb{R}^m$ via the first m coordinate for $m \leq n$	
ρ^n_m	tubular function of the standard tubular neighborhood T^n_m for $m \leq n$	p. 93
ξ, ζ	elements of a Lie algebra $\mathfrak{g}$	
ξ_M	fundamental vector field of $\xi \in \mathfrak{g}$	p. 153
$\Omega^k(\mathsf{M}), \Omega^k(\mathsf{X})$	space of k-forms on a manifold M resp. the stratified space X	p. 68
$\langle \cdot, \cdot \rangle$	dual pairing of two vector spaces, often Euclidean metric	
$\vert \cdot \vert, \Vert \cdot \Vert, \mathbf{\vert} \cdot \mathbf{\vert}$	norms or seminorms on a vector space	
$\partial \mathsf{M}$	boundary of a manifold-with-boundary M	
$\partial \mathsf{S}$	boundary $\overline{\mathsf{S}} \setminus \mathsf{S}$ of a stratum $\mathsf{S} \subset \mathsf{X}$, does not coincide in general with the topological boundary $\mathrm{bdr}\,(\mathsf{S}) = \overline{\mathsf{S}} \cap \overline{\mathsf{X} \setminus \mathsf{S}}$	p. 15
M°	interior of a manifold-with-boundary M	
A°	topological interior of a topological subspace $\mathsf{A} \subset \mathsf{X}$	

Chapter 1

Stratified Spaces and Functional Structures

1.1 Decomposed spaces

1.1.1 Let X be a paracompact Hausdorff space with countable topology, and $\mathcal{Z}$ a locally finite partition of X into locally closed subspaces $\mathsf{S} \subset \mathsf{X}$. Then one calls X or better the pair $(\mathsf{X}, \mathcal{Z})$ a *decomposed space* with *pieces* $\mathsf{S} \in \mathcal{Z}$ and $\mathcal{Z}$ a *decomposition* of X, if the following conditions are satisfied:

(DS1) Every piece $\mathsf{S} \in \mathcal{Z}$ is a smooth manifold in the induced topology.

(DS2) (*condition of frontier*) If $\mathsf{R} \cap \overline{\mathsf{S}} \neq \emptyset$ for a pair of pieces $\mathsf{R}, \mathsf{S} \in \mathcal{Z}$, then $\mathsf{R} \subset \overline{\mathsf{S}}$. We write in this case $\mathsf{R} \leq \mathsf{S}$ and call R *incident* to S, or a *boundary piece* of S.

One checks immediately that the incidence relation is an order relation on the set of pieces of X, hence the notation $\mathsf{R} \leq \mathsf{S}$ is justified.

1.1.2 Explanation The notion "locally closed" will appear more often in this work. Therefore let us briefly recall its meaning. By a *locally closed* subset of a topological space X we understand a subset $\mathsf{A} \subset \mathsf{X}$ such that every point of A has a neighborhood U in X with $\mathsf{A} \cap \mathsf{U}$ closed in U. Equivalently, A is the intersection of an open and a closed subset of X, or in other words A is open in its closure. Obviously, the finite intersection of locally closed subsets is again locally closed. Submanifolds lie always locally closed in their ambient manifold.

By the *boundary* $\partial \mathsf{A}$ of a locally closed subset $\mathsf{A} \subset \mathsf{X}$ we will understand the closed subspace $\overline{\mathsf{A}} \setminus \mathsf{A}$, which in general does not coincide with the *topological boundary* $\operatorname{bdr}(\mathsf{A}) = \overline{\mathsf{A}} \cap \complement \mathsf{A}$. If X is a decomposed space, and $\mathsf{S} \subset \mathsf{X}$ one of its pieces, then $\partial \mathsf{S}$ consists of all boundary pieces $\mathsf{R} < \mathsf{S}$.

Let us note that the notation $\partial \mathsf{A}$ will not lead to any confusion with the boundary $\partial \mathsf{M}$ of a manifold-with-boundary M. Namely, if M is embedded as a closed subspace of some Euclidean space $\mathbb{R}^n$, then the *interior* M° of M is locally closed in $\mathbb{R}^n$, and the boundary $\partial \mathsf{M}$ of the manifold M is just the boundary $\partial \mathsf{M}^\circ$ of the locally closed subset $\mathsf{M}^\circ \subset \mathbb{R}^n$ as defined above.

M.J. Pflaum: LNM 1768, pp. 15 - 62, 2001

1.1.3 Remark As X is separable, the decomposition $\mathcal{Z}$ contains at most countably many pieces.

1.1.4 Remark Instead of manifolds one can take in condition (DS1) any object of an arbitrary category $\mathfrak{T}$ of topological spaces. Thus one obtains the so-called $\mathfrak{T}$-*decomposed spaces.* As an example for $\mathfrak{T}$ let us name the category of real or complex analytic manifolds, or the category of polyhedra.

In this context we introduce the category Σ-$\mathfrak{Var}$ of Σ-*manifolds*, the objects of which consist of the topological sum of countably many connected smooth and separable manifolds. The morphisms of Σ-$\mathfrak{Var}$ are the continuous and on every component smooth functions between Σ-manifolds. If the dimension of the components of a Σ-manifold M is bounded, we will say that M has *finite dimension*, and denote the supremum of these dimensions by $\dim M$.

A decomposition of X into Σ-manifolds is called a Σ-decomposition.

1.1.5 The *dimension* of a decomposed space $(X, \mathcal{Z})$ is defined by

$$\dim X = \sup \left\{ \dim S \,\middle|\, S \in \mathcal{Z} \right\}.$$

In most applications we will consider only finitely dimensional decomposed spaces.

By the *k-skeleton* of $(X, \mathcal{Z})$, where $k \in \mathbb{N}$, we denote the decomposed space

$$X^k = \bigcup_{S \in \mathcal{Z},\, \dim S \leq k} S$$

with the topology induced by X.

For every element $x \in X$ one defines its *depth* by

$$\operatorname{dp}_{\mathcal{Z}}(x) = \sup \left\{ k \in \mathbb{N} \,\middle|\, \exists S_0, S_1, \ldots, S_k \in \mathcal{Z} \colon x \in S_0 < S_1 < \ldots < S_k \right\}.$$

By definition $\operatorname{dp}_{\mathcal{Z}}(x) = \operatorname{dp}_{\mathcal{Z}}(y)$ holds for all elements x, y of a piece S, hence the depth of a piece S is well-defined by $\operatorname{dp}_{\mathcal{Z}}(S) = \operatorname{dp}_{\mathcal{Z}}(x)$. Finally, the depth of X is given by

$$\operatorname{dp}_{\mathcal{Z}}(X) = \sup \left\{ \operatorname{dp}_{\mathcal{Z}}(S) \,\middle|\, S \in \mathcal{Z} \right\}.$$

1.1.6 A continuous mapping $f : X \to Y$ between decomposed spaces is called a *morphism of decomposed spaces*, if for every piece $S \in \mathcal{Z}$ there is a piece $R_S \in \mathcal{Y}$ such that the following holds:

(DS3) $f(S) \subset R_S$,

(DS4) the restriction $f_{|S} : S \to R_S$ is smooth.

By the condition of frontier and continuity a morphism f also has the following property:

(DS5) for all $S \leq S'$ the relation $R_S \leq R_{S'}$ is satisfied.

Obviously, the composition of two morphisms is again a morphism, hence the decomposed spaces together with their morphisms form a category $\mathfrak{Esp}_{\mathfrak{Dec}}$.

If a paracompact topological space X has two decompositions $\mathcal{Z}$ and $\mathcal{Y}$, we will say that $\mathcal{Z}$ is *coarser* than $\mathcal{Y}$ or that $\mathcal{Y}$ is *finer* than $\mathcal{Z}$, if the identity mapping is a morphism of decomposed spaces from $(X, \mathcal{Y})$ to $(X, \mathcal{Z})$. In most cases we are interested in a rather coarse decomposition of a paracompact topological space.

1.1.7 Given two decomposed spaces $(X, \mathcal{Z})$ and $(Y, \mathcal{Y})$ one can form their cartesian product $X \times Y$ and their topological sum $X \amalg Y$. The subspaces $S \times R$ with $S \in \mathcal{Z}$ and $R \in \mathcal{Y}$ then form a decomposition of $X \times Y$, the subspaces $T \subset X \amalg Y$ with $T \in \mathcal{Z}$ or $T \in \mathcal{Y}$ a decomposition of $X \amalg Y$. Thus $\mathfrak{Esp}_{\mathfrak{Dec}}$ becomes a category with (finite) products and sums.

A topological space $Y \subset X$ is called a *decomposed subspace*, if for all pieces $S \in \mathcal{Z}$ the intersection $S \cap Y$ is a submanifold of S, and the corresponding partition $\mathcal{Z} \cap Y$ of Y satisfies the condition of frontier. In this case $(Y, \mathcal{Z} \cap Y)$ is again a decomposed space.

To give the reader a first impression about the variety of all possible decomposed spaces, a long list of useful and instructive examples follows.

1.1.8 Manifolds Every smooth manifold M is a decomposed space in a canonical way with one single piece. It is easily possible to construct (infinitely) many different partitions of M which turn M into a decomposed space. All of these decompositions are coarser than the canonical decomposition and usually will not be considered further.

1.1.9 Intervals The most simple examples of (nontrivial) decomposed spaces are given by intervals of the form $[a, b[$, $]a, b]$ or $[a, b]$ with $-\infty \leq a < b \leq \infty$, and the obviously coarsest decomposition into the set $]a, b[$ and one respectively two boundary points.

1.1.10 Manifolds-with-boundary Let M be a manifold with boundary, $S_1 = \partial M$ its boundary and $S_2 = M^\circ = M \setminus \partial M$ its interior. Then M comprises a decomposed space with pieces S_1 and S_2.

1.1.11 Cones If X is a topological space, the *cone* over X is defined as the quotient space

$$CX = [0, 1[\times X / \{0\} \times X.$$

If now M is a manifold, the cone CM is a decomposed space with its pieces given by the *cusp* $o := [\{0\} \times M]$ and the set $]0, 1[\times M$. In case $M = S^1$ we obtain the well-known *standard cone* $X_{\text{Cone}} := CS^1$ (Fig. 1.1), and in case $M = S^0 = \{\pm 1\}$ the *edge* $X_{\text{Edge}} := CS^0$ (Fig. 1.2). Instead of M one can take a decomposed space X with finitely many pieces $S \in \mathcal{Z}$. The cone CX then is decomposed as well, where its pieces are given by the cusp o and the sets $]0, 1[\times S$. Hereby

$$\begin{aligned} \dim CX &= \dim X + 1, \\ \operatorname{dp} CX &= \operatorname{dp} X + 1. \end{aligned} \tag{1.1.1}$$

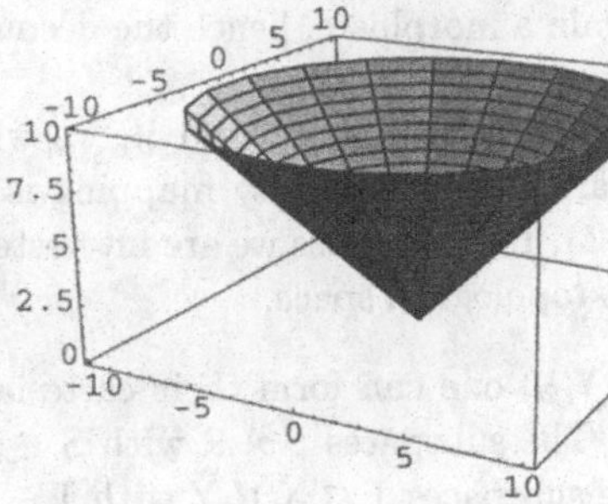

Figure 1.1: Standard Cone

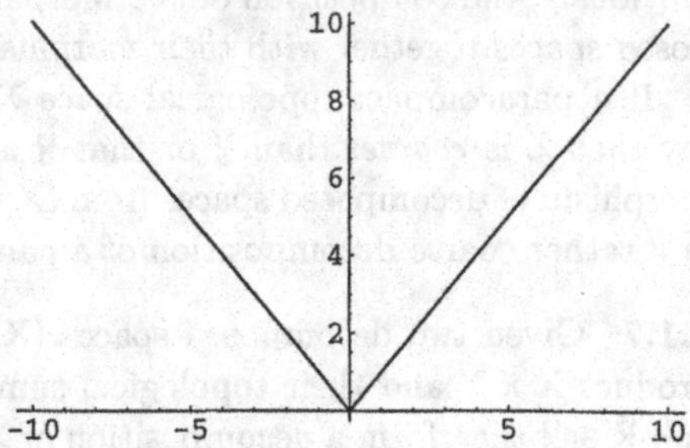

Figure 1.2: Edge

1.1.12 Pathological example The space $X = S_1 \cup S_2 \subset \mathbb{R}^2$ with

$$S_1 = \{0\} \times]-1, 1[\quad \text{and} \quad S_2 = \left\{ (x, y) \in \mathbb{R}^2 \,\middle|\, x > 0, y = \sin\left(\frac{1}{x}\right) \right\}$$

and the induced topology by $\mathbb{R}^2$ is not locally connected, but a a decomposed space with pieces S_1 and S_2. Hereby $S_1 < S_2$, but simultaneously $\dim S_1 = \dim S_2$. Such kind of spaces should not be included in our considerations. Therefore we will later impose further conditions on a decomposed space which will exclude such examples.

Even "more pathological" is the decomposed space $Y = R_1 \cup R_2 \subset \mathbb{R}^3$ with pieces

$$R_1 = \{(0, y, z) \in \mathbb{R}^3 \,|\, y^2 + z^2 < 1\} \quad \text{and}$$
$$R_2 = \left\{ (x, y, z) \in \mathbb{R}^3 \,\middle|\, x > 0, y = \sin\left(\frac{1}{x}\right), z = \sin\left(\frac{\alpha}{x}\right) \right\},$$

where $\alpha > 0$ is transcendental. In this case even $\dim R_1 > \dim R_2$ holds, though $R_1 < R_2$.

1.1.13 Spirals The *fast spiral* (Fig. 1.3)

$$X_{\mathrm{Spirr}} = \{0\} \cup \left\{ (r \sin\theta, r\cos\theta) \,\middle|\, r = e^{-\theta^2},\ \theta > 0 \right\} \subset \mathbb{R}^2$$

and the *slow spiral* (Fig. 1.4)

$$X_{\mathrm{Spir}} = \{0\} \cup \left\{ (r \sin\theta, r\cos\theta) \,\middle|\, r = e^{-\theta},\ \theta > 0 \right\} \subset \mathbb{R}^2$$

are both decomposed spaces with the origin as one piece and the rest as second piece. Note that both the slow as well as the fast spiral turn infinitely often around the origin.

1.1.14 Simplices and polyhedra By an *affine simplex* of dimension m one understands a point set $s \subset \mathbb{R}^n$ with $n \geq m$ of the form

$$s = s[v_0, v_1, \cdots, v_m] := \left\{ \sum_{j=0}^{m} \lambda_j v_j \,\middle|\, \sum_{j=0}^{m} \lambda_j = 1 \text{ and } \lambda_j \geq 0 \text{ for } j = 0, \cdots, m \right\},$$

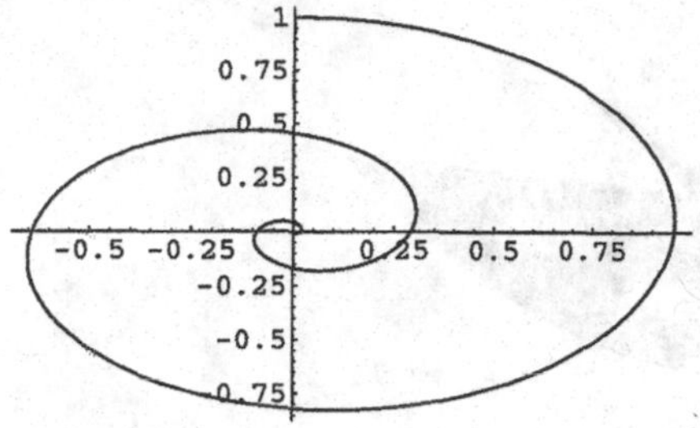

Figure 1.3: Fast Spiral

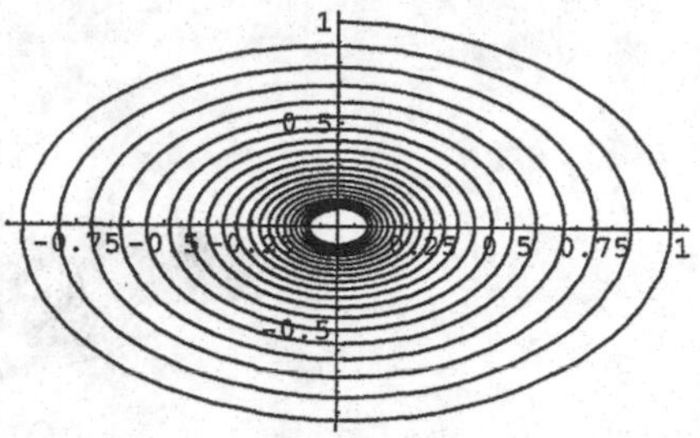

Figure 1.4: Slow Spiral

where $v_0, \cdots, v_m$ are affinely independent points of $\mathbb{R}^n$, and will be denoted as *vertices* of s. With the standard basis $(e_1, \cdots, e_m)$ of $\mathbb{R}^m$ one obtains the so-called m-th *standard simplex* $s_m := s[0, e_1, \cdots, e_m]$.

For every simplex s the $(m-k)$-dimensional manifolds

$$s_{j_0,\cdots j_k} := \left\{ \sum_{j=0}^{m} \lambda_j v_j \in s \,\middle|\, \lambda_{j_0}, \cdots, \lambda_{j_k} > 0 \text{ and } \lambda_j = 0 \text{ for } \lambda_j \neq \lambda_{j_0}, \cdots, \lambda_{j_k} \right\},$$

where k runs through the natural numbers from 0 to m and the $(k+1)$-tupel $(j_0, \cdots, j_k)$ through all sequences of the form $0 \leq j_0 < j_1 < \cdots < j_k \leq m$, comprise a decomposition of s. The sets will be called *open faces* of s, the closed hulls $\overline{s_{j_0,\cdots,j_k}} = s[v_{j_0}, \cdots, v_{j_k}]$ are simplices again and are named *faces* of s.

By gluing together simplices one obtains polyhedra which are decomposed spaces as well. A *finite simplicial complex* K consists of a nonempty set of simplices in $\mathbb{R}^n$, such that the following axioms hold:

(SC1) If the simplex s belongs to K, then every face of s belongs to K.

(SC2) For two simplices of K the intersection is either empty or a common face.

One associates to every finite simplicial complex K its *geometric realization* $|K|$, which is the subspace of $\mathbb{R}^n$ consisting of the union of all simplices of K. The partition of $|K|$ into the open faces of the simplices of K turns $|K|$ into a decomposed space. By a *polyhedron* or a *triangulizable space* one finally understands a topological space which is homeomorphic to such a space $|K|$. Given an explicit *triangulation* , that means a homeomorphism h from $|K|$ onto X, the canonical decomposition of $|K|$ can be carried over to X naturally via h.

1.1.15 Neil's parabola A well-known example for a decomposed space is (the real part of) *Neil's parabola* $X_{\text{Neil}} = \{(x, y) \in \mathbb{R}^2 \mid x^3 = y^2\}$. Its pieces are given by $S_0 = \{0\}$ and the union S_1 of the two legs $S_{11} = \{(x, y) \in X_{\text{Neil}} \mid y > 0\}$ and $S_{12} = \{(x, y) \in X_{\text{Neil}} \mid y < 0\}$.

1.1.16 Whitney umbrella The *Whitney umbrella* X_{WUmb} (Fig. 1.5) is the zero set of the real polynomial $x^2 - y^2 z$, i.e.

$$X_{\text{WUmb}} = \{(x, y, z) \in \mathbb{R}^3 \,|\, x^2 = y^2 z\}.$$

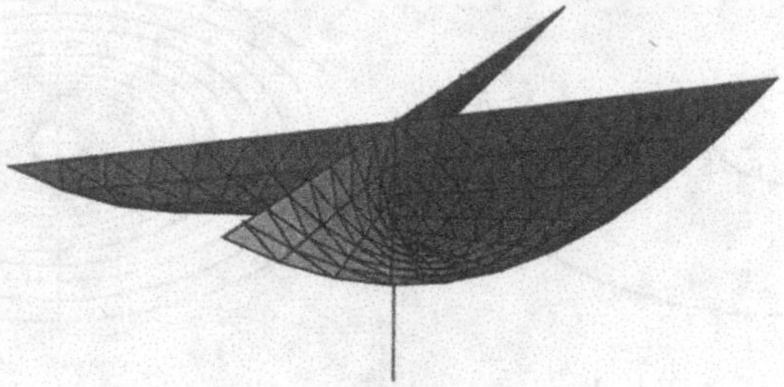

Figure 1.5: Whitney Umbrella

The standard decomposition of X_{WUmb} has the pieces $S_0 = \{0\}$, $S_1 = \{(0,0,z) | z < 0\}$, $S_2 = \{(0,0,z) | z > 0\}$ and $S_3 = \{(x,y,z) \in X_{WUmb} | y \neq 0\}$. Observe that the partition of X_{WUmb} into $T_0 = \{(0,0,z) | z \in \mathbb{R}\}$ and $T_1 = S_3$ is not a decomposition, because it does not satisfy axiom (DS2); instead one has $T_0 \cap \overline{T_1} = S_0 \cup S_2 \neq T_0$. In Figure 1.5 one can see the Whitney umbrella including the "handle", which is given by S_1.

1.1.17 Whitney cusp Another example for a decomposed space given by WHITNEY [192] is the so-called *Whitney cusp* (Fig. 1.6). It is defined as the zero set X_{WCsp} of the real polynomial $y^2 + x^3 - z^2 x^2$. The variety X_{WCsp} has two natural decomposition, namely first the decomposition into the z-axis S_1 and the complement $S_2 = X_{WCsp} \setminus S_1$ and secondly the decomposition by $R_0 = \{0\}$, $R_1 = S_1 \setminus \{0\}$ and $R_2 = S_2$. We will see in the section on the Whitney conditions which one of these decompositions is the "right" one.

1.1.18 Cone comb The decomposed spaces which have been presented up to now are all finite dimensional. An example of an infinite dimensional decomposed space is the *cone comb* X_{Cmb}, which arises from appropriately gluing the cones CS^n to the real half axis $\mathbb{R}^{\geq 0}$. More precisely, one first forms the topological sum of $\mathbb{R}^{\geq 0}$ and all CS^n. From this space one constructs X_{Cmb} by identifying every point $n \in \mathbb{N} \subset \mathbb{R}^{\geq 0}$ with the cusp o_n of the cone CS^n. Intuitively, one thus obtains a comb, the teeth of which are given by cones. The pieces of the natural decomposition of X_{Cmb} are given by $\mathbb{R}^{\geq 0} \setminus \mathbb{N}$, $]0,1[\times S^n$ and $\{n\}$, where n runs through the natural numbers.

1.1.19 Manifolds-with-corners Manifolds-with-corners are decomposed spaces, if they are defined in appropriate way, namely like in MELROSE [126, Sec. 2.1]. The definition by MELROSE entails that one can find a canonical partition of the boundary by embedded hypersurfaces. The usual definition of manifolds-with-corners via charts in model spaces $\mathbb{R}^{n,m} = (\mathbb{R}^{\geq 0})^m \times \mathbb{R}^{n-m}$ does not allow this in general. This problem is discussed in [127]. In the following we will introduce the definition by MELROSE [126, Sec. 2.1] in a slightly more general form and will give the canonical decomposition of manifolds-with-corners.

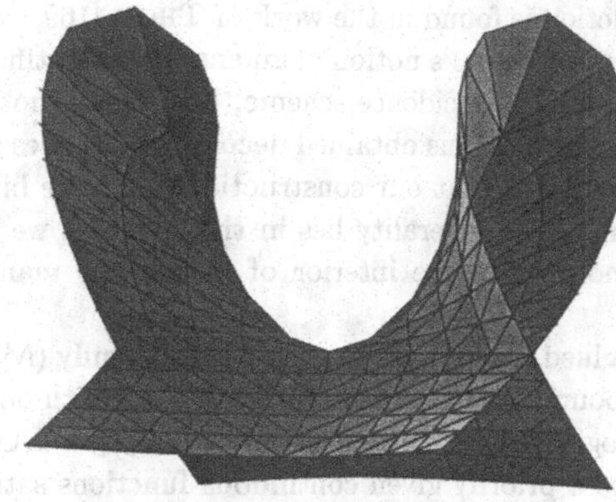

Figure 1.6: Whitney Cusp

Let M be a topological manifold with boundary of dimension m. By a *corner datum* for M we understand a topological embedding $\iota : \mathsf{M} \to \mathbb{R}^n$ together with a finite family $(\eta_j)_{j\in J}$ of smooth functions on $\mathbb{R}^m$ such that the following holds:

(CD1) $\iota(\mathsf{M}) = \{x \in \mathbb{R}^m \mid \eta_j(x) \geq 0 \text{ for all } j \in J\}$.

(CD2) For every $I \subset J$ define

$$\mathsf{M}_I = \{y \in \mathsf{M} \mid \eta_j(\iota(y)) = 0 \text{ for all } j \in I \text{ and } \eta_j(\iota(y)) > 0 \text{ for all } j \notin I\}.$$

Then the family $\left(d_{\iota(x)}\eta_j\right)_{j\in I}$ of cotangent vectors corresponding to the functions η_j, which define M_I, is linearly independent at $\iota(x)$ for all $x \in \mathsf{M}_I$.

Two corner data ι, $(\eta_j)_{j\in J}$ and $\tilde{\iota}$, $(\tilde{\eta}_j)_{j\in\tilde{J}}$ of M are called *equivalent*, if there is an $N \geq \max(n, \tilde{n})$, a diffeomorphism $\mathsf{H} : \mathsf{O} \to \tilde{\mathsf{O}}$ between open subsets of $\mathbb{R}^N$ with $\iota(\mathsf{M}) \subset \mathsf{O}$, and a bijective mapping $\alpha : J \to \tilde{J}$ such that

(CD3) $\tilde{\iota} = \mathsf{H} \circ \iota$.

(CD4) $\mathsf{M}_{\alpha(I)} = \mathsf{M}_I$ for all $I \subset J$.

Hereby we have embedded $\mathbb{R}^n$ resp. $\mathbb{R}^{\tilde{n}}$ into $\mathbb{R}^N$ via the first coordinates, and extended ι and $\tilde{\iota}$ correspondingly.

By a *manifold-with-corner* we now understand a topological manifold M with boundary together with an equivalence class of corner data. The family $\mathcal{Z} = (\mathsf{M}_I)_{I\subset J}$ then is independent of the special choice of the corner datum in the equivalence class and comprises a decomposition of M. By (CD2) every $\iota(\mathsf{M}_I)$ is a submanifold of $\mathbb{R}^n$, hence its manifold structure can be carried over to M_I via the embedding ι. By (CD3) this manifold structure is independent from the particular corner datum. Moreover, $\mathsf{M}_{I'} \cap \overline{\mathsf{M}_I} \neq \emptyset$ implies $I' \subset I$, hence $\mathsf{M}_{I'} \subset \overline{\mathsf{M}_I}$. So the condition of frontier is satisfied as well, and M is a decomposed space indeed. Examples of manifolds-with-corners are given by the simplices defined above.

1.1.20 Glued spaces A method for the construction of decomposed spaces is given by iteratively gluing together manifolds-with-boundary along their boundary. The corresponding construction is found in the work of THOM [169, Sec. C] and comprises one essential component of THOM's notion of an ensemble stratifié (the other essential component is the notion of an incidence scheme, but we will not discuss this further at this point). We will give the thus obtained decomposed spaces a new name, namely glued spaces. Let us mention that our construction is a little bit more general than the one of THOM. The larger generality lies in the fact that we allow as pieces even manifolds not diffeomorphic to the interior of a compact manifold-with-boundary (cf. also [75]).

As ingredients for glued spaces we first need a finite family $(M_i)_{1\leq i\leq k}$ of manifolds, where M_1 is without boundary, but all M_i with $i > 1$ are with boundary. Recursively one then defines topological spaces E_i by $E_1 = M_1$ and $E_{i+1} = E_i \cup_{h_{i+1}} M_{i+1}$, where the $h_{i+1} : \partial M_{i+1} \to E_i$ are a priorily given continuous functions satisfying the following gluing condition

(GC) $M_i^\circ \cap h_j(\partial M_i) = M_i^\circ$ for all $j > i$ with $M_i^\circ \cap h_j(\partial M_i) \neq \emptyset$.

Thus we obtain the *glued space* $X = E_k$. It is a decomposed space and possesses the pieces $S_i = M_i^\circ$. The functions h_i belonging to X are called its *gluing functions.*

PROOF: Axiom (DS11) is satisfied trivially by definition of X; hence it remains to show axiom (DS2). According to definition the sets $\bigcup_{j>i} M_j^\circ$ are open in X, that means $\overline{M_i} \subset \bigcup_{j\leq i} M_j^\circ$. Let $\overline{M_i} \cap M_j^\circ \neq 0$ for $j < i$, and $x \in M_j^\circ$ arbitrary. Moreover, let $U \subset X$ be an open neighborhood of x. By the gluing condition (GL) $h_i^{-1}(U \cap M_j^\circ)$ has to be nonempty, hence by construction of X there is an open $V \subset M_i$ with $h_i^{-1}(U \cap M_j^\circ) = V \cap \partial M_i$ and $\emptyset \neq V \cap M_i^\circ \subset U$. But that means $U \cap M_i^\circ \neq \emptyset$, hence $x \in \overline{M_i}$ and $M_j^\circ \subset \overline{M_i}$. This proves (DS2). □

1.1.21 Proposition *If* X *is a glued space, so is the cone* CX *over* X.

PROOF: Let $M_1, \cdots, M_k$ be manifolds-with-boundary which after glueing them together give the space X. Using Proposition C.4.2 of Appendix C one can furnish the products $[0, 1[\times M_1, \cdots, [0, 1[\times M_k$ with differentiable structures. The gluing functions h_{i+1} induce gluing functions $Ch_{i+1} : \partial([0, 1[\times M_{i+1}) \to CE_i$ by

$$Ch_{i+1}(t, m) = \begin{cases} [\{0\} \times E_i] & \text{if } t = 0, \\ (t, h_{i+1}(m)) & \text{if } t > 0. \end{cases}$$

Now, the reader will convince himself easily that $CE_1 = CM_1$ and $CE_{i+1} = CE_i \cup_{Ch_{i+1}} CM_{i+1}$ for $i = 1, \cdots, k-1$. Together with $CX = CE_k$ this implies that CX is a glued space. □

1.2 Stratifications

Let us consider a decomposed space $(X, \mathcal{Z})$. Then within the class of all decompositions of X one can find decompositions which differ from $\mathcal{Z}$ only slightly, as locally around every point they look like $\mathcal{Z}$, and decompositions which differ from $\mathcal{Z}$ in an essential way. As an example for that take $\mathbb{R}$ with the decomposition $\mathcal{Z}$ into $\{0\}$, $\mathbb{R}^{>0}$ and $\mathbb{R}^{<0}$. Other decompositions of $\mathbb{R}$ are given for example by $\mathcal{Z}_1 = \{\{0\}, \mathbb{R} \setminus \{0\}\}$ and $\mathcal{Y} = \{\mathbb{R}\}$. Intuitively it is clear that $\mathcal{Z}_1$ looks similar to $\mathcal{Z}$, where $\mathcal{Y}$ is really different from $\mathcal{Z}$.

Let us formulate this phenomenon in more precise mathematical terms and introduce the notion of a stratification according to MATHER [123]. Stratifications in MATHER's sense generate equivalence classes of decompositions of X. It will turn out that within every such equivalence classes there exists a coarsest decomposition; its pieces are the so-called strata of X.

1.2.1 Before we introduce in this section stratified spaces let us briefly recall the notion of a set germ.

Denote by X a topological space, and let $x \in X$. Two subsets A and B of X are called *equivalent at* X, if there is an open neighborhood $U \subset X$ of x, such that $A \cap U = B \cap U$. This relation comprises an equivalence relation on the power set of X. The class of all sets equivalent to $A \subset X$ at x will be denoted by $[A]_x$ and is called the *set germ* at x. If $A \subset B \subset X$ we sometimes say that $[A]_x$ is a subgerm of $[B]_x$, in signs $[A]_x \subset [B]_x$. One checks easily that two sets $A, B \subset X$ are equivalent at x, if and only if the function germs $[\chi_A]_x$ and $[\chi_B]_x$ of the characteristic functions of A and B at x are equivalent.

1.2.2 Definition By a *stratification* of a topological space X we understand a mapping $\mathcal{S}$ which associates to $x \in X$ the set germ $\mathcal{S}_x$ of a closed subset of X such that the following axiom is satisfied:

(ST1) For every $x \in X$ there is a neighborhood U of x and a decomposition $\mathcal{Z}$ of U such that for all $y \in U$ the germ $\mathcal{S}_y$ coincides with the set germ of the piece of $\mathcal{Z}$ of which y is an element.

The pair $(X, \mathcal{S})$ is called a *stratified space.*

Every decomposition $\mathcal{Z}$ of X defines a stratification by associating to $x \in X$ the germ $\mathcal{S}_x$ of the piece of which x is an element. In this case we say that $\mathcal{S}$ is *induced* by $\mathcal{Z}$. By definition a stratification is always induced at least locally by a decomposition $\mathcal{Z}$.

A continuous map $f: X \to Y$ between stratified spaces $(X, \mathcal{S})$ and $(Y, \mathcal{R})$ is called a *morphism of stratified spaces* or shortly a *stratified mapping,* if for every $x \in X$ there exist neighborhoods V of $f(x)$ and $U \subset f^{-1}(V)$ of x together with decompositions $\mathcal{Z}$ of U and $\mathcal{Y}$ of V inducing $\mathcal{S}_{|U}$ resp. $\mathcal{R}_{|V}$ in the sense of (ST1) such that the following holds:

(ST2) For every $y \in U$ there is an open neighborhood $O \subset U$ such that the map $f_{|S \cap O}$ restricted to the open subset $S \cap O$ of the piece $S \in \mathcal{Z}$ containing y has image in the piece $R \in \mathcal{Y}$ containing $f(y)$ and such that $f_{|S \cap O}$ is a smooth map from $S \cap O$ to R.

In particular $f(\mathcal{S}_x)$ then is a subgerm of $\mathcal{R}_{f(x)}$.

The stratified spaces and their morphisms form a category $\mathfrak{Esp}_{\mathrm{strat}}$.

1.2.3 Remark Similarly like for decomposed spaces one can form the notion of a $\mathfrak{T}$-stratification, where $\mathfrak{T}$ is a category of topological spaces. More precisely, by a *$\mathfrak{T}$-stratification* we understand a mapping $x \mapsto \mathcal{S}_x$ such that $\mathcal{S}_x$ is locally induced by a $\mathfrak{T}$-decomposition in the sense of (ST1). Now it is clear what to understand by a Σ-stratification.

1.2.4 The category $\mathfrak{Esp}_{\mathrm{strat}}$ has finite products and sums. Moreover, one can define subobjects in $\mathfrak{Esp}_{\mathrm{strat}}$. By a *stratified subspace* of $(X, \mathcal{S})$ we mean a topological subspace $Y \subset X$ such that for every $x \in Y$ there is an open neighborhood U in X and a decomposition $\mathcal{Z}$ inducing $\mathcal{S}_{|U}$ such that $(Y \cap U, \mathcal{Z} \cap Y)$ is a decomposed subspace of $(U, \mathcal{Z})$. In this case the pair$(Y, \mathcal{S} \cap Y)$ is again a stratified space.

We want to regard two decompositions $\mathcal{Z}_1$ and $\mathcal{Z}_2$ of X as essentially the same, if the stratifications induced by them are the same. In such a case we call $\mathcal{Z}_1$ and $\mathcal{Z}_2$ *equivalent.*

1.2.5 Lemma (MATHER [123, Lem. 2.1]) *If there are two equivalent decompositions $\mathcal{Z}_1$ and $\mathcal{Z}_2$ on X then for all $x \in X$*

$$\mathrm{dp}_{\mathcal{Z}_1}(x) = \mathrm{dp}_{\mathcal{Z}_2}(x). \tag{1.2.1}$$

PROOF: We show by induction on $\mathrm{dp}_{\mathcal{Z}_1}(x)$ that

$$\mathrm{dp}_{\mathcal{Z}_1}(x) \leq \mathrm{dp}_{\mathcal{Z}_2}(x).$$

After interchanging $\mathcal{Z}_1$ and $\mathcal{Z}_2$ the claim then follows. If $\mathrm{dp}_{\mathcal{Z}_1}(x) = 0$ nothing has to be shown. So let us assume that $\mathrm{dp}_{\mathcal{Z}_1}(x) = k + 1$ and that the claim holds for $\mathrm{dp}_{\mathcal{Z}_1}(y) \leq k$. Let $x \in S = S_0 < S_1 < \cdots < S_{k+1}$ be a maximal sequence of pieces from $\mathcal{Z}_1$ and R the piece of $\mathcal{Z}_2$ with $x \in R$. Then there exists an open neighborhood U of x with $S \cap U = R \cap U$ such that U meets only finitely many pieces of $\mathcal{Z}_2$. By $x \in \overline{S_1}$ there then exists a piece R_1 of $\mathcal{Z}_2$ such that $x \in \overline{S_1 \cap R_1 \cap U}$. Then $R < R_1$ follows. After the choice of an element $y \in S_1 \cap R_1$ the induction hypothesis entails

$$\mathrm{dp}_{\mathcal{Z}_1}(x) = \mathrm{dp}_{\mathcal{Z}_1}(y) + 1 \leq \mathrm{dp}_{\mathcal{Z}_2}(y) + 1 \leq \mathrm{dp}_{\mathcal{Z}_2}(x).$$

This proves the claim. □

By the lemma the depth $\mathrm{dp}_{\mathcal{Z}}(x)$ of a point x of a stratified space $(X, \mathcal{S})$ is independent of the specially chosen decomposition $\mathcal{Z}$ which defines $\mathcal{S}$ in a neighborhood of x. Therefore it makes sense, to define the *depth* of x with respect to the stratification $\mathcal{S}$ by $\mathrm{dp}_{\mathcal{S}}(x) = \mathrm{dp}_{\mathcal{Z}}(x)$.

1.2.6 Lemma (*cf.* MATHER [123, Lem. 2.2]) *Let T_k be the set of all points of X of depth k. Then the family $(T_k)_{k \in \mathbb{N}}$ is a Σ-decomposition of X inducing $\mathcal{S}$.*

PROOF: By (ST1) and the separability of X the set T_k has to be the countable topological sum of smooth manifolds, hence is a Σ-manifold. The local finiteness of the family $(T_k)_{k \in \mathbb{N}}$ follows immediately from the fact that every point possesses a neighborhood decomposed in finitely many pieces according to (ST1). As the points of a piece of a local decomposition have equal depth, the stratification $\mathcal{S}$ is induced by $(T_k)_{k\in\mathbb{N}}$ that means the set germ $[T_k]_x$ coincides for every $x \in X$ with $\mathcal{S}_x$. So it remains to show that the condition of frontier is satisfied. Let $x \in \overline{T_l} \setminus T_l$ be a point of depth k. Obviously then $k > l$. Let y be a further point of depth k. Then there exists an open neighborhood U of y, a decomposition $\mathcal{Z}$ of U defining $\mathcal{S}$ and a not extendable chain $y \in S_0 < S_1 < \cdots < S_k$ of pieces of $\mathcal{Z}$. As $y \in \overline{S_{k-l}}$ and $S_{k-l} \subset T_l$, the relation $y \in \overline{T_l}$ follows, hence $T_k \subset \overline{T_l}$. This finishes the proof. □

1.2.7 Proposition *Any stratified space* $(X, \mathcal{S})$ *has a decomposition* $\mathcal{Z}_{\mathcal{S}}$ *with the following maximal property: for every open subset* $U \subset X$ *and every decomposition* $\mathcal{Z}$ *inducing* $\mathcal{S}$ *over* U *the restriction of* $\mathcal{Z}_{\mathcal{S}}$ *is coarser than* $\mathcal{Z}$.

The unique decomposition $\mathcal{Z}_{\mathcal{S}}$ is called the *canonical decomposition* of the stratified space $(X, \mathcal{S})$ and will often be denoted by $\mathcal{S}$. The pieces of $\mathcal{S}$ are called the *strata* of X.

PROOF: We construct the decomposition $\mathcal{Z}_{\mathcal{S}}$ inductively. Let $S_{0,d}$ with $d \in \mathbb{N}$ be the union of all d-dimensional connected components of T_0 and $\mathcal{P}^0$. Then $\mathcal{P}^0$ is a decomposition of T_0 inducing $\mathcal{S}$ over T_0.

Now, let us suppose that we have for all T_l with $0 \le l \le k$ a partition P^l of T_l in open submanifolds $S \subset T_l$ such that the following holds:

(1) The family $\mathcal{Z}^k := (S)_{S\in\mathcal{P}^l, l\le k}$ is a decomposition of $\bigcup_{l\le k} T_l$ inducing $\mathcal{S}_{|\bigcup_{l\le k} T_l}$.

(2) For every open subset $U \subset X$ and every decomposition $\mathcal{Z}$ of U every piece $S \subset U$ of $\mathcal{Z}^k_{|U}$ of depth $l \le k$ is the union of pieces R of $\mathcal{Z}$.

Consider a point $x \in T_{k+1}$ and let S_x be the connected component of x in T_{k+1}. Then S_x is a smooth manifold of dimension d_x. Let S be a piece of $\mathcal{Z}^k$. We claim that either $S_x \subset \overline{S}$ or $S_x \cap \overline{S} = \emptyset$. This claim will follow from the fact that the set S'_x of all $y \in S_x$ with $y \in \overline{S}$ is open and closed in S_x. That S' is closed follows from the obvious fact that S_x is open in S_x. To prove that S' is open suppose that $y \in \overline{S}$ and choose an open neighborhood U of y in X together with a decomposition $\mathcal{Z}$ of U inducing $\mathcal{S}$. Let R be the piece of $\mathcal{Z}$, in which y lies. After shrinking U we can suppose that $R = U \cap S_x$. By induction hypothesis there is a countable family of pieces R_j, $j \in \mathbb{N}$ of $\mathcal{Z}$ with $\bigcup_{j\in\mathbb{N}} R_j = S \cap U$. As $y \in \overline{S}$ and $\mathcal{Z}$ satisfies the condition of frontier one has $R \subset \overline{\bigcup_{j\in\mathbb{N}} R_j}$, hence $S_x \cap U = R \subset \overline{S}$. This shows S' to be open in S_x.

For every subset $A \subset \mathcal{P}^k$ let us now define $S_{k+1,d,A}$ as the union of all d-dimensional S_x with $x \in T_{k+1}$ such that $S_x \subset \overline{S}$ for all $S \in A$ but $S_x \cap \overline{S} = \emptyset$ for all $S \in \mathcal{P}^k \setminus A$. Every nonempty $S_{k+1,d,A}$ is a d-dimensional manifold, and the family of $\mathcal{P}^{k+1}$ of all nonempty $S_{k+1,d,A}$ with $d \in \mathbb{N}$ and $A \subset \mathcal{P}^k$ is a partition of T_{k+1}. By definition of the $S_{k+1,d,A}$ one checks immediately that the corresponding partition $\mathcal{Z}^{k+1}$ satisfies condition (1). Condition (2) holds as well. To show this let U and $\mathcal{Z}$ be like in (2) and let $R \subset U$ be a piece of $\mathcal{Z}$ of depth $k+1$. By induction hypothesis every piece S of $\mathcal{Z}^k$ satisfies either $R \subset \overline{S}$ or $R \cap \overline{S} = \emptyset$, hence R lies in some piece $S_{k+1,\dim R,A}$. Therefore $S_{k+1,d,A} \cap U$ is the union of all d-dimensional pieces of depth $k+1$ of $\mathcal{Z}$ which intersect $S_{k+1,d,A} \cap U$. This completes the induction.

So finally, the family $(S)_{S \in \mathcal{P}^k, k \in \mathbb{N}}$ comprises a decomposition of X and satisfies the claim. □

1.2.8 Corollary *For every stratum S of X there are only finitely many strata R with $R > S$.*

PROOF: Assume that $(R_j)_{j \in \mathbb{N}}$ is a sequence of pairwise disjoint strata $R_j > S$. Choose $x \in R$. As $S \subset \overline{R_j}$ for all j, any neighborhood of x meets each of the R_j. This contradicts the local finiteness of the decomposition $\mathcal{Z}_S$. □

1.2.9 Now, as every stratified space $(X, \mathcal{S})$ has a canonical decomposition, its *k-skeleton* X^k can be defined as the union of all strata of X of dimension $\leq k$ and inherits the structure of a stratified space.

In this context we will denote the Σ-manifold T_0 of all points of depth 0 by X°. If X° is even a manifold, X° has to be a stratum, the so-called *top stratum* . In this case $X = \overline{X^\circ}$ follows. Note that not every connected stratified space needs to have a top stratum.

For a stratum $S \subset X$ one calls the union of all strata $R \geq S$ the *star* (or *étoile*) of S in X, in signs $\mathrm{Et}_X(S)$. Obviously, the star $\mathrm{Et}_X(S)$ is again a stratified space.

1.2.10 Given a morphism $f : X \to Y$ between stratified spaces $(X, \mathcal{S})$ and $(Y, \mathcal{R})$ one checks easily that for every connected component S_0 of a stratum S of X there exists a stratum R_{S_0} of Y with $f(S_0) \subset R_{S_0}$ and $f_{|S_0} \in \mathcal{C}^\infty(S_0)$. If all restrictions $f_{|S_0}$ are immersions (resp. submersions), we call f a *stratified immersion* (resp. *stratified submersion*).

1.2.11 Example At the end of this section let me give an example of a stratified subspace such that the canonical injection does not map strata into strata. Obviously such an example does not stand in contradiction to the fact that connected components of strata are mapped into strata. Consider the stratified space $X \subset \mathbb{R}^5$ given by the union of the cube $W = [0,1]^3 \times \{(0,0)\}$ and the square $Q = \{(0,0,0)\} \times [0,1]^2$. The set of all closed edges $K = [0,1] \times \{(0,0,0,0)\} \cup \cdots \cup \{(0,0,0,0)\} \times [0,1]$ lying on the coordinate axes is a stratified subspace of X and consists of a 0-dimensional stratum and a 1-dimensional stratum. Both are not subsets of a stratum of X, as the edges of the cube have depth 2, in contrast to the ones of the square which have depth 1. Analogously the vertices of the cube have depth 3, where the vertices of Q which do not lie on W have depth 2.

1.3 Smooth Structures

1.3.1 Let X be a stratified space, and $\mathcal{S}$ the family of its strata. A *singular chart of class* $\mathcal{C}^m$, $m \in \mathbb{N}^{>0} \cup \{\infty\}$ is a homeomorphism $x : U \to x(U) \subset \mathbb{R}^n$ from an open set $U \subset X$ to a locally closed subspace of $\mathbb{R}^n$ such that for every stratum $S \in \mathcal{S}$ the image $x(U \cap S)$ is a submanifold of $\mathbb{R}^n$ and the restriction $x_{|U \cap S} : U \cap S \to x(U \cap S)$ is a diffeomorphism of class $\mathcal{C}^m$. Sometimes we call the domain U of a singular chart shortly *chart domain.* Moreover, we often use for singular charts a notion of the form $\mathsf{x} : \mathsf{U} \to \mathsf{O} \subset \mathbb{R}^n$ to express that $\mathsf{O} \subset \mathbb{R}^n$ is open and $\mathsf{x}(\mathsf{U}) \subset \mathsf{O}$ is locally closed. Two singular charts $\mathsf{x} : \mathsf{U} \to \mathbb{R}^n$ and $\tilde{\mathsf{x}} : \tilde{\mathsf{U}} \to \mathbb{R}^{\tilde{n}}$ are called *compatible* , if for every $x \in \mathsf{U} \cap \tilde{\mathsf{U}}$ there exists an open neighborhood $\mathsf{U}_x \subset \mathsf{U} \cap \tilde{\mathsf{U}}$, an integer $N \geq \max(n, \tilde{n})$, open neighborhoods $\mathsf{O} \subset \mathbb{R}^N$ and $\tilde{\mathsf{O}} \subset \mathbb{R}^N$ of $\mathsf{x}(\mathsf{U}_x) \times \{0\}$ resp. $\tilde{\mathsf{x}}(\mathsf{U}_x) \times \{0\}$, and a diffeomorphism $\mathsf{H} : \mathsf{O} \to \tilde{\mathsf{O}}$ of class $\mathcal{C}^m$ such that $\iota^N_{\tilde{n}} \circ \tilde{\mathsf{x}}_{|\mathsf{U}_x} = \mathsf{H} \circ \iota^N_n \circ \mathsf{x}_{|\mathsf{U}_x}$. Hereby we have denoted by ι^N_m for $N \geq m$ the canonical embedding of $\mathbb{R}^m$ in $\mathbb{R}^N$ via the first m coordinates. We call the diffeomorphism H a *transition map* from x to $\tilde{\mathsf{x}}$ over the domain U_x. To keep notation reasonable we will identify singular charts $\mathsf{x} : \mathsf{U} \to \mathbb{R}^n$ in the following with their *extensions* $\iota^N_n \circ \mathsf{x} : \mathsf{U} \to \mathbb{R}^N$, $N \geq n$.

Like in differential geometry one defines the notion of a *singular atlas* on X of *class* $\mathcal{C}^m$ as a family $(\mathsf{x}_j)_{j \in J}$ of pairwise compatible singular charts $\mathsf{x}_j : \mathsf{U}_j \to \mathbb{R}^{n_j}$ of class $\mathcal{C}^m$ on X such that $\bigcup_{j \in J} \mathsf{U}_j = X$. Often we will denote such a singular atlas by $\mathfrak{U} = (\mathsf{U}_j, \mathsf{x}_j)_{j \in J}$; this will emphasize the domains U_j and will express that the U_j provide a covering of X. Sometimes we will say that $\mathfrak{U}$ is a *covering by charts.* Two atlases $\mathfrak{U}$ and $\tilde{\mathfrak{U}}$ of X are called *compatible,* if every singular chart of $\mathfrak{U}$ is compatible with every singular chart of $\tilde{\mathfrak{U}}$.

1.3.2 Lemma *The compatibility of singular atlases is an equivalence relation.*

PROOF: Obviously the compatibility of atlases is reflexive and symmetric. It remains to prove transitivity. Let $\mathfrak{U}$, $\tilde{\mathfrak{U}}$ and $\check{\mathfrak{U}}$ be three singular atlases such that $\mathfrak{U}$ and $\tilde{\mathfrak{U}}$ are compatible as well as $\tilde{\mathfrak{U}}$ and $\check{\mathfrak{U}}$. We have to show that every chart $\mathsf{x} : \mathsf{U} \to \mathbb{R}^n$ out of $\mathfrak{U}$ is compatible with every $\check{\mathsf{x}} : \check{\mathsf{U}} \to \mathbb{R}^{\check{n}}$ out of $\check{\mathfrak{U}}$. For $x \in \mathsf{U} \cap \check{\mathsf{U}}$ choose a sufficiently small open neighborhood $\mathsf{U}_x \subset \mathsf{U} \cap \check{\mathsf{U}}$ and a chart $\tilde{\mathsf{x}} : \mathsf{U}_x \to \mathbb{R}^{\tilde{n}}$ out of $\tilde{\mathfrak{U}}$. After shrinking U_x and enlarging n, $\tilde{n}$ and $\check{n}$ we can suppose $n = \tilde{n} = \check{n}$ and can find over U_x transition maps $\mathsf{H} : \mathsf{O} \to \tilde{\mathsf{O}} \subset \mathbb{R}^n$ from x to $\tilde{\mathsf{x}}$ and $\tilde{\mathsf{H}} : \tilde{\mathsf{O}} \to \check{\mathsf{O}} \subset \mathbb{R}^n$ from $\tilde{\mathsf{x}}$ to $\check{\mathsf{x}}$. But then $\check{\mathsf{H}} = \tilde{\mathsf{H}} \circ \mathsf{H} : \mathsf{O} \to \check{\mathsf{O}}$ is a transition map from x to $\check{\mathsf{x}}$ over the domain U_x, hence x and $\check{\mathsf{x}}$ are compatible. □

Like for differentiable manifolds the set theoretic inclusion induces an order relation between compatible singular atlases. Now, combining all charts of all atlases in a fixed equivalence class one obtains a *maximal atlas* containing all other atlases of the equivalence class as subsets. In particular the maximal atlas determines the equivalence class uniquely.

1.3.3 Definition A maximal atlas of singular charts of class $\mathcal{C}^m$ on a stratified space X is called a $\mathcal{C}^m$*-structure* on X, and for the case that $m = \infty$ a *smooth structure* on X.

1.3.4 Remark In the mathematical literature one can already find various approaches to define "differentiable" or "smooth" functional structures on singular, though not necessarily stratified spaces. Probably SIKORSKI [159, 160] was the first who worked in this direction and introduced the notion of a *differential space.* Mainly for the purpose to study singular complex spaces from a differential viewpoint SPALLEK developed in [163] his concept of *differenzierbare Räume.* Finally there are the *subcartesian spaces* which go back to the work of ARONSZAJN [3] and which have been used by him to consider analytical questions in a singular setting. For further information on subcartesian spaces see [4] or MARSHALL's paper [120]. The common aspect of these approaches and the one introduced here is that they all embed a singular space locally in some Euclidean space. The differences become apparent in the additional conditions imposed on these embeddings or on the transition maps.

In the context of orbit spaces (see for example SCHWARZ [156], BIERSTONE [14], SJAMAAR–LERMAN [162] and HUEBSCHMANN [93]) one can find a notion of a smooth structure for stratified spaces as well, but it is a weaker one than the notion introduced in this work. By a smooth structure the just named authors understand an algebra of smooth functions on a stratified space such that the restrictions to the strata are smooth. But nevertheless, the algebras of smooth functions constructed in [156, 14, 162, 93] always give rise to singular atlases in the sense as defined above. For a proof of this fact we refer the reader to Section 4.4.

1.3.5 Remark In the definition of a singular chart $x : U \rightarrow \mathbb{R}^n$ we have required that $x(U)$ is a locally closed subset of $\mathbb{R}^n$. This property is indispensable when we later want to apply the rich theory of Whitney functions to the study of stratified spaces. But for many applications, in particular for the definition of smooth functions and notions connected with that the local closedness is not absolutely necessary. To be able to allow a greater generality when needed we therefore will speak of a *weak singular chart* and correspondingly of a *weak smooth structure*, if all axioms besides the one of local closedness in $\mathbb{R}^n$ are satisfied.

1.3.6 With the help of a smooth structure on X represented by the maximal atlas $\mathcal{U}$, we can now construct the structure sheaf $\mathcal{C}^\infty_X$, the so-called sheaf of *smooth functions*. Let $U \subset X$ be open. Then one defines $\mathcal{C}^\infty_X(U)$ as the set of all continuous functions $g : U \rightarrow \mathbb{R}$ such that for all $x \in U$ and all singular charts $x : \tilde{U} \rightarrow \mathbb{R}^n$ from $\mathcal{U}$ with $x \in \tilde{U}$ there exists an open set $U_x \subset U \cap \tilde{U}$ and a smooth function $g : \mathbb{R}^n \rightarrow \mathbb{R}$ with $g_{|U_x} = g \circ x_{|U_x}$. One now checks easily that the $\mathcal{C}^\infty_X(U)$ are the sectional spaces of a sheaf $\mathcal{C}^\infty_X$. In case no confusion is possible we will often denote $\mathcal{C}^\infty_X$ by $\mathcal{C}^\infty$. Moreover, it follows immediately by definition that for every singular chart $x : U \rightarrow O \subset \mathbb{R}^n$ the algebra $\mathcal{C}^\infty(U)$ is canonically isomorphic to $\mathcal{C}^\infty(O)/\mathcal{J}$, where $\mathcal{J}$ is the ideal $\subset \mathcal{C}^\infty(O)$ of all smooth functions vanishing on $x(U)$.

On a stratified space X with a $\mathcal{C}^m$-structure one can define analogously for every $k \in \mathbb{N}$, $k \leq m$ the sheaf $\mathcal{C}^k_X$ of *k-times differentiable functions* on X by pullback of the corresponding sheaves on the $\mathbb{R}^n$. Obviously we then have $\mathcal{C}^0_X = \mathcal{C}_X$. In most cases we will restrict ourselves to consider only smooth structures.

Now let us come back to stratified spaces with a smooth structure. Every stalk $\mathcal{C}^\infty_x$ of the structure sheaf with footpoint $x \in X$ has a unique maximal ideal $\mathfrak{m}_x$, namely the ideal of functions vanishing at x. In other words $\mathfrak{m}_x$ is the set of all germs $[g]_x \in \mathcal{C}^\infty_x$ with $g(x) = 0$. Thus the pair $(X, \mathcal{C}^\infty)$ becomes a locally ringed space which we will also call – though formally not quite correct – a stratified space with a smooth structure.

1.3.7 Let $(X, \mathcal{C}^\infty_X)$ and $(Y, \mathcal{C}^\infty_Y)$ be two stratified spaces with smooth structure. A continuous map $f : X \rightarrow Y$ is called *smooth*, if $f_* \mathcal{C}^\infty_X \subset \mathcal{C}^\infty_Y$ or in other words if for all $g \in \mathcal{C}^\infty_Y(V)$ with $V \subset Y$ open the relation $g \circ f \in \mathcal{C}^\infty_X\,(f^{-1}(V))$ holds. Analogously, one calls the map f of *class* $\mathcal{C}^m$, $m \in \mathbb{N}$, if $f_*\mathcal{C}^m_X \subset \mathcal{C}^m_Y$. Note that a smooth map between stratified spaces need not be a stratified map, and that a stratified map between such spaces need not be smooth.

By definition the composition of smooth maps is again smooth. Therefore the stratified spaces with smooth structures with the smooth maps as morphisms form a category $\mathfrak{Esp}_{\mathfrak{lis}}$.

1.3.8 Proposition *A map $f : X \to Y$ between stratified spaces with smooth structure $(X, \mathcal{C}^\infty_X)$ and $(Y, \mathcal{C}^\infty_Y)$ is smooth if and only if for every $x \in X$ and singular charts $x : U \to O \subset \mathbb{R}^n$ around x and $y : \tilde{U} \to \tilde{O} \subset \mathbb{R}^N$ around $f(x)$ there exists an open neighborhood $U_x \subset U$ with $f(U_x) \subset \tilde{U}$ and a smooth mapping $\mathbf{f} : \mathbb{R}^n \to \mathbb{R}^N$ such that*

$$y \circ f_{|U_x} = \mathbf{f} \circ x_{|U_x}. \tag{1.3.1}$$

PROOF: As the problem is a local one, we can suppose without loss of generality that $X \subset O$ and $Y \subset \tilde{O}$ are locally closed subsets. By $y^1, \cdots, y^N$ we denote the coordinate functions of $\mathbb{R}^N$. If now $f : X \to Y$ is smooth, we can find functions $\mathbf{f}_i \in \mathcal{C}^\infty(O)$, $i = 1, \cdots, N$ such that $\mathbf{f}_{i|X} = y^i \circ f$. Now, choose a smooth function $\varphi : \mathbb{R}^n \to [0, 1]$ with $\operatorname{supp} \varphi \subset O$ and $\varphi_{|O_x} = 1$ on an open neighborhood $O_x \subset O$ of x. Then $\mathbf{f} : \mathbb{R}^n \to \mathbb{R}^N$ with

$$\mathbf{f}(x) = \begin{cases} \varphi(x)\,(\mathbf{f}_1(x), \cdots, \mathbf{f}_N(x)) & \text{for } x \in O, \\ 0 & \text{else,} \end{cases}$$

is a smooth map with the desired properties. As the inverse implication is obvious, the claim follows. □

The proposition just proven allows us to transfer the notions of immersions and submersions to smooth maps $f : X \to M$ between a stratified space and a manifold M. More precisely, we call f a *smooth immersion*, if there is a covering $(U_j)_{j \in J}$ of X by domains of singular charts $U_j \xrightarrow{x_j} O_j \subset \mathbb{R}^{n_j}$ and a family $(\mathbf{f}_j)_{j \in J}$ of immersions $\mathbf{f}_j : O_j \to M$ such that for all j

$$f_{|U_j} = \mathbf{f}_j \circ x_j.$$

In the case that all $\mathbf{f}_j$ can be chosen submersive and all restrictions $f_{|S} \to M$, $S \in \mathcal{S}$ are submersive, we call f a *smooth stratified submersion* from X to M.

1.3.9 Next we associate to every point $x \in X$ its *rank* $\operatorname{rk} x = \dim(\mathfrak{m}_x/\mathfrak{m}_x^2)$. The rank is finite indeed, because every chart $x : U \to \mathbb{R}^n$ of X around x induces a surjective homomorphism $x^* : \mathfrak{m}_{x(x)} \to \mathfrak{m}_x$ between the maximal ideals of the stalks $\mathcal{C}^\infty_{\mathbb{R}^n, x(x)}$ and $\mathcal{C}^\infty_x$, which implies $\dim(\mathfrak{m}_x/\mathfrak{m}_x^2) \leq \dim\left(\mathfrak{m}_{x(x)}/\mathfrak{m}_{x(x)}^2\right) = n$. The rank has the following interpretation.

1.3.10 Proposition *For every point x of a stratified space $(X, \mathcal{C}^\infty)$ with smooth structure there exists a chart around x of the form $x : U \to \mathbb{R}^{\operatorname{rk} x}$.*

PROOF: Let $x : U \to \mathbb{R}^n$ be a singular chart around x with minimal $n \in \mathbb{N}$. We already know that $\operatorname{rk} x \leq n$. Suppose that $\operatorname{rk} x < n$. Without restriction we can achieve after an affine transformation that $x(x) = 0$. As mentioned above the chart x induces a surjective homomorphism $x^* : \mathfrak{m}_0 \to \mathfrak{m}_x$, hence a surjective linear map $\overline{x^*} : \mathfrak{m}_0/\mathfrak{m}_0^2 \to \mathfrak{m}_x/\mathfrak{m}_x^2$. Now choose $y^1, \cdots, y^{\mathrm{rk}\,x} \in \mathcal{C}^\infty(U_x)$ such that $U_x \subset U$ is an open neighborhood of x and such that the elements $\overline{y^i} = [y_j]_x + \mathfrak{m}_x^2$, $i = 1, \cdots, \mathrm{rk}\,x$ form a basis of $\mathfrak{m}_x/\mathfrak{m}_x^2$. Then choose smooth functions $H_1, \cdots, H_{\mathrm{rk}\,x} \in \mathcal{C}^\infty(O)$ with $H_i \circ x_{|U_x} = y^i$, where O is an open neighborhood of the origin in $\mathbb{R}^n$ and U_x has been shrinked, if necessary. Every germ $[f]_0$ of a function $f \in \mathcal{C}^\infty(O)$ can then be written in the form

$$[f]_0 = c_0 + \sum_{j=1}^{\mathrm{rk}\,x} c_i\,[H_i]_0 + [h]_0 \mod \mathfrak{m}_0^2,$$

where the c_i are real numbers. The smooth function h can be chosen to lie in the ideal $\mathcal{J} = \{g \in \mathcal{C}^\infty(O) |\, g_{|x(U)\cap O} = 0\}$ of functions vanishing over $x(U)$, because $\ker \overline{x^*} = \mathfrak{m}_0 \cap \mathcal{J}_0 + \mathfrak{m}_0^2$. Now, by $\dim(\mathfrak{m}_0/\mathfrak{m}_0^2) = n$ there exist $H_{\mathrm{rk}\,x+1}, \cdots, H_n \in \mathcal{J}$ such that the germs $[H_1]_0, \cdots, [H_n]_0$ form a basis of $\mathfrak{m}_0/\mathfrak{m}_0^2$. Hence, after further shrinking O the map $H = (H_1, \ldots, H_n) : O \to \mathbb{R}^n$ is a diffeomorphism onto its image. By $H_i \circ x_{|U_x} = y^i$ for $i = 1, \cdots, \mathrm{rk}\,x$ and $H_i \circ x_{|U_x} = 0$ for $i = \mathrm{rk}\,x + 1, \cdots, n$ the map $y = (y^1, \cdots, y^{\mathrm{rk}\,x})$ has to be a singular chart over U_x and H a transition map from x to y. By $\mathrm{rk}\,x < n$ this contradicts the minimality assumption of n, hence $\mathrm{rk}\,x = n$ is true. Quod erat demonstrandum. □

The following two corollaries follow immediately from the proof just given.

1.3.11 Corollary *The function* $\mathrm{rk} : X \to \mathbb{N}$, $x \mapsto \mathrm{rk}\,x$ *is lower semicontinuous that means for every* $x \in X$ *there exists a neighborhood* U *with* $\mathrm{rk}\,y \le \mathrm{rk}\,x$ *for all* $y \in U$.

1.3.12 Corollary *For every pair of singular charts* $y : U \to \mathbb{R}^n$ *and* $x : U \to O \subset \mathbb{R}^{\mathrm{rk}\,x}$ *around* $x \in X$ *there exists after shrinking* U *and* O *an embedding* $H : O \to \mathbb{R}^n$ *such that* $y = H \circ x$.

By the proposition and the last corollary one can interpret the rank of x as the smallest natural number n such that a neighborhood of x can be embedded into $\mathbb{R}^n$ via a singular chart.

The sheaf of smooth functions on $\mathbb{R}^n$ is fine. Via singular charts this property can be carried over to a stratified space with smooth structure as is shown in the proof of the following theorem.

1.3.13 Theorem *The structure sheaf* $\mathcal{C}^m_X$ *of a stratified space* X *with* $\mathcal{C}^m$*-structure is fine.*

PROOF: First note that every stratified space with a $\mathcal{C}^m$-structure must be locally compact, as every locally closed subset of $\mathbb{R}^n$ is locally compact. To prove that $\mathcal{C}^m_X$ is a fine sheaf, it suffices by the paracompactness of X to construct for every locally finite open covering $\mathcal{U} = (U_j)_{j\in J}$ of X a subordinate partition of unity $(\varphi_j)_{j\in J}$ by $\mathcal{C}^m$-functions $\varphi_j : X \to \mathbb{R}$. After refinement of $\mathcal{U}$ we can assume without loss of generality that every U_j is the domain of a singular chart $x_j : U_j \to \mathbb{R}^{n_j}$. As X is normal as a topological space, there exists an open covering $(V_j)_{j\in J}$ of X with $\overline{V_j} \subset\subset U_j$ for every

$j \in J$. Now choose for every $j \in J$ an open subset $O_j \in \mathbb{R}^{n_j}$ with $\overline{x_j(U_j) \cap O_j} = x_j(V_j)$ and for every $x \in X$ an index j_x with $x \in V_{j_x}$. Next choose for every x a relatively compact open neighborhood $W''_x \subset\subset V_{j_x}$. By paracompactness of X we can then find two locally finite open coverings $(W_x)_{x\in X}$ and $(W'_x)_{x\in X}$ subordinate to $(W''_x)_{x\in X}$ such that $\overline{W_x} \subset W'_x \subset \overline{W'_x} \subset W''_x$. The $\overline{W_x}$ have to be compact, hence there exist smooth functions $\varphi_x : O_{j_x} \to [0,1]$ with compact support such that $\varphi_{x|x_{j_x}(W_x)} = 1$ and $\operatorname{supp} \varphi_x \cap x_{j_x}(V_{j_x}) \subset x_{j_x}(W'_x)$. Now let us set

$$\widetilde{\varphi}_j := \sum_{\{x\in X | j_x = j\}} \varphi_x \circ x_{j_x},$$

where $\varphi_x \circ x_{j_x}$ is set to 0 outside V_{j_x}. Then $\operatorname{supp} \widetilde{\varphi}_j \subset \overline{V_j} \subset U_j$ and $(\operatorname{supp} \widetilde{\varphi}_j)_{j\in J}$ is a covering of X. Hence $(\varphi_j)_{j\in J}$ with

$$\varphi_j(x) = \frac{\widetilde{\varphi}_j(x)}{\sum\limits_{j\in J} \widetilde{\varphi}_j(x)}, \qquad x \in X,$$

comprises a partition of unity by $\mathcal{C}^m$-functions subordinate to $(U_j)_{j\in J}$. □

1.3.14 If the situation occurs that X possesses a global chart $x : X \to \mathbb{R}^n$ of class $\mathcal{C}^\infty$, we will say that X is *Euclidean embeddable*. In most applications the regarded stratified space will be Euclidean embeddable. In the following we will provide criteria which guarantee the existence of a global singular chart. Hereby it will turn out useful to have a new name for injective smooth maps $f : X \to M$ between a stratified space X with smooth structure and a manifold M such that f is proper and such that the pullback

$$f^* : \mathcal{C}^\infty(M) \to \mathcal{C}^\infty(X)$$

is surjective. We will call such maps *proper embeddings*.

1.3.15 Proposition *Every proper embedding $f : X \to \mathbb{R}^n$ is a global singular chart for X.*

PROOF: As f is continuous, proper and injective, f is a homeomorphism onto its image, and the image is closed in $\mathbb{R}^n$. Therefore it only remains to show that f is compatible with all singular charts of an atlas of X. Let $x : U \to O \subset \mathbb{R}^{\operatorname{rk} x}$ be a singular chart around $x \in X$. By smoothness of f there exists after shrinking U and O a smooth function $\mathbf{f} : O \to \mathbb{R}^n$ such that $\mathbf{f} \circ x = f_{|U}$. As $f^*_x : \mathfrak{m}_{f(x)}/\mathfrak{m}^2_{f(x)} \to \mathfrak{m}_x/\mathfrak{m}^2_x$ is surjective by assumption on f, and $\overline{x^*} : \mathfrak{m}_{x(x)}/\mathfrak{m}^2_{x(x)} \to \mathfrak{m}_x/\mathfrak{m}^2_x$ an isomorphism, the derivative $D_{x(x)}\mathbf{f}$ of $\mathbf{f}$ at the point $x(x)$ is injective. Hence, after shrinking O further $\mathbf{f}$ is an immersion, and $\mathbf{f}(O)$ a submanifold of $\mathbb{R}^n$. After shrinking O a last time one can find a diffeomorphism $H : O \times V \to \tilde{O} \subset \mathbb{R}^n$, where $V \subset \mathbb{R}^{n-\operatorname{rk} x}$ and $\tilde{O} \subset \mathbb{R}^n$ are open subsets, such that $H_{|O\times\{0\}} = \mathbf{f}$. Therefore f is compatible with x, hence the claim follows. □

1.3.16 Proposition *Assume that X has a countable atlas of singular charts $x_j : U_j \to \mathbb{R}^{N_j}$, $j \in \mathbb{N}$, of class $\mathcal{C}^\infty$ such that $N_j \leq N$ for a number $N \in \mathbb{N}$ and all $j \in \mathbb{N}$.*

Then X can be embedded into $\mathbb{R}^{2N+1}$ via a proper singular chart. In particular, every compact stratified space with a smooth structure is Euclidean embeddable.

PROOF: We divide the proof in two steps, and will prove the claim in the first step for the case, where X is compact. Then we will extend this result in the second step to the general case. In the course of the argumentation the reader will notice that the proof is very close to the one of the WHITNEY Embedding Theorem.
1. STEP: By compactness of X there exists a finite covering $(\mathsf{U}_j)_{j=1}^k$ of X by chart domains $\mathsf{U}_j \xrightarrow{\mathsf{x}_j} \mathsf{O}_j \subset \mathbb{R}^{N_j}$. Let $(\mathsf{V}_j)_{j=1}^k$ be an open covering subordinate to $(\mathsf{U}_j)_{j=1}^k$. Then first there exist smooth functions $\varphi_j : \mathsf{X} \to \mathbb{R}$ with support in U_j and identical to 1 over $\overline{\mathsf{V}_j}$. Secondly fix maps $\mathsf{y}_j \in \mathcal{C}^\infty(\mathsf{X}, \mathbb{R}^{n_j})$ by $\mathsf{y}_j(\mathsf{x}) = \varphi_j(\mathsf{x})\,\mathsf{x}_j(\mathsf{x})$ for $\mathsf{x} \in \mathsf{U}_j$ and $= 0$ for $\mathsf{x} \notin \mathsf{U}_j$. Using these functions we can now define a map $\mathsf{x} : \mathsf{X} \to \mathbb{R}^n$ with $n = N_1 + \cdots + N_k + k$ by the following:

$$\mathsf{x} = \left(\mathsf{y}_1, \cdots, \mathsf{y}_k, \varphi_1, \cdots, \varphi_k\right).$$

Obviously x is smooth and injective. Moreover, the map x is proper, hence a homeomorphism onto its image. As for every j the restriction of y_j to V_j comprises a singular chart, $\mathsf{x}^*_\mathsf{x} : \mathcal{C}^\infty_{\mathsf{x}(\mathsf{x})} \to \mathcal{C}^\infty_\mathsf{x}$ is surjective for all $\mathsf{x} \in \mathsf{X}$. By the proof of Proposition 1.3.15 then x is a global singular chart of X.

By Corollary 1.3.12 and the compactness of X there exist submanifolds $\mathsf{M}_1, \cdots, \mathsf{M}_l \subset \mathbb{R}^n$ of dimension $\leq N$ such that $\mathsf{X} \subset \mathsf{M}_1 \cup \cdots \cup \mathsf{M}_l$. Let us now suppose there exists a vector $v \in \mathbb{R}^n$ which for every pair $y, z \in \mathsf{x}(\mathsf{X})$, $y \neq z$ is not parallel to $y - z$ and which is not tangent to any of the submanifolds M_j. The composition $\overline{\mathsf{x}}$ of x with the projection from $\mathbb{R}^n$ to the hyperplane $v^\perp \cong \mathbb{R}^{n-1}$ then is injective and induces surjective morphisms $\overline{\mathsf{x}}^*_\mathsf{x} : \mathcal{C}^\infty_{\overline{\mathsf{x}}(\mathsf{x})} \to \mathcal{C}^\infty_\mathsf{x}$, $\mathsf{x} \in \mathsf{X}$, hence is a global singular chart of X. If we can now prove that for $n > 2N + 1$ there exists such a vector v, we will obtain after a finite recursion a global singular chart of X with values in $\mathbb{R}^{2N+1}$.

Consider the maps $\sigma_j : T\mathsf{M}_j \setminus \mathsf{M}_j \to \mathbb{RP}^{n-1}$, $j = 1, \cdots, l$ which result from assigning to every nonvanishing tangent vector of M_j its equivalence class in the projective space $\mathbb{RP}^{n-1}$. Denote the diagonal in $\mathbb{R}^n \times \mathbb{R}^n$ by the symbol Δ and consider the maps $\tau_{ij} : (\mathsf{M}_i \times \mathsf{M}_j) \setminus \Delta \to \mathbb{RP}^{n-1}$, $i, j = 1, \cdots, l$, which assign to every pair (y, z) with $y \neq z$ the line through $y - z$. Both the σ_j as well as the τ_{ij} are smooth and defined on manifolds of dimension $\leq 2N$. As long as $2N < n - 1$ holds, the images of the σ_j and τ_{ij} are of first category in $\mathbb{RP}^{n-1}$ by the theorem of SARD, and so is their union. Consequently its complement is nonempty, hence there is a vector $v \in \mathbb{R}^n$ with the desired properties. This proves the claim for compact X.

2. STEP: Now we drop the assumption that X is compact. Without loss of generality we can assume that all chart domains U_j have compact closure. Then we choose a locally finite smooth partition of unity $(\varphi_j)_{j\in\mathbb{N}}$ subordinate to $(\mathsf{U}_j)_{j\in\mathbb{N}}$. and set $\lambda(\mathsf{x}) = \sum_{j\in\mathbb{N}} j\varphi_j(\mathsf{x})$. Then $\lambda : \mathsf{X} \to \mathbb{R}$ is smooth and proper. Let $\mathsf{V}_j = \lambda^{-1}\left(]j - 1/4, j + 5/4[\right)$ and $\mathsf{K}_j = \lambda^{-1}\left([j - 1/3, j + 4/3]\right)$. Then V_j is open, K_j is compact and $\overline{\mathsf{U}_j} \subset \mathsf{K}_j^\circ$. Moreover, all K_{2j} are pairwise disjoint, just as the K_{2j+1}. According to the 1. Step we can choose smooth functions $\mathsf{g}_j : \mathsf{X} \to \mathbb{R}^{2N+1}$ with bounded image such that the restriction of g_j to $\overline{\mathsf{V}_j}$ is a singular chart and such that $\operatorname{supp} \mathsf{g}_j \subset \mathsf{K}_j^\circ$. We now define $\mathsf{x}_e := \sum_{j\in\mathbb{N}} \mathsf{g}_{2j}$, $\mathsf{x}_o := \sum_{j\in\mathbb{N}} \mathsf{g}_{2j+1}$ and set

$x := (x_e, x_o, \lambda) : \mathbb{R}^{2N+1} \times \mathbb{R}^{2N+1} \times \mathbb{R}$. Then x is smooth, proper and injective. By the fact that the g_j are singular charts on $\overline{V_j}$ and by definition of x every one of the induced maps $x_x^* : \mathcal{C}^\infty_{x(x)} \to \mathcal{C}^\infty_{x(x)}$, $x \in X$ has to be surjective, hence by the proof of Proposition 1.3.15 x is a singular chart of X. Analogously to the arguments in the 1. Step one can now find by the theorem of SARD a $(2N+1)$-dimensional hyperplane $H \subset \mathbb{R}^{4N+3}$ such that the composition of x with the orthogonal projection π_H onto H is a again a proper singular chart of X. To guarantee the properness of $\pi_H \circ x$ one has to choose H such that subspace $\mathbb{R}_{4N+3}$ generated by the last coordinate of $\mathbb{R}^{4N+3}$ does not lie in the kernel of π_H. But this is possible indeed, as $\mathbb{R}_{4N+3}$ is of first category in $\mathbb{R}^{4N+3}$. So finally we obtain a global singular and proper chart of X with values in a $(2N+1)$-dimensional vector space. □

For the case that X is not Euclidean embeddable one has special atlases at ones disposal which in many cases achieve almost the same like global charts. But before we can explain this in more detail let us briefly recall the notion of a *compact exhaustion* of X: this is a family $(K_j)_{j\in\mathbb{N}}$ of compact subsets of X such that $K_j \subset K^\circ_{j+1}$ and $\bigcup_{j\in\mathbb{N}} K_j = X$. Such a compact exhaustion of X exists, as X is locally compact Hausdorff with countable topology. If one now chooses a compact exhaustion $(K_j)_{j\in\mathbb{N}}$ of X, a singular atlas $(x_j)_{j\in\mathbb{N}}$ of X consisting of charts of the form $x_j : K^\circ_{j+1} \to O_j \subset \mathbb{R}^{n_j}$ is called *inductively embedding* with respect to $(K_j)_{j\in\mathbb{N}}$, if $n_{j+1} \geq n_j$ for all j, and if there are relatively compact open neighborhoods $U_j \subset\subset K^\circ_{j+1}$ of K_j such that $x_{j+1}(x) = \iota^{n_{j+1}}_{n_j} \circ x_j(x)$ for all $x \in U_j$.

1.3.17 Lemma *For every compact exhaustion there is an inductively embedding atlas.*

PROOF: As all K_j are compact, there exists by Proposition 1.3.16 an atlas of singular charts $y_j : K^\circ_{j+2} \to \mathbb{R}^{m_j}$. Let now V_j, W_j and U_j be relatively compact open neighborhoods of K_{j-1} in K°_{j+1} with

$$K_{j-1} \subset U_{j-1} \subset\subset V_j \subset\subset W_j \subset\subset K^\circ_j \subset\subset U_j \subset\subset K^\circ_{j+1}.$$

Then there exist smooth functions $\varphi_j : X \to [0,1]$ and $\psi_j : X \to [0,1]$ with $\operatorname{supp}\varphi_j \subset\subset K^\circ_j$, $\varphi_{j|W_j} = 1$, $\operatorname{supp}\psi_j \subset K^\circ_{j+1} \setminus \overline{U_{j-1}}$ and $\psi_{j|X\setminus\overline{V_j}} = 1$. Define $x_0 : K^\circ_1 \to \mathbb{R}^{n_0}$ by $x_0 = y_0$. If for some index j all x_i with $i < j$ are already determined, then fix $x_j : K^\circ_{j+1} \to \mathbb{R}^{n_j} := \mathbb{R}^{m_j} \times \mathbb{R}^{n_{j-1}} \times \mathbb{R}^2$ recursively by

$$x_j(x) = \begin{cases} (\varphi_j(x)\, x_{j-1}(x),\ \psi_j(x)\, y_j(x),\ 1-\varphi_j(x),\ \psi_j(x)) & \text{for } x \in K^\circ_j, \\ (0,\ y_j(x),\ 1-\varphi_j(x),\ \psi_j(x)), & \text{for } x \in K^\circ_{j+1} \setminus K^\circ_j. \end{cases}$$

Then by definition of x_j the relation $x_{j+1|U_j} = \iota^{n_{j+1}}_{n_j} \circ x_{j|U_j}$ holds. Moreover, one checks easily that x_j is injective, a homeomorphism onto its image, and compatible with the singular charts x_{j-1} and y_j. By induction one thus obtains a singular atlas with the desired properties. □

1.3.18 Example Manifolds-with-corners possess a smooth structure induced by their corner data. Moreover, manifolds-with-corners are Euclidean embeddable by definition. The smooth functions with respect to this smooth structure coincide with the smooth functions on manifolds-with-corners in the usual sense.

1.3.19 Example Examples 1.1.12 to 1.1.17 inherit a canonical smooth structure as locally closed subspaces of Euclidean space.

1.3.20 Example Every triangulation of a polyhedron provides a smooth structure. Note that the smooth structures defined by two different triangulations need not be compatible with each other.

1.3.21 Example The absolute value $|\cdot| : \mathbb{R} \to \mathbb{R}$ is not a smooth function, if $\mathbb{R}$ carries the ordinary smooth structure. But it is possible to interpret $\mathbb{R}$ as a stratified space with the decomposition $\mathcal{R}$ into $R_0 = \{0\}$ and $R_1 = \mathbb{R} \setminus \{0\}$, and then embed this space by $x \mapsto (x, |x|)$ into $\mathbb{R}^2$. The stratified space $(\mathbb{R}, \mathcal{R})$ then inherits from $\mathbb{R}^2$ a smooth structure with respect to which the absolute value is a smooth map. Incidentally $(\mathbb{R}, \mathcal{R})$ is diffeomorphic in a canonical way with the edge X_{Edge}.

1.3.22 Example The cone comb possesses a natural smooth structure, but as an infinite dimensional stratified space it is not Euclidean embeddable. Starting from the cone comb one can even construct an example for a stratified space with smooth structure having only two strata, hence being finite dimensional, but which is not Euclidean embeddable. Let us explain this in more detail.

By definition X_{Cmb} arises by appropriately gluing the cones CS^n to the half line $\mathbb{R}^{\geq 0}$. Now set $U_n := CS^n \cup]n - 3/4, n + 3/4[$, and note that

$$CS^n = \{(ty, t) \in \mathbb{R}^{n+2} \mid y \in S^n\}$$

can be regarded as a stratified subspace of $\mathbb{R}^{n+2}$. Then

$$x_n : U_n \to \mathbb{R}^{n+3} \cong \mathbb{R} \times \mathbb{R}^{n+2}, \quad x \mapsto \begin{cases} (x, 0) & \text{if } x \in]n - 3/4, n + 3/4[, \\ (n, x) & \text{if } x \in CS^n \setminus \{o_n\}, \end{cases}$$

comprises a singular chart of X_{Cmb}, and the family $(x_n)_{n \in \mathbb{N}}$ is a singular atlas. On the other hand the set

$$CS^n_F := \{(ty, t) \in \mathbb{R}^{n+1} \times [0, 1[\mid y \in \{e_1, \cdots, e_{n+1}\}\},$$

where "F" stands for frame, is for every n canonically a stratified subspace of CS^n of dimension 1. Hence

$$X_{\text{FCmb}} := \{x \in X_{\text{Cmb}} \mid x \in \mathbb{R} \text{ or } x \in CS^n_F \text{ for an } n \in \mathbb{N}\}$$

becomes a stratified subspace of X_{Cmb} of dimension 1, has only two strata, and inherits from X_{Cmb} a smooth structure. Obviously, X_{FCmb} together with this smooth structure is not Euclidean embeddable.

1.4 Local Triviality and the Whitney conditions

Several of the decomposed spaces introduced in Section 1.1 have properties which seem unnatural, like for example the space Y from 1.1.12, which satisfies $\dim S_1 > \dim S_2$, though S_1 is a boundary piece of S_2. Such and other "pathological" stratified spaces should not be admitted for further consideration, so in the course of the formation of stratification theory people have tried to find criteria which exclude such unwanted spaces. Usually the conditions on stratifications appearing in the mathematical literature impose further restrictions to the behavior of a stratum near a boundary stratum. This should guarantee that the remaining stratified spaces have nice properties which admit further topological, geometric or analytic considerations.

First in this section we will introduce topological local triviality of a stratified space. As already explained in the introduction this condition says that a locally trivial space is locally around each of its points isomorphic to a trivial fiber bundle over the stratum of the point. Often it is supposed additionally that the typical fiber is given by the cone over a compact stratified space, but the literature is not uniform in this point. As cones are important nontrivial and well to study examples of stratified spaces we have named locally trivial spaces with cones as typical fibers "cone spaces". In Section 3.10 we will treat such spaces in detail.

In a certain sense one can regard topological local triviality as a minimal requirement to a reasonable stratified space. Therefore some authors require local triviality in the definition of a stratified space (cf. e.g [64, 162]).

For a stratified space given (locally) as subspaces of manifolds we will afterwards introduce in 1.4.3 the famous Whitney conditions (A) and (B). These conditions essentially impose restrictions on the behavior of the limit tangent spaces of a higher stratum when approaching a boundary stratum. The Whitney conditions have far reaching implications, in particular condition (B) guarantees that the considered stratified space is locally trivial (see Corollary 3.9.3). The corresponding proofs are quite involved and have led to the control theory of J. MATHER which will be explained in Chapter 3.

In the evolution of stratification theory many more conditions have been imposed on "good" stratified spaces. The goal is to formulate criteria, which are as easy as possible to prove and which entail essential but more difficult properties like locally triviality, a particular metric structure or even geometric features. At the end of this section we will introduce some of these further criteria and explain their meaning.

1.4.1 Topological local triviality A stratified space X is called *topologically locally trivial*, if for every $x \in X$ there exists a neighborhood U, a stratified space F with stratification $\mathcal{S}^F$, a distinguished point $o \in F$ and an isomorphism of stratified spaces

$$h : U \to (S \cap U) \times F$$

such that $h^{-1}(y, o) = y$ for all $y \in S \cap U$ and such that $\mathcal{S}^F_o$ is the germ of the set $\{o\}$. Hereby, S is the stratum of X with $x \in S$. Sometimes we call F the *typical fiber* over x. In many cases F is given by a cone $F = CL$ over a compact stratified space L. Then L is called the *link* of x, and one says that X is *locally trivial* with *cones as typical fibers.* If L is locally trivial with cones as typical fibers and if that holds again for the links of the points of the link and so on, we obtain a class of stratified spaces called *cone spaces* of *class* $\mathcal{C}^0$. For a precise definition of the notion of a cone space and some explanations we refer the reader to Section 3.10.

1.4.2 Example It is relatively easy to prove that manifolds-with-boundary or manifolds-with-corners are locally trivial, in particular so are simplices and polyhedra. It is much more difficult to see that (real or complex) algebraic varieties possess topologically locally trivial stratifications, more generally even all semialgebraic, semianalytic and subanalytic sets. This follows from the fact that all these spaces have an essentially unique Whitney stratification (see Example 1.4.10 for references) and that Whitney stratifications are locally trivial according to THOM [169] and MATHER [122]. In the course of this monograph we will show explicitly local triviality for Whitney stratifications in Corollary 3.9.3 and for orbit spaces in 4.4.6.

Local triviality alone does not automatically imply that the fibers or links are locally trivial as well. In case all the different trivializations are compatible with each other one can show that local triviality does also hold for the fibers and links. The right axiomatics for that, an appropriate definition of compatibility and the corresponding implications is given by the control theory of MATHER (see Chap. 3).

1.4.3 The Whitney conditions In the following we will consider a manifold M as well as submanifolds R and S. On says that the pair (R, S) fulfills the *Whitney condition* (A) at $x \in R$, or that (R, S) is (A)-*regular* at x, if the following axiom is satisfied:

(A) Let $(y_k)_{k\in\mathbb{N}}$ be a sequence of points $y_k \in S$ converging to x such that the sequence of tangent spaces $T_{y_k}S$ converges in the Graßmannian of $\dim S$-dimensional subspaces of TM to $\tau \subset T_xM$. Then $T_xR \subset \tau$.

If $x : U \to \mathbb{R}^n$ is a smooth chart of M around $x \in R$ such that the following axiom (B) is fulfilled, one says that (R, S) satisfies the *Whitney condition* (B) at x with respect to the chart x.

(B) Let $(x_k)_{k\in\mathbb{N}}$ and $(y_k)_{k\in\mathbb{N}}$ be two sequences of points $x_k \in R \cap U$, $y_k \in S \cap U$ fulfilling the following three conditions:

(B1) $x_k \neq y_k$ and $\lim_{k\to\infty} x_k = \lim_{k\to\infty} y_k = x$.

(B2) The sequence of connecting lines $\overline{x(x_k)\,x(y_k)} \subset \mathbb{R}^n$ converges in projective space to a line ℓ.

(B3) The sequence of tangent spaces $T_{y_k}S$ converges in the Graßmannian to a subspace $\tau \subset T_xM$.

Then $(T_x x)^{-1}(\ell) \subset \tau$.

Now the question arises how Whitney's condition (B) transforms under a coordinate change.

1.4.4 Lemma *If* (R, S) *satisfies the Whitney condition* (B) *at* x *with respect to the chart* $x : U \to \mathbb{R}^n$, *and if* $y : U \to \mathbb{R}^n$ *is a further chart of* M *around* x, *then* (B) *is satisfied as well with respect to* y.

PROOF: Let $(x_k)_{k\in\mathbb{N}}$ and $(y_k)_{k\in\mathbb{N}}$ be two sequences of points $x_k \in R\cap U$, $y_k \in S\cap U$ converging to x such that the sequence of secants $\ell_k = \overline{y(x_k)\,y(y_k)}$ converges to the line ℓ and such that the condition (B3) is satisfied. Let further $H : O \longrightarrow \mathbb{R}^n$ be an open embedding such that (after possibly shrinking U) the convex hull K of the set $x(U)$ is compactly contained in O and such that $H\circ y = x$. After transition to subsequences we can suppose that the sequence of unit vectors $v_k := \frac{y(y_k)-y(x_k)}{\|y(y_k)-y(x_k)\|}$ converges to a unit vector $v_\infty \in \ell$. For $k\in\mathbb{N}\cup\{\infty\}$ now define curves $\gamma_k :]-1,1[\to \mathbb{R}^n$ by $t\mapsto y(x_k)+tv_k$, where $x_\infty := x$. Next consider the transformed curves $\eta_k = H\circ\gamma_k :]-1,1[\to\mathbb{R}^n$. By applying the triangle inequality and Taylor's formula we obtain for all $k\in\mathbb{N}$ and all $t\in]-1,1[$ the following estimate:

$$\begin{aligned}
&\left\| \eta_k(t) - x(x_k) - t\frac{x(y_k)-x(x_k)}{\|y(y_k)-y(x_k)\|} \right\| \leq \\
&\quad \leq \|\eta_k(t) - x(x_k) - tDH(y(x_k)).v_k\| + \\
&\quad + \frac{t}{\|y(y_k)-y(x_k)\|}\|x(y_k)-x(x_k) - DH(y(x_k)).(y(y_k)-y(x_k))\| \\
&\quad \leq \frac{1}{2}t(t+\|y(y_k)-y(x_k)\|)\sup_{z\in K}\|D_z^2H\|.
\end{aligned} \tag{1.4.1}$$

Note that $C := \sup_{z\in K}\|D_z^2H\| < \infty$ and that C is independent of k and t. After transition to further subsequences we can suppose that $\frac{x(y_k)-x(x_k)}{\|y(y_k)-y(x_k)\|}$ converges to a nonvanishing vector w_∞. The estimate (1.4.1) then entails

$$w_\infty = \dot{\eta}_\infty(0) = T_{y(x)}H(\dot{\gamma}_\infty(0)) = T_{y(x)}H(v_\infty).$$

By hypothesis on the singular chart x the relation $(T_x x)^{-1}(w_\infty)\in\tau$ is true, so altogether

$$(T_x y)^{-1}(v_\infty) = (T_x x)^{-1}(w_\infty)\in\tau$$

follows. This proves the claim. □

By the lemma just proven the validity of the Whitney condition (B) is independent of the chosen chart. Hence it is clear what to understand by the sentence "(R, S) satisfies the Whitney condition (B) at x" or equivalently by "(R, S) is (B)-regular at x".

1.4.5 Lemma *If the pair* (R, S) *is* (B)*-regular at* $x\in R$, *then* (R, S) *satisfies Whitney's condition* (A) *at* x.

PROOF: As the claim is a local one, we can suppose that R and S are submanifolds of Euclidean space $\mathbb{R}^n$.

Let $(y_k)_{k\in\mathbb{N}}$ be a sequence of points of S converging to x such that the sequence of tangent spaces $T_{y_k}S$ converges to $\tau\subset\mathbb{R}^n$. Let further $v\in T_xR$ be a nonvanishing tangent vector and $\gamma(t)$, $t\in[-1,1]$ a smooth path with $\dot\gamma(0)=v$. Let $t_k = \|y_k - x\|$ and $w_k = \frac{y_k - x}{t_k}$. Then $(t_k)_{k\in\mathbb{N}}$ converges to 0 and, after a transition to a subsequence, $(w_k)_{k\in\mathbb{N}}$ to some $w\in\mathbb{R}^n$. On the other hand the sequence of the vectors $v_k := \frac{\gamma(t_k)-x}{t_k}$ converges to v by assumption on γ. Hence the sequence $(z_k)_{k\in\mathbb{N}}$ with $z_k = w_k - v_k = \frac{y_k-\gamma(t_k)}{t_k}$ converges to $z\in\mathbb{R}^n$. By Whitney (B) the vectors w and z lie in τ, hence so does $v = w - z$. This proves the claim. □

Let us agree on some further notation. If the condition (A) resp. (B) is satisfied at every point $x \in R$, we will say that the pair (R, S) satisfies the *Whitney condition* (A) resp. (B), or that S is (A) resp. (B)-regular over R. A stratified space with smooth structure such that for every pair (R, S) of strata Whitney's condition (A) holds is called a *Whitney* (A) *space* or an (A)-*stratified space*. As a *Whitney space* or a (B)-*stratified space* we will denote a stratified space with smooth structure such that for every pair (R, S) of strata Whitney's condition (B) holds.

1.4.6 Remark If $m \in \mathbb{N}^{>0}$, one can formulate for $\mathcal{C}^m$-manifolds M, R, S the Whitney conditions (A) and (B). But for the proof given in Lemma 1.4.4 of the invariance of the Whitney condition (B) under a chart transition one has to assume that $m \geq 2$. Nevertheless the condition (B) is a $\mathcal{C}^1$-invariant. A proof of this fact can be found in TROTMAN [172] (see 3.4.2 as well).

1.4.7 Example One can construct a stratified space which is not (A)-regular by starting from Whitney's umbrella X_{WUmb}. Intuitively we fold down one leaf of the umbrella and obtain in mathematically more precise terms the following topological space:

$$X = \{(x, y, z) \in \mathbb{R}^3 \,|\, x^2 = y^2\,|z| \;\&\; \mathrm{sgn}(x) = \mathrm{sgn}(yz)\}.$$

As stratification of X choose the one generated by $S_0 = \{(0, 0, z) \mid z \in \mathbb{R}\}$ and $S_1 = X \setminus S_0$. In the origin, the pair (S_0, S_1) is not (A)-regular. To see this, consider the limit τ of the tangent spaces $T_{x_k}S_1$ with $x_k = (0, 1/k, 0)$. Then τ is given by the xy-plane, but the z-axis, which is the tangent space of S_0 in the origin, is not contained in the xy-plane.

In general it is rather difficult to find examples of not (A)-regular stratifications in particular of not (A)-regular stratified varieties. A source of such examples is given by the Trotman varieties [173, 10].

1.4.8 Example The fast spiral X_{Spirr} of example 1.1.13 is a Whitney stratified space, the slow spiral X_{Spir} on the other hand not. Let us show this in some more detail. The top stratum of the fast spiral can be parametrized by $\gamma(\theta) = e^{-\theta^2}(\sin\theta, \cos\theta)$, $\theta \in \mathbb{R}^{>0}$, the one of the slow spiral by $\eta(\theta) = e^{-\theta}(\sin\theta, \cos\theta)$. This gives

$$\dot{\gamma}(\theta) = e^{-\theta^2}\big((\cos\theta, -\sin\theta) - 2\theta(\sin\theta, \cos\theta)\big).$$

Besides that $\frac{\dot{\gamma}(\theta)}{\|\dot{\gamma}(\theta)\|}$ is a unit tangent vector with footpoint $\gamma(\theta)$, and $(\sin\theta, \cos\theta)$ spans the secant connecting the origin and $\gamma(\theta)$. As

$$\lim_{\theta\to\infty}\left(\frac{\dot{\gamma}(\theta)}{\|\dot{\gamma}(\theta)\|} + (\sin\theta, \cos\theta)\right) = 0,$$

the space X_{Spirr} satisfies the Whitney condition (B). Let us now consider the slow spiral and calculate:

$$\dot{\eta}(\theta) = e^{-\theta}\big((\cos\theta, -\sin\theta) - (\sin\theta, \cos\theta)\big).$$

For $\theta_k = \frac{\pi}{4} + 2\pi k$ this implies $\frac{\dot{\eta}(\theta_k)}{\|\dot{\eta}(\theta_k)\|} = (0, -1)$ and $\eta(\theta_k) = \frac{1}{2}e^{-\theta_k}(\sqrt{2}, \sqrt{2})$. Now the sequence of the points $\eta(\theta_k)$ converges to the origin, the sequence of tangent spaces converges to the subspace spanned by $(0,1)$, and finally the sequence of secants $\overline{\eta(\theta_k)0}$ converges to the line generated by $(1,1)$. Hence X_{Spir} cannot satisfy Whitney (B). Moreover this argument shows as well that no finer decomposition of the slow spiral exists which makes X_{Spir} into a Whitney stratified space.

1.4.9 Example Consider the two decompositions of the Whitney cusp X_{WCsp} given in Example 1.1.17. One can prove easily that the stratification induced by the decomposition into $R_0 = \{0\}$, $R_1 = \{(x,y,z) \in \mathbb{R}^3 \mid x = 0,\ y = 0,\ z \neq 0\}$ and $R_2 = X_{\mathrm{WCsp}} \setminus (R_0 \cup R_1)$ is a Whitney stratification. One the other hand the decomposition of X_{WCsp} into the z-axis S_1 and its complement S_2 fulfills Whitney (A), but not Whitney (B). Let us explain this in more detail. Consider the sequence of points $w_k = (1/k^2, 0, 1/k) \in X_{\mathrm{WCsp}}$ converging to the origin. Now, if (x,y,z) is an element of X_{WCsp}, then the point $(x,-y,z)$ is one as well. Hence the tangent space of S_2 with footpoint w_k is spanned by the vectors $(0,1,0)$ and $(2/k, 0, 1)$. Thus for $k \to \infty$ the sequence of tangent spaces converges to the hyperplane τ spanned by the vectors $(0,1,0)$ and $(0,0,1)$. But the connecting secants $\overline{w_k w_k'}$ with $w_k' = (0,0,1/k) \in X_{\mathrm{WCsp}}$ converge to the line ℓ spanned by $(1,0,0)$. As obviously ℓ does not lie in τ, Whitney (B) does not hold for the decomposition (S_1, S_2).

1.4.10 Example Since the emergence of stratification theory one could show for more and more general classes of spaces that they possess Whitney stratifications. The beginnings of this go back to WHITNEY [191], who showed first that every real or complex analytic variety has a Whitney stratification. ŁOJASIEWICZ succeeded in [115] to prove that every semianalytic subset of a real analytic manifold possesses a Whitney stratification by analytic manifolds, and that the strata are *strong analytic*, i.e. they comprise analytic manifolds which are semianalytic. For subanalytic sets HARDT [77, 78] and HIRONAKA [86] could show that they are Whitney stratifiable. But it should not remain unmentioned that the first ideas for a proof of this fact goes back to THOM. In his work [169] THOM had already worked out some of the fundamental properties of subanalytic sets to which he gave the name *PSA* for *Projection d'ensemble Semi-Analytique.* In the book by SHIOTA [158] one can find a detailed and modern account of the theory of semialgebraic and subanalytic sets.

1.4.11 Thom's Condition (T) One of the first regularity conditions imposed on a stratified space has been introduced in 1964 by THOM [168]. In the mathematical literature THOM's condition is often called condition (T). Using our notation we call a pair (R, S) of disjoint submanifolds (T)-*regular*, if every smooth function $g : \mathbb{R}^N \to M$ transversal to R is also transversal to S in a neighborhood of R.

The condition (T) is a relatively weak requirement to a stratified space. WHITNEY proved in 1964 in his article [192] that his condition (A) implies THOM's transversality condition (T). The significance of (T) lies mainly in the stability theory of differentiable mappings.

1.4.12 Verdier's condition (W) A pair (R, S) of submanifolds of $\mathbb{R}^n$ is called (W)-*regular* at $x_0 \in R$, , if there exists a neighborhood U of x_0 and a constant $C > 0$ such that for all $x \in R \cap U$ and all $y \in S \cap U$

$$d_{Gr}(T_xR, T_yS) < C\,\|y - x\|,$$

where d_{Gr} is the vector space distance defined in Appendix A.1. Similarly to Whitney's condition (B) one shows that the condition (W) is invariant under diffeomorphisms of class $\mathcal{C}^2$, so the notion of (W)-*regularity* at $x_0 \in R$ of a pair (R, S) of $\mathcal{C}^2$-submanifolds of a manifold M is well-defined.

For every $\mathcal{C}^2$-stratified subspace of a manifold M condition (W) implies according to [175] always topological local triviality of the considered stratified space. By the work of KUO [105] and VERDIER [175] it follows that for a subanalytic stratification of a subanalytic space, which means that all strata are subanalytic, condition (W) implies condition (B). In the category of complex analytic stratifications of a complex analytic variety (W) and (B) are even equivalent (see TROTMAN [174]). In the real algebraic case the situation is different; here the two conditions are not equivalent. But there is more. Using HIRONAKA's desingularization theorem VERDIER [175] could prove that every subanalytic set (resp. semialgebraic set, resp. complex analytic variety) has a (W)-regular subanalytic (resp. semialgebraic, resp. complex analytic) stratification. Meanwhile there exist proofs of this result by ŁOJASIEWICZ–STASICA–WACHTA [116] and DENKOWSKA–WACHTA [51] which use only elementary methods and do not need HIRONAKA's resolution of singularities.

1.4.13 Bekka's condition (C) Suppose that $R, S \subset M$ are disjoint and that on an open neighborhood T of $R \subset M$ there is given a $\mathcal{C}^1$-mapping $\rho : T \to \mathbb{R}^{\geq 0}$ such that $R = \rho^{-1}(0)$. BEKKA [8, 9] calls the pair (R, S) (C)-*regular* at $x \in R \cap \overline{S}$, if

(C1) there exists an open neighborhood U of x such that $\rho_{|S\cap U} : S \cap U \to \mathbb{R}$ is submersive, and

(C2) for every sequence $(y_k)_{k\in\mathbb{N}}$ of points $y_k \in S$ converging to x such that the sequence of kernels $\ker T_{y_k}(\rho_{|S})$ converges to a subspace $\lambda \subset T_xM$, the relation $T_xR \subset \lambda$ is satisfied.

Now, the chain of implications (B) $\Rightarrow$ (C) $\Rightarrow$ (A) holds, where the second implication (C) $\Rightarrow$ (A) follows immediately by definition of (C). The reader will find a proof of (B) $\Rightarrow$ (C) in Corollary 3.4.3. Note that the inverse implications are not true in general.

According to the results of BEKKA [8, 9] an essential feature of (C)-regular stratified spaces is that (C)-regularity already allows to construct control data on the underlying stratified in the sense of MATHER (as they are defined in 3.6.4). A consequence of this is that (C)-regular stratified spaces are topologically locally trivial.

1.4.14 An important ingredient for the formulation of the following condition as well as for the later defined curvature moderate stratifications is the notion of a projection valued section.

A $\mathcal{C}^m$-mapping $P : T \to \mathrm{End}(TM)$ from a submanifold $T \subset M$ to the endomorphism bundle of TM is called *projection valued section* (of *class* $\mathcal{C}^m$), if for every $x \in T$ the image P_x is a projection in $\mathrm{End}(T_xM)$, that means if $P_x^2 = P_x$.

Every submanifold $S \subset \mathbb{R}^n$ induces a canonical projection valued section $P_S : S \to \mathrm{End}(\mathbb{R}^n)$ by mapping to every $x \in S$ the orthogonal projection of $\mathbb{R}^n$ onto T_xS.

1.4.15 The Bekka–Trotman condition (δ) We say that a pair (R, S) of submanifolds of $\mathbb{R}^n$ satisfies the condition (δ) at $x \in R$, if there exists an open neighborhood U of x in $\mathbb{R}^n$ and a $\delta > 0$ such that for all $y \in S \cap U$ and $z \in R \cap U$

$$\|P_{S,y}(y - z)\| \geq \delta \|y - z\|. \tag{1.4.2}$$

A stratified space $(X, \mathcal{C}^\infty)$ with smooth structure is called *(δ)-stratified*, if for every pair (R, S) of strata and every point $x \in R$ there exists a singular chart $x : U \to \mathbb{R}^n$ around x such that $(x(R \cap U), x(S \cap U))$ satisfies condition (δ) at $x(x)$.

The condition (δ) has been introduced originally by Bekka–Trotman [11] and Bekka [8]. It is very useful for any considerations of metric properties of a stratified space X. In particular, the condition (δ) guarantees that the geodesic distance on X is locally finite (see Section 1.6) and, in case X has no strata of dimension $\leq \dim X - 2$ that X is volume regular and has finite Hausdorff measure (see Bekka [8] and Ferrarotti [55, 58]).

One can somewhat weaken the Bekka–Trotman-condition (δ) in such a way that the local finiteness of the geodesic length is preserved. We say that the pair (R, S) satisfies at x the condition (δ_l) with $l \in \mathbb{R}^{\geq 1}$, if there is an open neighborhood U of x in $\mathbb{R}^n$ and a $\delta > 0$ such that for all $y \in S \cap U$ and $z \in R \cap U$

$$\|P_{S,y}(y - z)\| \geq \delta \|y - z\|^{2-1/l}. \tag{1.4.3}$$

1.4.16 Proposition *If the pair (R, S) of submanifolds of Euclidean space satisfies Whitney's condition* (B) *at $x \in R$, then (R, S) is* (A)+(δ)*-regular at x.*

Proof: That (R, S) satisfies (A) at x, has been shown already shortly after the introduction of the Whitney conditions. Let us now suppose that (δ) does not hold at x. In other words this means that there are sequences $(y_k)_{k\in\mathbb{N}}$ and $(z_k)_{k\in\mathbb{N}}$ of points $y_k \in S$ and $z_k \in R$ with $\lim_{k\to\infty} y_k = \lim_{k\to\infty} z_k = x$ and

$$\frac{1}{k}\|y_k - z_k\| \geq \|P_{S,y_k}(y_k - z_k)\|.$$

After selection of a subsequence $\left(\frac{y_k - z_k}{\|y_k - z_k\|}\right)_{k\in\mathbb{N}}$ then converges to a unit vector $v \in \mathbb{R}^n$ and $(P_{S,y_k})_{k\in\mathbb{N}}$ to the orthogonal projection onto a subspace $\tau \subset \mathbb{R}^n$. By assumption on the sequences $(y_k)_{k\in\mathbb{N}}$ and $(z_k)_{k\in\mathbb{N}}$ the relation $\|Pv\| \leq \frac{1}{k}$ holds for all $k \in \mathbb{N}$, hence v cannot be an element of τ. But this contradicts Whitney (B). □

1.4.17 Remark Stratified sets which satisfy conditions (A)+(δ) have been considered in detail in a recent work by Bekka–Trotman [12] and have been named *weakly Whitney stratified sets.* These spaces form an intermediate class between Whitney stratified spaces and (C)-regular spaces. We have to postpone the proof that (A)+(δ) implies (C) till 3.4.4.

1.5 The sheaf of Whitney functions

A smooth structure on a stratified space X generates besides the sheaf $\mathcal{C}^\infty$ of smooth functions via a fixed covering by chart domains the sheaf of so-called Whitney functions. This sheaf will play an important role for the extension theory of smooth functions (Sections 1.7 and 3.8) as well as for cohomological considerations of X (Section 5.4).

1.5.1 For the definition of Whitney functions let us suppose first that X has a global singular chart $x : X \to \mathbb{R}$. Then $A := x(X)$ is a locally closed stratified subspace of $\mathbb{R}^n$ and there exists an open subset O of Euclidean space such that $A \cap O$ is closed in O. Furthermore let $m \in \mathbb{N}^{>0} \cup \{\infty\}$. For every in X locally closed set U the subset $x(U)$ of $\mathbb{R}^n$ then is locally closed, hence by Appendix C there exist the spaces $J^m(x(U))$ and $\mathcal{E}^m(x(U))$ of m-jets resp. Whitney functions of class $\mathcal{C}^m$ on X. Via the chart x we now pull back these spaces to X, that means in other words we set

$$J^m_{X,x}(U) := J^m(x(U)) \quad \text{and} \quad \mathcal{E}^m_{X,x}(U) := \mathcal{E}^m(x(U)).$$

Now, if U runs through all open subsets of X, then we obtain two sheaves $J^m_{X,\mathbb{R}^n}$ and $\mathcal{E}^m_{X,\mathbb{R}^n}$, where the details of the corresponding argumentation are left to the reader. The first sheaf is called the sheaf of *m-jets* on X with respect to the chart x, the second one the sheaf of *Whitney functions* of *class* $\mathcal{C}^m$ on X with respect to the chart x. According to the construction in Section C.3 $\mathcal{E}^m_{X,x}$ is a subsheaf of $J^m_{X,x}$.

For the following a new representation for $\mathcal{E}^m_{X,x}$ will prove to be useful. It is well-known that a sheaf is determined uniquely by its espace étalé (see GODEMENT [60] for the necessary sheaf theoretic notions). This suggests to determine the espace étalé $\acute{E}(\mathcal{E}^m_{X,\mathbb{R}^n})$ for Whitney functions. To shorten notion we will often write $\mathcal{E}^m$ instead of $\mathcal{E}^m_{X,x}$. As a set $\acute{E}(\mathcal{E}^m)$ is the union of all stalks $\mathcal{E}^m_x$, $x \in X$. How do the stalks $\mathcal{E}^m_x$ look like? To answer this question, let $z = x(x) \in \mathbb{R}^n$. Then let us recall that $\mathcal{C}^m_{\mathbb{R}^n,z}$ denotes the stalk of all germs of smooth functions on $\mathbb{R}^n$ at z and $\mathcal{J}^m(A, \mathbb{R}^n)_z \subset \mathcal{C}^m_{\mathbb{R}^n,z}$ the ideal of of function germs which are *flat* on A of *order* m (see Section C.3). That means $\mathcal{J}^m(A, \mathbb{R}^n)_z$ consists of all germs $[f]_z \in \mathcal{C}^m_{\mathbb{R}^n,z}$ such that all partial derivatives of f up to order m vanish on A. According to WHITNEY's extension theorem C.3.2 one can identify the stalk $\mathcal{E}^m_x$ with the quotient $\mathcal{C}^m_{\mathbb{R}^n,z}/\mathcal{J}^m(X, \mathbb{R}^n)_z$. The topology of $\acute{E}(\mathcal{E}^m_{X,\mathbb{R}^n})$ is given as follows. For every smooth function $f : V \to \mathbb{R}$ with $V \subset O$ open denote by $\overline{f} : A \cap V \to \acute{E}(\mathcal{E}^m)$ the mapping $z \mapsto [f]_z + \mathcal{J}^m(X, \mathbb{R}^n)_z$. Then $\acute{E}(\mathcal{E}^m)$ is equipped with the finest topology such that all $\overline{f}$ are continuous. Thus $\mathcal{E}^m(U)$ with $U \subset X$ open consists of all sections $F : U \to \acute{E}(\mathcal{E}^m)$ such that for every $x \in U$ there exists an open set $V \subset O$ with $x \in V$ and a function $f \in \mathcal{C}^m(V)$ such that $[F]_y = [f]_{x(y)} + \mathcal{J}^m(X, \mathbb{R}^n)_{x(y)}$ for all $y \in x^{-1}(V)$. Now, setting $F^{(0)}(x) := f(x)$ the value $F^{(0)}(x)$ does obviously not depend from the special choice of f, that means we can assign to F a function $F^{(0)} : X \to \mathbb{R}$ which lies in $\mathcal{C}^m(U)$ by definition of $\mathcal{E}^m$. Altogether one thus obtains a canonical epimorphism $\mathcal{E}^m \to \mathcal{C}^m$ of sheaves of commutative algebras.

The definition of the sheaf of Whitney functions does depend on the special choice of the global chart x, but in most applications this dependence does not play an essential role. Therefore we will often not denote it.

1.5.2 Let us now consider the general case of a stratified space X with a smooth or even only a $\mathcal{C}^m$-structure. Then choose a (countable) singular atlas of X consisting of singular charts $x_j : U_j \to O_j \subset \mathbb{R}^{n_j}$, $j \in J$. Moreover, let us abbreviate and write J_j^m instead of $\mathcal{E}^m_{U_j, x_j}$. Then we define for every open (or even locally closed) subset $U \subset X$ the space $J^m_{X,\mathcal{U}}(U)$ as the set of all families $F = (F_j)_{j\in J}$ of m-jets $F_j \in J_j^m(U \cap U_j)$ such that for every pair of indices j, i the relation

$$F_j^{(0)}{}_{|U\cap U_j\cap U_i} = F_i^{(0)}{}_{|U\cap U_j\cap U_i} \tag{1.5.1}$$

is satisfied. The subspace of all families $F = (F_j)_{j\in J}$ with $F_j \in \mathcal{E}_j^m$ will be denoted by $\mathcal{E}^m_{X,\mathcal{U}}(U)$. The restriction morphisms $r^U_{\tilde{U}} : J^m_{X,\mathcal{U}}(U) \to J^m_{X,\mathcal{U}}(\tilde{U})$ with $\tilde{U} \subset U \subset X$ open and analogously for $\mathcal{E}^m_{X,\mathcal{U}}(U)$ result immediately from those of J_j^m: for $F = (F_j) \in J^m_{X,\mathcal{U}}(U)_{j\in J}$ let $r^U_{\tilde{U}}(F) = \left(F_{j|\tilde{U}\cap U_j}\right)$. Then $J^m_{X,\mathcal{U}}$ and $\mathcal{E}^m_{X,\mathcal{U}}$ are presheaves, but fulfill the sheaf axioms as well, as one derives from the corresponding properties of J_j^m and $\mathcal{E}_j^m$. Finally, we thus have obtained the desired sheaves $J^m_{X,\mathcal{U}}$ and $\mathcal{E}^m_{X,\mathcal{U}}$. We will call them the sheaf of *m-jets* respectively the sheaf of *Whitney functions* on X of *class* $\mathcal{C}^m$. Let us mention explicitly that both sheaves depend on the special choice of the atlas $\mathcal{U}$.

Like in the Euclidean embeddable case there exists a natural sheaf morphism $\mathcal{E}^m_{X,\mathcal{U}} \to \mathcal{C}^m$ by assigning to $F \in \mathcal{E}^m_{X,\mathcal{U}}$ the function $F^{(0)} \in \mathcal{C}^m(U)$ with $F^{(0)}_{|U\cap U_j} = F_j^{(0)}$.

1.5.3 Example Let X be an n-dimensional manifold. Then, on the one hand there exists an embedding $x : X \to \mathbb{R}^N$ into some Euclidean space of large enough dimension and on the other hand an open covering $(U_j)_{j\in J}$ together with differentiable charts $x_j : U_j \to \mathbb{R}^n$. The sheaves of Whitney functions $\mathcal{E}^m_{X,x}$ and $\mathcal{E}^m_{X,\mathcal{U}}$ associated to these two initial situations are different in general, and comprise in a certain sense the two extreme examples of such sheaves. If $N > n$, then $\mathcal{E}^m_{X,x}(X)$ is equal to the space of m-jets of $x(X)$ in $\mathbb{R}^N$, and $\mathcal{E}^m_{X,\mathcal{U}}(X)$ is canonically isomorphic to the algebra of m-times continuously differentiable functions on X.

1.5.4 Proposition *The sheaf $\mathcal{E}^m_{X,\mathcal{U}}$ of Whitney functions of class $\mathcal{C}^m$ associated to a stratified space $(X, \mathcal{C}^m)$ with $\mathcal{C}^m$-structure and an atlas $\mathcal{U}$ is a fine sheaf. The same holds for the sheaf $J^m_{X,\mathcal{U}}$ of m-jets.*

PROOF: The proof can be performed analogously to the one for Theorem 1.3.13. □

1.5.5 The well-known norms defined in Appendix C on spaces of jets and Whitney functions can be carried over to $J^m_{X,\mathcal{U}}(U)$ and $\mathcal{E}^m_{X,\mathcal{U}}(U)$. To see this let $\mathcal{K} = (K_j)_{j\in J}$ be a family of compact sets $K_j \subset U_j \cap U$ and $I \subset J$ a finite family of indices. Then define seminorms $|\cdot|_{\mathcal{K},m,I}$ and $\|\cdot\|_{\mathcal{K},m,I}$ on $J^m_{X,\mathcal{U}}(U)$ resp. $\mathcal{E}^m_{X,\mathcal{U}}(U)$ by

$$|F|_{\mathcal{K},m,I} := \sum_{j\in I} |F_j|_{K_j,m}, \quad F \in J^m_{X,\mathcal{U}}(U), \quad \text{and}$$

$$\|G\|_{\mathcal{K},m,I} := \sum_{j\in I} \|G_j\|_{K_j,m}, \quad G \in \mathcal{E}^m_{X,\mathcal{U}}(U).$$

If now K_j runs through a compact exhaustion of U_j and I through all finite index sets, then $J^m_{X,\mathcal{U}}(U)$ and $\mathcal{E}^m_{X,\mathcal{U}}(U)$ become Fréchet algebras with seminorms $|\cdot|_{\mathcal{K},m,I}$ resp. $\|\cdot\|_{\mathcal{K},m,I}$.

1.6 Rectifiable curves and regularity

Any two points of a connected differentiable manifold can be connected by a curve of finite length. In a stratified space with smooth structure this need not be the case anymore. Moreover, for such spaces it is not immediately clear what to understand by a curve of finite length or in other words under a rectifiable curve. The goal of this section is to introduce the appropriate notions. Besides that we will introduce different regularity notions which locally in singular charts relate the length of a rectifiable curve to the Euclidean distance. These regularity notions will serve to better understand the metric properties of a singular space, but also to formulate and prove extension theorems for smooth functions on an (A)-stratified space.

1.6.1 First let us consider the following situation. Let (Y, d) be a path connected metric space. Then for every pair of points $x, y \in Y$ the *geodesic distance*

$$\delta(x,y) = \inf\left\{|\gamma| \mid \gamma \in \mathcal{C}([0,1];Y),\ \gamma(0) = x \text{ and } \gamma(1) = y\right\} \in [0,\infty]$$

is well-defined, where $\mathcal{C}([t^-,t^+],Y)$ is the set of all curves in Y defined on the interval $[t^-,t^+]$, and the *length* of the curve $\gamma : [t^-,t^+] \to Y$ is given by

$$|\gamma| = \sup\left\{\sum_{1\leq j\leq k} d(\gamma(t_j),\gamma(t_{j-1})) \mid k\in\mathbb{N},\ t^- = t_0 < t_1 < \cdots < t_{k-1} < t_k = t^+\right\}.$$

1.6.2 Explanation To avoid any confusion with respect to the used notation let us recall that by a *curve* or a *path* in a topological space Y we always understand a continuous mapping $\gamma : [t^-,t^+] \to Y$ with real numbers $t^- \leq t^+$. By an *arc* we understand the image of a curve.

The geodesic distance need not always be finite, as the following example from [59] shows.

1.6.3 Example Let $X \subset \mathbb{R}^3$ be the stratified subspace

$$X = \{0\} \cup \left\{(t, r_t\cos\theta, r_t\sin\theta) \in \mathbb{R}^3 \mid t > 0,\ \theta\in[0,2\pi],\ r_t = 2t + t\sin(1/t)\right\}.$$

Then X inherits from $\mathbb{R}^3$ a metric by restriction of the Euclidean distance. Any two points $x, y \in X\setminus\{0\}$ can obviously be connected by a rectifiable curve. But $\delta(0,x) = \infty$ holds for every point $x \in X\setminus\{0\}$, as the following calculation shows with $x = (t, r_t\cos\theta, r_t\sin\theta)$, $s_n := \frac{1}{2\pi n}$ and $s'_n := \frac{1}{2\pi n + \frac{\pi}{6}}$:

$$\begin{aligned}
\delta(0,x) &= \int_0^t \sqrt{1 + |\dot r_s|^2}\, ds \geq \int_0^t |\dot r_s|\, ds \\
&\geq -3 + \int_0^t \frac{1}{s}\left|\cos(1/s)\right| ds \geq -3 + \sum_{n\in\mathbb{N}^{>0}} \int_{s'_n}^{s_n} \frac{1}{s}\left|\cos(1/s)\right| ds \\
&\geq -3 + \sum_{n\in\mathbb{N}^{>0}} \frac{1}{12n+1} = \infty.
\end{aligned}$$

1.6.4 In case two points of a metric space (Y, d) can be connected by a *rectifiable* curve that means by a curve γ of finite length, then $\delta(x, y) < \infty$ holds for all $x, y \in Y$. In this case we say that Y is a *finitely path connected* space. If every point of (Y, d) has a basis of finitely path connected neighborhoods, then (Y, d) is called *locally finitely path connected.* A connected locally finitely path connected metric space is finitely path connected.

A curve $\gamma : [t^-, t^+] \to X$ in a stratified space $(X, \mathcal{C}^\infty)$ with smooth structure is called *rectifiable,* if there exists a neighborhood $U \subset X$ of $\gamma([t^-, t^+])$ and a singular chart $x : U \to \mathbb{R}^n$, such that the curve $x \circ \gamma$ is rectifiable with respect to the Euclidean metric. In analogy to the metric case one calls a stratified space $(X, \mathcal{C}^\infty)$ (*locally*) *finitely path connected,* if any two points of X can be connected by a rectifiable curve, respectively if every point has a basis of finitely path connected neighborhoods.

One now checks easily that the rectifiability of a curve in X is independent of the special choice of the singular chart that means if $x \circ \gamma$ is a rectifiable curve with $x : U \to \mathbb{R}^n$ a singular chart, then $y \circ \gamma$ is rectifiable as well for every chart $y : V \to \mathbb{R}^N$ compatible with x and which satisfies $\mathrm{im}\,\gamma \subset V \subset U$.

1.6.5 Let us suppose that A is a subset of the Euclidean space $\mathbb{R}^n$ and that d is the Euclidean distance restricted to A. Obviously the geodesic distance $\delta(x, y)$ on A has a lower bound from below given by the Euclidean distance, that means

$$d(x, y) = \|x - y\| \leq \delta(x, y), \qquad x, y \in A.$$

Vice versa this need not be the case anymore, so the two metrics are in general not equivalent. Therefore we make the following definition.

1.6.6 Definition (Tougeron [170, Def. 3.10]) A compact set $K \subset \mathbb{R}^n$ is called l-*regular* with $l \in \mathbb{R}^{\geq 1}$, if K is finitely path connected and there exists a constant $C > 0$, such that for all $x, y \in K$

$$\delta(x, y) \leq C d(x, y)^{1/l}.$$

A locally closed connected and locally finitely path connected set $A \subset \mathbb{R}^n$ is called l-*regular,* if each of its points has a compact l-regular neighborhood. If for every point z of A there is an $l \in \mathbb{R}^{\geq 1}$ depending on z and an l-regular compact neighborhood $K \subset \mathbb{R}^n$ we will say that A is *Whitney–Tougeron regular* or briefly that A is *regular.*

Finally one calls a connected stratified set $(X, \mathcal{C}^\infty)$ with smooth structure *Whitney–Tougeron regular* or *regular,* (resp. l-*regular*), if there exists a covering of X by singular charts $x : U \to O \subset \mathbb{R}^n$ such that $x(U)$ is a regular (resp. l-regular) set in Euclidean space.

By a simple calculation one shows that l-regularity is invariant under diffeomorphisms between open subsets of $\mathbb{R}^n$. This implies for the stratified case that l-regularity of $x(U)$ entails l-regularity of $y(U)$ for any further singular chart compatible with x.

1.6.7 Example Every subanalytic set $X \subset \mathbb{R}^n$ is Whitney–Tougeron regular. For a proof of this fact see Kurdyka–Orro [107, Cor. 2].

1.6.8 Proposition *Every* (δ_l)*-stratified space* $(X, \mathcal{C}^\infty)$ *is locally finitely path connected and* l*-regular in the sense of Tougeron. In particular, every Whitney stratified space is* 1*-regular.*

For the case $l = 1$ this result has been proved in BEKKA–TROTMAN [11]. There, one can also find further methods, how to check whether a space composed by manifolds (but which need not be a stratified space) is finitely path connected.

PROOF: Without loss of generality we can assume that X is a connected subset of $\mathbb{R}^n$. Let $x \in X$ be a point, S the stratum of x and U an open ball around x such that U meets only finitely many strata of X and such that for every stratum R with $R \cap U \neq \emptyset$ the relation $R \geq S$ holds. After shrinking U and choosing $\delta > 0$ appropriately the following relation holds for all $z \in S \cap U$ and $y \in (X \setminus S) \cap U$

$$\|P_y(z-y)\| \geq \delta \|z-y\|^{2-1/l}.$$

Hereby P_y is the orthogonal projection onto the tangent space T_yR of the stratum of y. Next we supply S with the Riemannian metric induced by the Euclidean scalar product and consider the corresponding exponential function exp. After further shrinking U one can achieve that $U \cap S = \exp B_r$, where B_r is the ball around the origin with radius $r < 1$ and exp shall be injective on a neighborhood of $\overline{B_r}$. For every $y \in S \cap U$ define the path $\gamma_y : [0,1] \to S$ by $\gamma_y(t) = \exp(tw)$, where $w = \exp_x^{-1}(y)$ and $t \in [0,1]$. Then there exists $C_1 > 0$ such that

$$|\gamma_y| < C_1 \|y - x\| \leq C_1 \|y - x\|^{1/l}, \tag{1.6.1}$$

where the second inequality follows from $\|y-x\| < r < 1$ and $l \geq 1$. Now we consider the strata $R > S$. For every such stratum define a vector field $V_R : R \to TR$ by

$$V_R(y) = \frac{\|x-y\|}{\|P_y(x-y)\|^2} P_y(x-y).$$

For every $y \in R$ let $t_y^+ > 0$ be the positive escape time of y with respect to V_R, that means there is a smooth curve $\gamma_y : [0, t_y^+[\to R$ fulfilling

$$\dot{\gamma}_y(t) = V_R(\gamma_y(t)), \quad t \in [0, t_y^+[,$$

and t_y^+ is maximal with this property. As

$$\begin{aligned} \frac{d}{dt}\|\gamma_y(t) - x\| &= \frac{d}{dt}\sqrt{\langle \gamma_y(t) - x, \gamma_y(t) - x\rangle} = \frac{\langle \dot{\gamma}_y(t) - x, \gamma_y(t) - x\rangle}{\sqrt{\langle \gamma_y(t) - x, \gamma_y(t) - x\rangle}} \\ &= -\frac{1}{\|P_{\gamma_y(t)}(x - \gamma_y(t))\|^2} \langle P_{\gamma_y(t)}(x - \gamma_y(t)), x - \gamma_y(t)\rangle = -1 \end{aligned}$$

holds, one has for $0 \leq t < t_y^+$

$$\|\gamma_y(t) - x\| = -t + \|y - x\|. \tag{1.6.2}$$

We calculate further

$$\begin{aligned} \int_0^t \|\dot{\gamma}_y(s)\|\, ds &= \int_0^t \frac{\|\gamma_y(s) - x\|}{\|P_{\gamma_y(s)}(x - \gamma_y(s))\|} ds \leq \frac{1}{\delta}\int_0^t \|\gamma_y(s) - x\|^{1/l - 1} ds \\ &= \frac{1}{\delta l}\left(\|y - x\|^{1/l} - (\|y - x\| - t)^{1/l}\right), \end{aligned} \tag{1.6.3}$$

hence γ_y has finite length. As U is a relatively compact subset of $\mathbb{R}^n$, the set $Z := \overline{\gamma_y([0,t_y^+[)} \setminus \gamma_y([0,t_y^+[)$ cannot be empty. Moreover, by (1.6.2) the set Z is contained in U. Suppose that Z has two different elements z and z'. Then there exist two disjoint balls B and B' around z resp. z' and γ has to oscillate infinitely often between B and B'. In particular γ cannot have finite length. Consequently Z consists of exactly one point y_1 contained in in the stratum $R_1 \leq R$. By the maximality of t_y^+ the stratum R_1 cannot be equal to R, hence $R_1 < R$. If $R_1 \neq S$ we continue and obtain as end point of the path γ_{y_1} a point y_2 of a stratum $R_2 < R_1$. Recursively we thus obtain a sequence of points $y = y_0, y_1, \cdots, y_k$ and of strata $R > R_1 > \cdots > R_k$ with $y_l = \lim_{t \to t_{l-1}^+} \gamma_{y_{l-1}}(t) \in R_l$, where $1 \leq l \leq k$ and $t_{l-1}^+ = t_{y_{l-1}}^+$. As U contains only finitely many strata, the recursion terminates after finitely many steps, that means $R_k = S$ for an appropriate k. Now, we assign to the point $y_k \in S$ the path $\gamma_{y_k} \subset S$ defined above; it connects y_k with x within S. The total length of the path γ composed by $\gamma_y, \gamma_{y_1}, \cdots, \gamma_{y_k}$ then sums up by (1.6.3), (1.6.2) and (1.6.1) to

$$\begin{aligned} |\gamma| &\leq C\left(\left(\|y-x\|^{1/l} - \|y_1 - x\|^{1/l}\right) + \left(\|y_1 - x\|^{1/l} - \|y_2 - x\|^{1/l}\right) + \cdots + \|y_k - x\|^{1/l}\right) \\ &= C\,\|y-x\|^{1/l}, \end{aligned}$$

where $C = \max\{\frac{1}{\delta l}, C_1\}$. Consequently, every point y of $X \cap U$ can be connected with x by a rectifiable path in X, and

$$\delta(x,y) \leq C\,\|y-x\|^{1/l}.$$

holds. This proves the claim. □

In general it is not possible to connect any two points of a connected regular (A)-stratified space by a piecewise differentiable path, but at least it is possible to connect them by a so-called weakly piecewise differentiable path.

1.6.9 Definition Let X be a stratified space with smooth structure. A curve γ in X is called *weakly piecewise* $\mathcal{C}^1$, there exists a relatively open subset $I \subset [t^-, t^+]$ with countable complement and a singular chart $x : U \to X$ around $\gamma([t^-, t^+])$ such that $x \circ \gamma_{|I}$ is continuously differentiable, such that $\overline{I} = [t^-, t^+]$ holds and

$$\int_I \|Tx(\dot\gamma(t))\|\,dt < \infty. \tag{1.6.4}$$

The set I then is called the *differentiability set* of γ. If I has the shape $I = [t^-, t^+] \setminus \{t_0, \cdots, t_k\}$ with $t^- = t_0 < \cdots < t_k = t^+$, and if $x \circ \gamma_{|[t_j, t_{j+1}]}$ is continuously differentiable for $j = 0, \cdots, k-1$, then one calls γ *piecewise* $\mathcal{C}^1$.

The space of all piecewise-differentiable $\mathcal{C}^1$-curves from $[t^-, t^+]$ to X will be denoted by $\mathcal{C}^{w1}([t^-, t^+]; X)$.

A weakly piecewise $\mathcal{C}^1$-curve is obviously rectifiable, where the length of $x \circ \gamma$ is given by the integral in (1.6.4).

By Proposition 1.3.16 every curve composed of finitely many rectifiable resp. weakly piecewise $\mathcal{C}^1$-curves is rectifiable resp. weakly piecewise again.

1.6.10 Lemma *Let $\lambda : [t^-, t^+] \to X$ be a rectifiable curve and $x : U \to \mathbb{R}^n$ a singular chart around $\lambda([t^-, t^+])$. Then there exists a weakly-piecewise $\mathcal{C}^1$-curve $\gamma : [t^-, t^+] \to X$ fulfilling the following estimation:*

$$|x \circ \gamma| \leq |x \circ \lambda|. \tag{1.6.5}$$

PROOF: Without loss of generality we can assume $X \subset \mathbb{R}^n$. As every stratum $S \subset X$ is locally closed, $I_S := \{t \in [t^-, t^+] \mid \lambda(t) \in S\}$ has to be a locally closed subset of $[t^-, t^+]$. With the help of the fact that every open subset of $\mathbb{R}$ is the union of countably many open intervals one checks easily that I_S is the disjoint union of countably many open, closed or semiclosed intervals. In particular, I_S has countably many connected components. Explicitely

$$I_S = \bigcup_{k \in \mathbb{N}} I_{S,k},$$

where $I_{S,k} \cap I_{S,l} = \emptyset$ for $k \neq l$ and $I_{S,k}$ is either empty or one of the intervals $]t^-_{S,k}, t^+_{S,k}[$, $[t^-_{S,k}, t^+_{S,k}]$, $]t^-_{S,k}, t^+_{S,k}]$, or $[t^-_{S,k}, t^+_{S,k}[$ with $t^-_{S,k} < t^+_{S,k}$. We now define an open set

$$\tilde{I} := \bigcup_{(S,k) \in J} I^\circ_{S,k},$$

where J consists of all pairs $(S, k) \in \mathcal{S} \times \mathbb{N}$ with $I^\circ_{S,k} \neq \emptyset$. The complement $[t^-, t^+] \setminus \tilde{I}$ contains at most countably many points, and $\overline{\tilde{I}} = [t^-, t^+]$ holds. Next we choose for $(S, k) \in J$ a strictly monotone decreasing sequence $(t^-_{S,k,-l})_{l \in \mathbb{N}} \subset I^\circ_{S,k}$ with $\lim_{l \to \infty} t^-_{S,k,-l} = t^-_{S,k}$ and a strictly monotone increasing sequence $(t^+_{S,k,l})_{l \in \mathbb{N}} \subset I^\circ_{S,k}$ with $t^-_{S,k,0} < t^+_{S,k,0}$ and $\lim_{l \to \infty} t^+_{S,k,l} = t^+_{S,k}$. Then define for $l \in \mathbb{N}^{>0}$ further points $t^+_{S,k,-l} := t^-_{S,k,-l+1}$ and $t^-_{S,k,l} := t^+_{S,k,l-1}$. Finally set for $l \in \mathbb{Z}$

$$I_{S,k,l} := [t^-_{S,k,l}, t^+_{S,k,l}].$$

According to construction $\lambda(I_{S,k,l})$ then lies in the stratum S, and $\bigcup_{l \in \mathbb{Z}} I_{S,k,l} = I^\circ_{S,k}$ holds. The times $t^\pm_{S,k,l}$ can obviously chosen in such a way that the points $\lambda(t^-_{S,k,l})$ and $\lambda(t^+_{S,k,l})$ can be connected by differentiable curves of $\gamma_{S,k,l} : I_{S,k,l} \to S$ of minimal length, where S is equipped with the Riemannian metric induced by the Euclidean scalar product. Then

$$|\gamma_{S,k,l}| \leq |\lambda_{|I_{S,k,l}}|$$

holds. We now put together the curves $\gamma_{S,k,l}$ to a path $\gamma : [t^-, t^+] \to X$ in the following way:

$$t \mapsto \gamma(t) = \begin{cases} \gamma_{S,k,l}(t) & \text{if } t \in I_{S,k,l}, \\ \lambda(t) & \text{if } t \notin \tilde{I}. \end{cases}$$

According to construction γ then becomes a weakly piecewise $\mathcal{C}^1$-curve with differentiability set $I = \tilde{I} \setminus \bigcup_{(S,k,l) \in J \times \mathbb{Z}} \{t^\pm_{S,k,l}\}$ and satisfies inequality (1.6.5). □

Next we will introduce some first implications from Whitney–Tougeron regularity for Whitney functions on the considered space.

1.6.11 Lemma (Tougeron [170, Rem. 2.5]) *Let $x, y \in \mathbb{R}^n$ be two points and γ a rectifiable curve connecting x with y. Furthermore, let $g \in \mathcal{C}^m(\mathbb{R}^n)$, $m \in \mathbb{N}^{>0}$ be a function flat on x of order $(m-1)$ that means $D^\alpha g(x) = 0$ holds for all $\alpha \in \mathbb{N}^n$ with $|\alpha| \leq m-1$. Then $g(y)$ satisfies the following estimate:*

$$|g(y)| \leq n^{m/2}\, |\gamma|^m \sup_{\substack{\xi \in \operatorname{im}\gamma \\ |\alpha|=m}} |D^\alpha g(\xi)|. \tag{1.6.7}$$

Proof: According to the mean value theorem one has for every function $g \in \mathcal{C}^m(\mathbb{R}^n)$

$$|g(y) - g(x)| \leq \sqrt{n}\, |x-y| \sup_{\substack{\xi \in [x,y] \\ |\alpha|=1}} |D^\alpha g(\xi)|,$$

where $[x, y] \subset \mathbb{R}^n$ is the segment connecting x with y. Then we have for every piecewise linear path γ connecting x with y:

$$|g(y) - g(x)| \leq \sqrt{n}\, |\gamma| \sup_{\substack{\xi \in \operatorname{im}\gamma \\ |\alpha|=1}} |D^\alpha g(\xi)|. \tag{1.6.8}$$

Passing to the limit one shows that (1.6.8) is true for arbitrary rectifiable paths γ. If now g is flat over x of order $(m-1)$, then an easy induction argument proves the estimate (1.6.7). □

Now let $K \subset \mathbb{R}^n$ be an l-regular compact set, and F a Whitney function of class $\mathcal{C}^m$, $m < \infty$ over K. According to Whitney's extension theorem C.3.2 there exists a function $f \in \mathcal{C}^m(\mathbb{R}^n)$ such that $F = J^m(f)$. For $\alpha \in \mathbb{N}^n$ with $|\alpha| \leq m$ and $x \in K$ the function $g = D^\alpha(f - T_x^m F)$ is an element of $\mathcal{C}^{m-|\alpha|}(\mathbb{R}^n)$. Moreover, g is flat on x of order $m - |\alpha|$, and equal to the rest term $(R_x^m F)^{(\alpha)}$ over K. Hence by Lemma 1.6.11 we have for all $x, y \in K$ the following estimate

$$\begin{aligned} |(R_x^m F)^{(\alpha)}(y)| &\leq n^{\frac{m-|\alpha|}{2}}\, \delta(x,y)^{m-|\alpha|} \sup_{\substack{\xi \in K \\ |\beta|=m}} |F^{(\beta)}(\xi) - F^{(\beta)}(x)| \\ &\leq 2n^{\frac{m-|\alpha|}{2}}\, \delta(x,y)^{m-|\alpha|}\, |F|_{K,m}. \end{aligned} \tag{1.6.9}$$

This proves the first part of the following proposition.

1.6.12 Proposition (Whitney [187], Tougeron [170, Prop. 2.6]) *Let $K \subset \mathbb{R}^n$ be compact. If the Euclidean and geodesic distance on K define equivalent metrics, that means in other words if K is a 1-regular set, then for all $m \in \mathbb{N}$ the seminorms $\|\cdot\|_{K,m}$ and $|\cdot|_{K,m}$ are equivalent on $\mathcal{E}^m(A)$.*

(Tougeron [170, Prop. 3.11]) *If K is l-regular with $l \in \mathbb{R}^{\geq 1}$, then there exists a constant $C_m > 0$ such that for all $F \in \mathcal{E}^\infty(K)$*

$$\|F\|_{K,m} \leq C_m\, |F|_{K,ml}.$$

Proof: Let $\alpha \in \mathbb{N}^n$, $|\alpha| \leq m$. Then there exists a constant $C > 0$ such that for all $x, y \in K$ and all $F \in \mathcal{E}^\infty(K)$

$$|(R_x^m F)^{(\alpha)}(y)| \leq |(R_x^{ml} F)^{(\alpha)}(y)| + C|x-y|^{m-|\alpha|}\, |F|_{K,ml}.$$

On the other hand there exists by assumption and (1.6.9) a constant $D > 0$ such that for all $F \in \mathcal{E}^\infty(K)$

$$\begin{aligned}\left|(R_x^{ml}F)^{(\alpha)}(y)\right| &\leq 2n^{\frac{m-|\alpha|}{2}}\delta(x,y)^{ml-|\alpha|}\,|F|_{K,ml}\\ &\leq D|x-y|^{m-|\alpha|}|F|_{K,ml}.\end{aligned}$$

The claim now follows from these two estimates. □

1.6.13 Corollary *For every regular stratified space* X *with smooth structure and every atlas* $\mathfrak{U}$ *of* X *the space* $\mathcal{E}^\infty_{X,\mathfrak{U}}(X)$ *is a closed subspace of* $J^\infty_{X,\mathfrak{U}}(X)$.

The notion of regularity according to TOUGERON given in definition 1.6.6 for a locally closed subset $A \subset \mathbb{R}^n$ does not make a statement about the behavior of the geodesic distance of two points $x, y \in A$ with respect to their Euclidean distance, when the two points x and y approach the boundary ∂A of A. FERRAROTTI and WILSON have introduced in their work [59] a new notion of regularity which considers the behavior of the geodesic distance near the boundary. The important application of this new kind of regularity is the extension theory of smooth functions which will be the topic of the next section. To be compatible in our notation with l-regularity as defined above we have appropriately adapted the regularity notion of [59]. The (r, l)-regularity defined in the following corresponds to (r, l^{-1})-regularity in [59].

1.6.14 Definition (cf. FERRAROTTI–WILSON [59, Sec. I]) Let $A \subset \mathbb{R}^n$ be a connected, locally closed and locally finitely path connected set, and $Z \subset A$ a locally connected finitely path connected *boundary set*, that means $Z \subset \partial(A \setminus Z)$. The one calls A (r, l)-*regular relative* Z, where $l, r \in \mathbb{R}^{\geq 1}$, if for every point $z \in Z$ there exists a neighborhood $V \subset \mathbb{R}^n$ of z such that $V \cap A$ is finitely path connected and such that the following axioms (RA1) and (RA2) or (RB1) and (RB2) hold for some constants $C_1, C_2, D > 0$:

(RA1) $\delta(x, Z) \leq C_1\, d(x, Z)^{1/l}$ for every $x \in V \cap A \setminus Z$.

(RA2) $d(x, y) \leq D \max\{d(x, Z), d(y, Z)\}^r$ with $x, y \in V \cap A \setminus Z$ implies

$$\delta(x,y) \leq C_2\, d(x,y) \max\{d(x,Z), d(y,Z)\}^{1/l-r}.$$

(RB1) $\delta(x, y) \leq C_1\, d(x, y)^{1/l}$ for all $x, y \in V \cap A \setminus Z$.

(RB2) $d(x, y) \leq D \max\{d(x, Z), d(y, Z)\}^r$ with $x, y \in V \cap A \setminus Z$ implies

$$\delta(x,y) \leq C_2\, d(x,y) \max\{d(x,Z), d(y,Z)\}^{r(1/l-1)}.$$

In the first case one says that A is (r, l)-regular of *type* RA, in the second of *type* RB. In case there exists for every $z \in Z$ a neighborhood in A satisfying axiom (RA1) (resp. (RB1)) we will say briefly that A satisfies (RA1) (resp. (RB1)) relative Z with exponent l.

A stratum S of a stratified space X with smooth structure, or more precisely the closure $\overline{S}$ is called (r,l)-*regular* (of *type* RA resp. *type* RB) relative a relatively ∂S open subset $Z \subset \partial S$, if there exists a covering of X by singular charts $x : U \to \mathbb{R}^n$ such that $x(S \cap U)$ is an (r,l)-regular set (of type RA resp. type RB) with respect to $x(Z \cap U)$.

Finally we call A (resp. S) *Ferrarotti–Wilson regular* or briefly *FW regular* with respect to Z, if for every point z of Z there exists a neighborhood in A which is (r,l)-regular with respect to $Z \cap U$ for appropriate constants $l, r \geq 1$ depending on z.

1.6.15 Example Every 1-regular space $A \subset \mathbb{R}^n$ is $(1,1)$-regular (of type RA and of type RB) with respect to any closed, finitely path connected boundary set $Z \subset A$.

1.6.16 Example For every subanalytic set $X \subset \mathbb{R}^n$ the closure $\overline{X}$ is FW regular with respect to the boundary ∂X (see [59]).

Next we assign for $m \in \mathbb{N}^{>0}$ to every (r,l)-regular space A resp. to every stratum S a *critical constant* $c_A^m(r,l)$ resp. $c_S^m(r,l)$:

$$c_{A/S}^m(r,l) = \begin{cases} m(lr-1) & \text{if A (resp. S) is of type RA,} \\ mr(l-1)\frac{l'}{l} & \text{if A (resp. S) is of type RB.} \end{cases}$$

Hereby $l' \leq l$ is in the type RB case an exponent such that (RA1) is satisfied for A resp. X. Such an exponent exists according to the following lemma.

1.6.17 Lemma (Ferrarotti–Wilson [59, Lem. I.1]) *Let $V \subset \mathbb{R}^n$ be a neighborhood of point $z \in Z$ such that $V \cap A$ is finitely path connected and $l \in \mathbb{R}^{>0}$. Then* (RA1) *is equivalent to*

(RA1)' *There is an estimate*

$$\hat{\delta}(x,y) := \inf\{\delta(x,y), \delta(x,Z) + \delta(y,Z)\} \leq C \max\{d(x,Z), d(y,Z)\}^{1/l}$$

for all $x, y \in V \cap A \setminus Z$ and an appropriate $C > 0$.

Moreover, (RA1) *implies the estimate $l \geq 1$.*

If finally the axiom (RB1) *is satisfied over $V \cap A$, then after shrinking V the axiom* (RA1) *holds over $V \cap A$.*

Proof: If (RA1) holds, then (RA1)' follows immediately, as

$$\begin{aligned} \hat{\delta}(x,y) &\leq \delta(x,Z) + \delta(y,Z) \\ &\leq C_1\big(d(x,Z)^{1/l} + d(y,Z)^{1/l}\big) \leq 2C_1 \max\{d(x,Z)^{1/l}, d(y,Z)^{1/l}\}. \end{aligned}$$

If conversely (RA1)' is satisfied, then choose for $x \in V \cap A \setminus Z$ points $y_j \in V \cap A \setminus Z$, $j \in \mathbb{N}$ with $y_0 = x$ and $d(y_j, z) \leq \frac{1}{2^{jl}} d(x,Z)$ for all $j \geq 1$. Afterwards choose rectifiable curves γ_j from y_j to y_{j+1} such that the length μ_j of the part passing through $V \cap A \setminus Z$ is $\leq 2\hat{\delta}(y_j, y_{j+1})$. Composing the γ_j to a path γ in $V \cap A$, then γ is continuous by construction and the part $(\operatorname{im}\gamma)\setminus Z$ lying in $V\cap A\setminus Z$ has finite length. By assumption there exists a $C > 0$ with

$$\begin{aligned}\delta(x,Z) &\le \mu \le 2\hat{\delta}(x,y_1) + \sum_{j\ge 1} 2\hat{\delta}(y_j,y_{j+1}) \\ &\le 2C\Big(d(x,Z)^{1/l} + \sum_{j\ge 1} d(y_j,Z)^{1/l}\Big) \\ &\le 2C\Big(d(x,Z)^{1/l} + \sum_{j\ge 1} \frac{1}{2^j} d(x,Z)^{1/l}\Big) \le 4C d(x,Z)^{1/l},\end{aligned}$$

that means (RA1) is satisfied.

If $l < 1$, then (RA1) would entail for x close enough to Z the following chain of inequalities:

$$d(x,Z) \le \delta(x,Z) \le C_1 d(x,Z)^{1/l} < d(x,Z),$$

which is impossible.

Now let us come to the last claim. First shrink V to a neighborhood $\tilde{V}$ of z such that for every $x \in \tilde{V}\cap A\setminus Z$ there exists $y \in V\cap Z$ with $d(x,Z) = d(x,y)$. Then one can find for $\varepsilon > 0$ a sequence of points $y_j \in V\cap A\setminus Z$ converging to y such that $y_0 = x$, $d(y_0,y_1)^{1/l} \le d(x,y)^{1/l} + \varepsilon$ and $d(y_j,y_{j+1})^{1/l} \le \frac{\varepsilon}{2^j}$ for all $j \ge 1$. Additionally there are rectifiable curves γ_j from y_j to y_{j+1} such that $|\gamma_j| \le \delta(y_j,y_{j+1}) + \frac{\varepsilon}{2^j}$. The path γ composed by the γ_j connects x with y and has finite length which by assumption has the following upper bound:

$$\begin{aligned}|\gamma| &\le \sum_{j\in\mathbb{N}} |\gamma_j| \le \sum_{j\in\mathbb{N}} \delta(y_j,y_{j+1}) + \frac{\varepsilon}{2^j} \\ &\le C_1\Big(\sum_{j\in\mathbb{N}} d(y_j,y_{j+1})^{1/l}\Big) + 2\varepsilon \\ &\le C_1 d(x,b)^{1/l} + 2(C_1+1)\varepsilon.\end{aligned}$$

As ε was arbitrary, $\delta(x,Z) \le C_1 d(x,Z)$ follows, hence the claim. □

Finally in this section we will introduce a notion which rules the behavior of the distance of two subsets of Euclidean space or an (A)-stratified space while approaching the common intersection.

1.6.18 Definition (cf. [170, Def. 4.4], [118, Sec. I.5]) Let A, Z be two closed subsets of $O \subset \mathbb{R}^n$ open. Then one says that A, Z are *regularly situated* (*in* O), if for every point $z \in A\cap Z$ there exists a neighborhood V as well as constants $c \in \mathbb{N}$ and $C > 0$ such that

(RS) $$d(x,A) + d(x,Z) \ge C d(x, A\cap Z)^c \qquad \text{for all } x \in V.$$

If A, Z are closed subsets of an (A)-stratified space X then A, Z are called *regularly situated* (*relative* Z), if there exists a singular atlas $\mathfrak{U} = (U_j, x_j)_{j\in J}$ of X such that for every singular chart x_j the sets $\overline{x_j(A\cap U_j)}$ and $\overline{x_j(Z\cap U_j)}$ are regularly situated in $\mathbb{R}^{n_j}$.

An open neighborhood T of a submanifold $S \subset O$ or of an open submanifold S of a stratum of X is called *regularly situated*, if the relatively closed sets $O \setminus T$ and $\overline{S} \cap O \subset O$ resp. $X \setminus T$ and $\overline{S} \subset X$ are regularly situated.

The following rather useful result follows by a simple argument.

1.6.19 Lemma *Two relatively closed subsets* $A, Z \subset O$ *are regularly situated, if and only if for every point* $z \in A \cap Z$ *there exists a neighborhood* V *as well as constants* $c' \in \mathbb{N}$ *and* $C' > 0$ *such that*

(RS') $$d(x, Z) \geq C' d(x, A \cap Z)^{c'} \quad \text{for all } x \in V.$$

1.7 Extension theory for Whitney functions on regular spaces

A function $f \in \mathcal{C}^\infty(X)$ on an (A)-stratified space X induces over every stratum S a smooth restricted function $f_{|S} \in \mathcal{C}^\infty(S)$. If on the other hand a $\mathcal{C}^\infty$-function $g : S \to \mathbb{R}$ is given, then it is rather easy to see whether g can be extended to a continuous function on X. As X is a normal topological space this is exactly then the case, if for every $x \in \partial S$ and every convergent sequence $x_k \to x$ of points of S the limit $\lim_{k \to \infty} g(x_k)$ exists and if this limit is independent of the special choice of the sequence $(x_k)_{k \in \mathbb{N}}$. Now the much further reaching question arises, in particular in view of analytic applications, whether it is possible to give reasonable criteria to a smooth $g : S \to \mathbb{R}$ which guarantee that g has a smooth extension to X. Though it seems to be impossible to find such a condition as easy as in the continuous case, the theory of jets and Whitney functions will give us good tools in our hands which can help in many situations. In particular we then will be able to find criteria, when a Whitney function $G : S \to \mathbb{R}$ of $\mathcal{C}^m$ falling fast enough at the boundary of S can be extended to a Whitney function F on X. But before we come to the details let us explain what to understand by a Whitney function "flat at the boundary".

To simplify notation let us fix for the rest of this section a singular atlas $\mathcal{U} = (U_j, x_j)_{j \in J}$ on the given (A)-stratified space X with smooth structure $\mathcal{C}^\infty$.

1.7.1 Definition Let $m \in \mathbb{N} \cup \{\infty\}$, $A \subset \mathbb{R}^n$ be a locally closed set, and $Z \subset A$ relatively closed. Then an m-jet $F \in J^m(A)$ on A is called *flat* of *order* $c \in \mathbb{R}^{\geq 0}$ over Z, if the following conditions are satisfied:

(FJ1) $F_{|Z} = 0$.

(FJ2) For every point $z \in Z$ and all $\alpha \in \mathbb{N}^n$, $|\alpha| \leq m$ the following relation holds:

$$\lim_{\substack{x \to z \\ x \in A \setminus Z}} \frac{F^{(\alpha)}(x)}{d(x, Z)^c} = 0.$$

An m-jet $F \in J^m_{\mathcal{U}}(\overline{S})$ on the closure of a stratum S of X is called *flat* of *order* c over a closed set $Z \subset \partial S$, if for every singular chart x_j the m-jet $F_j \in J^m(x_j(\overline{S} \cap U_j))$ is flat over $x_j(Z \cap \mathcal{U}_j)$ of order c. Th space of m-jets on A (resp. $\overline{S}$) which are flat of order c over Z will be denoted by $\mathcal{J}^{m,c}(Z; A)$ (resp. $\mathcal{J}^{m,c}_{\mathcal{U}}(Z; \overline{S})$).

Finally we set

$$\mathcal{J}^m(Z; A) = \{F \in \mathcal{E}^m(A) \mid F^{(\alpha)}_{|Z} = 0 \text{ for all } |\alpha| \le m\},$$
$$\mathcal{J}^m_{\mathcal{U}}(Z; \overline{S}) = \{F \in \mathcal{E}^m_{\mathcal{U}}(\overline{S}) \mid F^{(\alpha)}_{|Z} = 0 \text{ for all } |\alpha| \le m\}.$$

Note that this definition coincides with the one given in Appendix C, if A is an open subset of $\mathbb{R}^n$. For such A this means $\mathcal{J}^m(Z; A) = \ker J^m$, where $J^m : \mathcal{C}^m(A) \to \mathcal{E}^m(Z)$ is the jet map.

Now we have all necessary ingredients to formulate and proof the announced extension result.

1.7.2 Generalized Lemma of Hesténès (Ferrarotti–Wilson [59]) *Let $A \subset \mathbb{R}^n$ be locally closed, $Z \subset A$ a relatively closed, locally finitely path connected boundary set with respect to which A is (r, l)-regular, and F a jet of order m on A. If the restriction $F_{|A \setminus Z}$ is a $\mathcal{C}^m$-Whitney function on $A \setminus Z$, and F flat over Z of order $c \ge c^m_A(r, l)$, then F is a Whitney function of class $\mathcal{C}^m$ on A.*

The generalized lemma of Hesténès immediately entails the following main result of this section.

1.7.3 Extension Theorem *Let X be a stratified space with smooth structure, S an (r, l)-regular stratum, and $\mathcal{U}$ a covering of X by charts. Then every m-jet $G \in J^m_{X,\mathcal{U}}(\overline{S})$ which is flat over ∂S of order $c \ge c^m_S(r, l)$ can be extended to a Whitney function $F \in \mathcal{E}^m_{X,\mathcal{U}}(X)$ with $F_{|S} = G$ and $F_{|X \setminus S} = 0$.*

Proof of 1.7.2: We follow the presentation given in [59]. By Whitney's extension theorem C.3.2 and by the fact that $F_{|A \setminus Z}$ is a $\mathcal{C}^m$-Whitney function it suffices to show that for every $z \in Z$ and all $\alpha \in \mathbb{N}^n$, $|\alpha| \le m$ the relation

$$\frac{(R^m_x F)^{(\alpha)}(y)}{|x - y|^{m-|\alpha|}} \to 0, \tag{1.7.1}$$

holds, if x and y converge to z. Now let $V \cap A$ with $V \subset \mathbb{R}^n$ be an open finitely connected neighborhood of z according to Definition 1.6.14. Obviously V can be chosen such that $V \cap Z$ is finitely path connected. Let furthermore $\gamma : [0, 1] \to V \cap A$ be a rectifiable curve with $\gamma(0) = x$ and $\gamma(1) = y$. The path $\Delta := \operatorname{im} \gamma$ possesses a unique partition of the form $\Delta = \Delta_1 \cup \Delta_2$ with $\Delta_1 \subset X \setminus Z$ and $\Delta_2 \subset Z$. Obviously the Δ_i are rectifiable. As $V \cap Z$ is finitely path connected, we can choose γ such that $|\Delta_1| \le 2\hat{\delta}(x, y)$ (with $\hat{\delta}$ given in Lemma 1.6.17) and such that

$$\sup_{\xi \in \Delta} d(\xi, Z) \le \max\{d(x, Z), d(y, Z)\} + 2\hat{\delta}(x, y) \le 3 \max\{\delta(x, Z), \delta(y, Z)\}. \tag{1.7.2}$$

By assumption $F_{A \setminus Z}$ is a Whitney function, hence there exists $f \in \mathcal{C}^m(O)$ with $F_{A \setminus Z} = J^m(f)$, where $O \subset \mathbb{R}^n$ is open and $A \setminus Z \subset O$ closed. The function $g = D^\alpha(f - T^m_x F) \in \mathcal{C}^{m-|\alpha|}(O)$ with $\alpha \in \mathbb{N}^n$, $|\alpha| \le m$ then is flat on x of order $m - |\alpha|$, and

$(R_x^m F)^{(\alpha)}(y) = g(y)$ holds for all $y \in A \setminus Z$. By $F_{|Z} = 0$ and the compactness of Δ an argumentation analogous to the one in the proof of Lemma 1.6.11 shows the following estimate:

$$\begin{aligned} \left|(R_x^m F)^{(\alpha)}(y)\right| &\leq n^{\frac{m-|\alpha|}{2}}\, \hat{\delta}(x,y)^{m-|\alpha|} \sup_{\substack{\xi\in\Delta_1 \\ |\beta|=m}} \left|F^{(\beta)}(\xi) - F^{(\beta)}(x)\right| \\ &\leq C\,\hat{\delta}(x,y)^{m-|\alpha|}\, |F|_{\Delta,m}. \end{aligned} \tag{1.7.3}$$

On the other hand the flatness assumption on F and Eq. (1.7.2) entail

$$|F|_{\Delta,m} = o\left(\sup_{\xi\in\Delta} d(\xi,Z)^c\right) = o\left(\max\left\{\delta(x,Z),\delta(y,Z)\right\}^c\right), \tag{1.7.4}$$

where we have used LANDAU's notation (see C.2.1 in the Appendix). Consequently

$$\frac{\left|(R_x^m F)^{(\alpha)}(y)\right|}{|x-y|^{m-|\alpha|}} = o\left(\left(\frac{\hat{\delta}(x,y)}{|x-y|}\right)^{m-|\alpha|} \max\left\{\delta(x,Z),\delta(y,Z)\right\}^c\right) \tag{1.7.5}$$

holds, hence the relation

$$\frac{\left|(R_x^m F)^{(\alpha)}(y)\right|}{|x-y|^{m-|\alpha|}} = o\left(\left(\frac{\hat{\delta}(x,y)}{|x-y|}\right)^{m-|\alpha|} \max\left\{d(x,Z),d(y,Z)\right\}^{c/l'}\right) \tag{1.7.6}$$

follows, where $l' \leq l$ is the exponent in (RA1) according to Lemma 1.6.17. To derive (1.7.1) we will from now on treat the type RA and the type RB cases separately.

TYPE RA: First note that in this case $l = l'$. Let D be the constant in axiom (RA2). If $|x-y| \leq D \max\left\{d(x,Z),d(y,Z)\right\}^r$, then (RA2) and $\hat{\delta} \leq \delta$ imply

$$\frac{\hat{\delta}(x,y)}{|x-y|} \leq C_2 \max\left\{d(x,Z),d(y,Z)\right\}^{1/l-r}.$$

If on the other hand $|x-y| \geq D \max\left\{d(x,Z),d(y,Z)\right\}^r$, then we have with (RA1)

$$\frac{\hat{\delta}(x,y)}{|x-y|} \leq \frac{C_1}{D} \max\left\{d(x,Z),d(y,Z)\right\}^{1/l-r}.$$

By (1.7.6) in either case

$$\begin{aligned} \frac{\left|(R_x^m F)^{(\alpha)}(y)\right|}{|x-y|^{m-|\alpha|}} &= o\left(\left(\max\left\{d(x,Z),d(y,Z)\right\}^{1/l-r}\right)^{m-|\alpha|} \max\left\{d(x,Z),d(y,Z)\right\}^{c/l}\right) \\ &= o\left(\max\left\{d(x,Z),d(y,Z)\right\}^{(1/l-r)(m-|\alpha|)+c/l}\right) \\ &= o\left(\max\left\{d(x,Z),d(y,Z)\right\}^{|\alpha|(r-1/l)}\right), \end{aligned}$$

holds, where we have used the fact that $c \geq c_A^m(r,l) = m(lr-1)$. As $(r-1/l) \geq 0$ by $l, r \geq 1$ the relation (1.7.1) follows.

TYPE RB: First let $|x-y| \le D \max\{d(x,Z), d(y,Z)\}^{r}$. Then by (RB2) and $\hat{\delta} \le \delta$

$$\frac{\hat{\delta}(x,y)}{|x-y|} \le C_2 \max\{d(x,Z), d(y,Z)\}^{r(1/l-1)}.$$

If on the other hand $|x-y| \ge D \max\{d(x,Z), d(y,Z)\}^{r}$, then by (RB1)

$$\frac{\hat{\delta}(x,y)}{|x-y|} \le \frac{C_1}{D} |x-y|^{(1/l-1)} \le \{d(x,Z), d(y,Z)\}^{r(1/l-1)}.$$

Using $c \ge c_A^m(r,l) = rm(l-1)\frac{l'}{l}$ one concludes

$$\begin{aligned}\frac{\left|(R_x^m F)^{(\alpha)}(y)\right|}{|x-y|^{m-|\alpha|}} &= o\left(\max\{d(x,Z), d(y,Z)\}^{r(1/l-1)(m-|\alpha|)+c/l'}\right) \\ &= o\left(\max\{d(x,Z), d(y,Z)\}^{|\alpha| r(1-1/l)}\right).\end{aligned}$$

This proves (1.7.1) also for the type RB case. □

1.7.4 Remark The classical lemma of HESTÉNÈS (see [84] or [170, Lem. 4.3]) follows from the generalized lemma of HESTÉNÈS, if one takes for A an open subset of $\mathbb{R}^n$. Then A is 1-regular, in particular $(1,1)$-regular with respect to $Z \subset A$. Therefore any m-jet F over A which vanishes over Z and which is a Whitney function over $A \setminus Z$ must be a $\mathcal{C}^m$-function over A.

1.7.5 Remark One might ask the question, whether the seemingly rather complicated (r,l)-regularity is really necessary to prove a generalized HESTÉNÈS lemma. In their article [59] FERRAROTTI–WILSON have shown by a counter example that Whitney–Tougeron regularity alone does not suffice to identify flat jets like in 1.7.2 as Whitney functions, and that a further notion of regularity which rules the behavior near the boundary is necessary for an extension result à la HESTÉNÈS.

1.7.6 Remark The first extension result for smooth functions on a special class of stratified space, namely manifolds-with-corners, originates from the work of SEELEY [157]. SEELEY's result says that for every manifold-with-corners X which is embedded in some $\mathbb{R}^n$ as a closed subset there exists a continuous extension operator

$$e : \mathcal{C}^\infty(X) \to \mathcal{C}^\infty(\mathbb{R}^n).$$

For an arbitrary stratified subspace $Y \subset \mathbb{R}^n$ an analogous statement does in general not hold.

Multiplying a jet F over A which is flat over Z of order c with a Whitney function on A the product FG is again flat over Z of order c. But if G is only a Whitney function over $A \setminus Z$, the product need not be flat over Z of order c. In the following we will give criteria on G which imply that the product FG is flat of order $d \le c$ relative Z.

1.7.7 Definition Let $A \subset \mathbb{R}^n$ be locally closed, $m \in \mathbb{N}$, and $Z \subset \overline{A}$ a locally closed subset such that $A \setminus Z$ is dense in A. A Whitney function $G \in \mathcal{E}^m(A \setminus Z)$ is called *tempered relative* Z of *class* $\mathcal{C}^m$ and *order* $c \in \mathbb{R}^{\geq 0}$, if either $Z = \emptyset$ or if for every $z \in Z$ and $\alpha \in \mathbb{N}^n$, $|\alpha| \leq m$ there exists a neighborhood $V \subset \mathbb{R}^n$ such that

$$\sup_{y \in V \cap A \setminus Z} G^{(\alpha)}(y)\, d(y, Z)^c < \infty.$$

A Whitney function $G \in \mathcal{E}^m_{\mathfrak{U}}(S)$ over a stratum S of X is called *tempered relative* ∂S of *class* $\mathcal{C}^m$ and *order* $c \in \mathbb{R}^{\geq 0}$, if in every chart x_j the Whitney function $G_j \in \mathcal{E}^m(x_j(S \cap U_j))$ is tempered relative $x_j(\partial S \cap U_j)$ of class $\mathcal{C}^m$ and order c.

The obviously linear space of all Whitney functions over $A \setminus Z$ (resp. S) which are tempered relative Z (resp. ∂S) of class $\mathcal{C}^m$ and order c will be denoted by $\mathcal{M}^{m,c}(Z; A)$ (resp. $\mathcal{M}^{m,c}_{\mathfrak{U}}(\partial S; \overline{S})$). Moreover we set $\mathcal{M}^m(Z; A) := \bigcup_{c \geq 0} \mathcal{M}^{m,c}(Z; A)$ and $\mathcal{M}^m_{\mathfrak{U}}(\partial S; \overline{S}) := \bigcup_{c \geq 0} \mathcal{M}^{m,c}_{\mathfrak{U}}(\partial S; \overline{S})$.

A Whitney function $G \in \mathcal{E}^\infty(A \setminus Z)$ is called *tempered relative* Z of *class* $\mathcal{C}^\infty$, if for every $x \in A$ there exists an open neighborhood O such that for every $m \in \mathbb{N}$ one can choose $c_m \in \mathbb{R}^{\geq 0}$ such that G is tempered relative $O \cap Z$ of class $\mathcal{C}^m$ and order c_m. This space will be denoted $\mathcal{M}^\infty(Z; A)$. Analogously we define $\mathcal{M}^\infty_{\mathfrak{U}}(Z; \overline{S})$, and call it the space of *relative* Z *tempered* Whitney functions of *class* $\mathcal{C}^\infty$ over S.

1.7.8 Proposition *Under the prerequisites of the preceding definition and the additional assumption that Z is closed in A let c, d be two nonnegative real numbers. If then $F \in J^m(A)$ is a jet flat of order $c + d$ and $G \in \mathcal{E}^m(A \setminus Z)$ tempered of order d, then the product $F \cdot G \in \mathcal{E}^m(A \setminus Z)$ can be continued to a jet $\in J^m(A)$ which is flat over Z of order c. In signs:*

$$\mathcal{M}^{m,d}(Z; A) \cdot \mathcal{J}^{m,c+d}(Z; A) \subset \mathcal{J}^{m,c}(Z; A).$$

Moreover, the space $\mathcal{M}^\infty(Z; A)$ is a multiplier algebra for the ideal $\mathcal{J}^\infty(Z; A) \subset \mathcal{E}^\infty(A)$ that means $\mathcal{M}^\infty(Z; A)$ is an algebra, and

$$\mathcal{M}^\infty(Z; A) \cdot \mathcal{J}^\infty(Z; A) \subset \mathcal{J}^\infty(Z; A).$$

In case Z is a closed subset of the boundary ∂S of a stratum S of X one has analogously

$$\mathcal{M}^{m,d}_{\mathfrak{U}}(Z; \overline{S}) \cdot \mathcal{J}^{m,c+d}_{\mathfrak{U}}(Z; \overline{S}) \subset \mathcal{J}^{m,c}_{\mathfrak{U}}(Z; \overline{S}).$$

Likewise $\mathcal{M}^\infty_{\mathfrak{U}}(Z; \overline{S})$ is a multiplier algebra for $\mathcal{J}^\infty_{\mathfrak{U}}(Z; \overline{S}) \subset \mathcal{E}^\infty_{\mathfrak{U}}(\overline{S})$.

PROOF: The first part of the claim follows easily from Definitions 1.7.1 and 1.7.7. The second part is essentially a consequence of the theorem of TAYLOR. To see this in detail first choose an open subset $O \subset \mathbb{R}^n$ such that A is closed on O. According to WHITNEY's extension theorem C.3.2 there exists for $F \in \mathcal{E}^\infty(A)$ a smooth function f on O such that $F = f \mod \mathcal{J}^\infty(A; O)$. If F is even an element of $\mathcal{J}^\infty(Z; A)$, all partial derivatives of f vanish on Z. Consequently the product of F with an element $G \in \mathcal{M}^\infty(Z; A)$ can be extended by 0 to a jet over A. It remains to show that the thus defined jet is again a Whitney function. More precisely one has to show that for all $z \in Z$, $m \in \mathbb{N}$ and $\alpha \in \mathbb{N}^n$ with $|\alpha| \leq m$

$$\frac{(R_x^m(F \cdot G))^{(\alpha)}(y)}{|x-y|^{m-|\alpha|}} \to 0, \tag{1.7.11}$$

if $x, y \in A$ converges to z. But as all partial derivatives of f vanish over Z, and G is tempered relative Z of class $\mathcal{C}^\infty$, relation 1.7.11 has to be true. This proves the claim. □

In the following considerations of this section we will construct nontrivial tempered functions. To this end we first provide in Lemma 1.7.9 a special partition of unity for the complement of a compact set $K \subset \mathbb{R}^n$. The corresponding result plays an essential role for the proof of WHITNEY's extension theorem and will be introduced here in a somewhat more general form than usual (cf. MALGRANGE [118, Lem. 3.1] or TOUGERON [170, Lem. 2.1]). Afterwards we will show that two regularly situated sets can be separated by a tempered function, and finally that the solution of an ordinary differential equation given by tempered functions of class $\mathcal{C}^m$ with $m \in \mathbb{N}^{>0}$ is again tempered of class $\mathcal{C}^m$.

1.7.9 Lemma *Let $d \in \mathbb{N}^{>0}$ and $\delta > 0$ be real, where we assume $\delta = 1$ in case $d = 1$. Then there exists for every given compact set $K \subset \mathbb{R}^n$ a countable family of smooth functions $\phi_j \in \mathcal{C}^\infty(\mathbb{R}^n \setminus K)$, $j \in J$, with the following properties:*

(1) *For all $j \in J$ the relation $\phi_j \geq 0$ holds.*

(2) *The family of supports $\operatorname{supp}\phi_j$ is locally finite, where the number $N(x)$ of the sets $\operatorname{supp}\phi_j$ with $x \in \operatorname{supp}\phi_j$ is bounded by 4^n that means $N(x) \leq 4^n$.*

(3) $\sum_{j\in J} \phi_j(x) = 1$ *for all $x \in \mathbb{R}^n \setminus K$.*

(4) *For all j the relation $2\delta\, d(\operatorname{supp}\phi_j, K)^d \geq \operatorname{diam}(\operatorname{supp}\phi_j)$ holds.*

(5) *For every $m \in \mathbb{N}$ there exists a constant D_m depending only on d, δ, n and m such that for all $x \in \mathbb{R}^n \setminus K$*

$$|\partial^\alpha \phi_j(x)| \leq D_m \left(1 + \frac{1}{d(x,K)^{d|\alpha|}}\right), \qquad |\alpha| = m,\ \alpha \in \mathbb{N}^n.$$

PROOF: For every $p \in \mathbb{N}$ let us decompose $\mathbb{R}^n$ into closed cubes of side length $\frac{1}{(\delta 2^p)^d}$ via the hyperplanes $x_k = \frac{j_k}{(\delta 2)^{pd}}$, where $k = 1, \cdots, n$ and j_k run through all integers. Let σ_p be the set of all such cubes.

Let J_0 be the set of all cubes W of Σ_0 such that $d(K, W) \geq \frac{\sqrt{n}}{\delta^d}$. We now define J_p recursively. The set J_p consists of all those cubes $W \in \Sigma_p$ which are not contained in a cube of $J_0, \cdots, J_{p-1}$ and which satisfy the estimate $d(K, W) \geq \frac{\sqrt{n}}{\delta^d 2^p}$. Finally we set $J = \bigcup_{p\in\mathbb{N}} J_p$.

If a cube $W \in \Sigma_p$ meets a cube W' of $J_0, \cdots, J_{p-1}$, then

$$d(K, W) \geq d(K, W') - \frac{\sqrt{n}}{\delta^d 2^{pd}} \geq \frac{\sqrt{n}}{\delta^d 2^{p-1}} - \frac{\sqrt{n}}{\delta^d 2^{pd}} \geq \frac{\sqrt{n}}{\delta^d 2^p},$$

hence W is either contained in a cube $J_0, \cdots, J_{p-1}$ or is an element of J_p. Thus the family J forms a covering of $\mathbb{R}^n \setminus K$, and the cubes of J_p meet at most cubes of J_{p-1}, J_p and J_{p+1}.

Next let $\psi : \mathbb{R}^n \to [0,1]$ be a $\mathcal{C}^\infty$-function such that $\psi(x) = 1$, if $|x_k| \leq \frac{1}{2}$ for every $k = 1, \cdots, n$, and $\psi(x) = 0$, if there exists an index k with $|x_k| \geq \frac{1}{2} + \frac{1}{4d}$. For $W \in J$ now set $\psi_W(x) = \psi\left(\frac{x - x_W}{l_W}\right)$, where x_W denotes the mid point of W and l_W the side length. Obviously the family $(\operatorname{supp} \psi_W)_{W \in J}$ then is locally finite and covers $\mathbb{R}^n \setminus K$ by compact sets. Therefore we set

$$\phi_W(x) = \frac{\psi_W(x)}{\sum_{j \in J} \psi_j(x)}, \qquad x \in \mathbb{R}^n \setminus K.$$

One checks easily that the family consisting of the ϕ_W $(W \in J)$ satisfies conditions (1) to (3).

If $W \in j_p$, then the following chain of equalities holds:

$$d(K, \operatorname{supp} \phi_W) \geq d(K, W) - \frac{\sqrt{n}}{\delta^d 2^{(p+2)d}} \geq \frac{3\sqrt{n}}{\delta^d 2^{p+2}} \geq \frac{\sqrt{n}}{2\sqrt[d]{n}\delta^{d-1}} \sqrt[d]{\operatorname{diam}(\operatorname{supp} \phi_W)},$$

hence (4) follows as well.

Obviously for all α with $|\alpha| \leq m$

$$|D^\alpha \psi_W(x)| \leq \left| \frac{1}{l_W^{|\alpha|}} D^\alpha \psi \left(\frac{x - x_W}{l_W} \right) \right| \leq \frac{D}{l_W^{|\alpha|}},$$

where D is constant depending only on m (and the choice of ψ). On the other hand we have for all $x \in \mathbb{R}^n \setminus K$

$$1 \leq \sum_{W \in J} \psi_W(x) \leq 4^n,$$

hence the LEIBNIZ rule and the preceding inequalities entail the existence of a constant $D_1 > 0$, which depending only on m and n, such that

$$|D^\alpha \phi_W(x)| \leq \frac{D_1}{l_W^{|\alpha|}}.$$

For $W \in J_0$ the inequality $|D^\alpha \phi_W(x)| \leq \frac{D_1}{\delta^{d|\alpha|}}$ holds. Now let $W \in J_p$, $p \geq 1$. Then we have for every cube W' of Σ_{p-1} containing W

$$d(K, W') < \frac{\sqrt{n}}{\delta^d 2^{p-1}},$$

hence for all $x \in W$

$$d(x, K) \leq \frac{\sqrt{n}}{\delta^d 2^{p-1}} + \operatorname{diam}(W') \leq \frac{\sqrt{n}}{\delta^d 2^{p-2}},$$

and for all $x \in \operatorname{supp} \phi_W$

$$d(x, K) \leq \frac{\sqrt{n}}{\delta^d 2^{p-2}} + \frac{\sqrt{n}}{\delta^d 2^{d(p+2)}} \leq D_2 \sqrt[d]{l_W},$$

where $D_2 > 0$ depends only on c, δ and n. So finally for all $x \in \mathbb{R}^n \setminus K$

$$|D^\alpha \phi_W(x)| \leq D_1 \left(1 + \frac{D_2^{d|\alpha|}}{d(x,K)^{d|\alpha|}}\right),$$

which proves the last claim of the lemma. □

1.7.10 Lemma (*cf.* TOUGERON [170, Lem. 4.5]) *Let A, Z be two regularly situated subsets of* $O \subset \mathbb{R}^n$ *open. Then there exists a function* $\phi \in \mathcal{M}^\infty(A \cap Z; O)$ *such that* $\phi = 0$ *on a neighborhood of* $A \setminus (A \cap Z)$ *and* $\phi = 1$ *on a neighborhood of* $B \setminus (A \cap Z)$.

PROOF: We proceed like in [170]. For the proof of the claim it suffices to assign to every point $z \in O$ an open neighborhood $V_z \subset O$ and a function $\phi_z \in \mathcal{E}(V_x \setminus (A \cap Z))$ such that $\phi_z = 0$ on a neighborhood of $(V_z \cap A) \setminus (A \cap Z)$, $\phi_z = 1$ on a neighborhood of $(V_z \cap Z) \setminus (A \cap Z)$ and such that for every $\alpha \in \mathbb{N}^n$ there exist constants $c' \in \mathbb{N}^{>0}$ and $C' > 0$ such that the following estimate holds:

$$\left|D^\alpha \phi_z(x)\right| \leq C' \left(1 + \frac{1}{d(x, A \cap Z)^{c'}}\right) \quad \text{for all } x \in V_z \setminus (A \cap Z).$$

Namely, if one has constructed such ϕ_z, then one can set $\phi = \sum_{j \in \mathbb{N}} \psi_j \, \phi_{x_j}$, where $(\psi_j)_{j \in \mathbb{N}}$ is a locally finite family of smooth functions on O such that $\sum_{j \in \mathbb{N}} \psi_j = 1$ and $\operatorname{supp} \psi_j \subset V_{z_j}$ for every $j \in \mathbb{N}$, where the points $z_j \in O$ have been chosen appropriately.

We can suppose $z \in A \cap Z$, the case $z \notin A \cap Z$ is trivial. Then there exists an open neighborhood V_z of z, relatively compact in O, such that over V_z an inequality of the form (RS) (see p. 52) holds. Under these prerequisites let $K \subset A$ be compact such that $d(x, A) = d(x, K)$ for all $x \in V_z$.

Now we consider for K and $d = \delta = 1$ the construction done in the preceding lemma and associate to every $W \in J$ an integer λ_W in the following way: if there exists an $x \in \operatorname{supp} \phi_W$ with $d(x, K) < \frac{1}{2} d(x, Z)$, then let $\lambda_W = 0$, otherwise let $\lambda_W = 0$. Consequently, the function ϕ_z can be extended to a $\mathcal{C}^\infty$-function on $\mathbb{R}^n \setminus (K \cap Z)$ (denoted by ϕ_z as well) which vanishes on a neighborhood of $K \setminus (K \cap Z)$. On the other hand $\phi_z = 1$ on a neighborhood of $Z \setminus (K \cap Z)$, because if W runs through all indices such that $\operatorname{supp} \phi_W \cap Z$ is nonempty, then for every $x \in \operatorname{supp} \phi_W$

$$d(x, K) \geq d(\operatorname{supp} \phi_W, K) \geq \frac{1}{2} \operatorname{diam}(\operatorname{supp} \phi_W) \geq \frac{1}{2} d(x, Z),$$

hence $\lambda_W = 1$. Let $x \in V_z \setminus (A \cap Z)$, in particular $d(x, K) = d(x, A)$. If $d(x, A) < \frac{1}{2} d(x, Z)$, then one has $\phi_z(x) = 0$, if otherwise $d(x, A) \geq \frac{1}{2} d(x, Z)$, then according to Lemma 1.7.9 and (RS') the inequality

$$|D^\alpha \phi_z(x)| \leq C' \, d(x, A \cap Z)^{-c|\alpha|},$$

holds, where C' is an appropriate constant. This prove the lemma. □

To formulate the last result in this section we will need the notion of a tubular neighborhood. Though this notion will be explained in detail only later in Section

3.1, we already introduce it here in its most simple form and refer to Chapter 3 for more details.

By a *tubular neighborhood* of a submanifold $S \subset \mathbb{R}^n$ we understand an open neighborhood $T \subset \mathbb{R}^n$ of S such that for every $x \in T$ there exists exactly one point $\pi(x) \in S$ with $d(x, \pi(x)) = d(x, S)$, and such that $\pi(x) + t(x - \pi(x)) \in T$ for all $t \in [0, 1]$. The mapping π will be called the *projection* of the tubular neighborhood.

1.7.11 Theorem *Let $m \in \mathbb{N}^{>0}$, $S \subset \mathbb{R}^n$ be a submanifold of class $\mathcal{C}^m$, and $T \subset \mathbb{R} \times \mathbb{R}^n$ a regularly situated tubular neighborhood of $\{0\} \times S \subset \mathbb{R} \times \mathbb{R}^n$. Moreover, let $f : T \to \mathbb{R}^n$ be a relatively $\{0\} \times \partial S$ tempered function of class $\mathcal{C}^m$. Then there exists a regularly situated tubular neighborhood $T^m \subset T$ of $\{0\} \times S$ together with a unique solution $\gamma : T^m \to \mathbb{R}^n$, $(t, x) \mapsto \gamma(t, x) = \gamma_x(t)$ of the initial value problem*

$$\dot{\gamma}_x(t) = f(t, \gamma_x(t)), \qquad \gamma_x(0) = x.$$

Moreover, T^m and γ can be chosen such that γ is tempered relative $\{0\} \times \partial S$ of class $\mathcal{C}^m$.

PROOF: By the existence and uniqueness of solutions of an ordinary differential equation we can glue together local solutions to global ones, hence we can suppose without loss of generality that S is a bounded subset of $\mathbb{R}^n$. By assumption on T there exist continuous functions $\varepsilon : S \to \mathbb{R}^{>0}$ and $\delta : S \to \mathbb{R}^{>0}$ as well as constants $c_1, c_2 \in \mathbb{N}$ and $C_1, C_2 > 0$ such that for all $x \in S$ first

$$\varepsilon(x) \geq C_1\, d(x, \partial S)^{c_1} \quad \text{and} \quad \delta(x) \geq C_2\, d(x, \partial S)^{c_2}$$

and secondly $B^1_{4\delta(x)}(0) \times B^n_{4\varepsilon(x)}(x) \subset T$. Hereby $B^k_r(y)$ denotes the ball in $\mathbb{R}^k$ around y with radius r. Next we set for $x \in S$

$$\begin{aligned} L(x) &:= 1 + \sup \left\{ \|\partial_y f(t, y)\| \;\middle|\; (t, y) \in B^1_{3\delta(x)}(0) \times B^n_{3\varepsilon(x)}(x) \right\}, \\ M(x) &:= \sup \left\{ \|f(t, y)\| \;\middle|\; (t, y) \in B^1_{3\delta(x)}(0) \times B^n_{3\varepsilon(x)}(x) \right\}, \end{aligned}$$

and define $\tau(x) := \min\left\{1, \delta(x), \frac{\varepsilon(x)}{M(x)}\right\}$. Then $\tau(x) \geq C_3\, d(x, \partial S)^{c_3}$ holds with appropriate constants $c_3 \in \mathbb{N}$ and $C_3 > 0$, so

$$T^0 := \bigcup_{x \in S} B^1_{\tau(x)}(0) \times \left(B^n_{\varepsilon(x)}(x) \cap \pi^{-1}(x) \right)$$

is a regularly situated neighborhood of S in $\mathbb{R} \times \mathbb{R}^n$. Finally define the Banach space $\mathcal{E}$ by

$$\mathcal{E} := \left\{ \varphi \in \mathcal{C}(T^0, \mathbb{R}^n) \;\middle|\; \varphi(t, y) \in \overline{B^n_{2\varepsilon(\pi(y))}(\pi(y))} \text{ for } (t, y) \in T^0 \right\},$$

where the norm is given by $\|\varphi\|_{\mathcal{E}} := \sup\left\{ \|\varphi(t, y)\|\, e^{-2L(y)|t|} \;\middle|\; (t, y) \in T^0 \right\}$. Then the operator

$$K : \mathcal{E} \to \mathcal{C}(T^0, \mathbb{R}^n), \; \varphi \mapsto K\varphi(t, y) = y + \int_0^t f(s, \varphi(s, y))\, ds$$

maps the Banach space $\mathcal{E}$ into itself, because

$$\|K\varphi(t, y) - y\| \leq \left| \int_0^t \|f(s, \varphi(s, y))\|\, ds \right| \leq \tau(\pi(y))\, M(\pi(y)) \leq \varepsilon(\pi(y)), \quad (t, y) \in T^0,$$

hence $\|K\varphi(t,y) - \pi(y)\| \leq 2\varepsilon(\pi(y))$. Like in the standard proof of the theorem of PICARD–LINDELÖF one now shows that K is a contraction. For $\varphi, \psi \in \mathcal{E}$ we namely have

$$\|K\varphi(t,y) - K\psi(t,y)\| \leq \left|\int_0^t \|f(s,\varphi(s,y)) - f(s,\psi(s,y))\| \, ds\right|$$
$$\leq L(y) \left|\int_0^t \|\phi(s,y) - \psi(s,y)\| \, ds\right| \leq L(y) \, \|\phi - \psi\|_E \left|\int_0^t e^{2L(y)|s|} \, ds\right| \leq \frac{1}{2} e^{2L(y)|t|},$$

hence $\|K\varphi - K\psi\|_{\mathcal{E}} \leq \frac{1}{2}\|\phi - \psi\|_{\mathcal{E}}$. So K has exactly one fixed point in $\mathcal{E}$, which we denote by γ and which is a solution of the above initial value problem. Moreover, by definition of $\mathcal{E}$ the solution γ is tempered relative $\{0\} \times \partial S$ of class $\mathcal{C}^0$.

Next shrink T^0 to a regularly situated tubular neighborhood T^1 of $\{0\} \times S$ such that for every $\nu \in \mathbb{R}^n$ with $\|\nu\| = 1$ the integral operator

$$K_\nu : \mathcal{E}_\nu \to \mathcal{C}(T^1, \mathbb{R}^n), \ \varphi \mapsto K_\nu\varphi(t,y) = \nu + \int_0^t Df(s,\gamma(s,y)).(0,\varphi(s,y)) \, ds$$

has exactly one fixed point γ_ν in the Banach space

$$\mathcal{E}_\nu := \left\{\varphi \in \mathcal{C}(T^1, \mathbb{R}^n) \mid \varphi(t,y) \in \overline{B^n_{2\varepsilon(\pi(y))}(\nu)} \text{ for } (t,y) \in T^1\right\}.$$

By definition of K_ν the fixed point γ_ν coincides with the derivative $D\gamma(t,y).(0,\nu)$, and the restriction $\gamma_{|T^1} : T^1 \to \mathbb{R}^n$ becomes a relatively $\{0\} \times \partial S$ tempered function of class $\mathcal{C}^1$. Recursively one thus obtains a sequence of regularly situated tubular neighborhoods $T^m \subset \cdots \subset T^1 \subset T^0 \subset T$ of $\{0\} \times S$ such that $\gamma_{|T^m} : T^m \to \mathbb{R}^n$ is tempered relative $\{0\} \times \partial S$ of class $\mathcal{C}^m$. This proves the claim. □

Chapter 2

Differential Geometric Objects on Singular Spaces

2.1 Stratified tangent bundles and Whitney's condition (A)

In this section we will construct for every stratified space X with stratification $\mathcal{S}$ and smooth structure $\mathcal{C}^\infty$ the stratified tangent bundle TX. The Whitney condition (A) crystallizes hereby in a natural way as the property which guarantees that the stratified tangent bundle is again a stratified space.

2.1.1 As a set we define the *stratified tangent bundle* of a stratified space X by

$$\mathsf{TX} = \bigcup_{\mathsf{S} \in \mathcal{S}} \mathsf{TS}.$$

Like in the case of manifolds one has a canonical projection,

$$\pi = \pi_{\mathsf{TX}} : \mathsf{TX} \to \mathsf{X}, \quad \mathsf{TS} \ni \nu \mapsto \pi_{\mathsf{TS}}(\nu) \in \mathsf{S},$$

where $\pi_{\mathsf{TS}} : \mathsf{TS} \to \mathsf{S}$ denotes the canonical projection of the tangent bundle of the stratum $\mathsf{S} \in \mathcal{S}$. If X possesses a smooth structure, then we can topologize TX in the following way. Let $\mathsf{U} \subset \mathsf{X}$ be open, and $\mathsf{x} : \mathsf{U} \to \mathsf{O} \subset \mathbb{R}^n$ be a singular chart. Then let us set $\mathsf{TU} := \pi^{-1}(\mathsf{U})$ and define $\mathsf{Tx} : \mathsf{TU} \to \mathsf{TO} \subset \mathbb{R}^{2n}$ by requiring that $(\mathsf{Tx})_{|\mathsf{TS} \cap \mathsf{TU}} = \mathsf{T}(\mathsf{x}_{|\mathsf{S} \cap \mathsf{U}})$ for all strata S. Now we can supply TX with the coarsest topology such that all $\mathsf{TU} \subset \mathsf{TX}$ are open and all Tx are continuous. The projection π then becomes a continuous mapping which is differentiable on all the spaces TS. Moreover, one concludes easily from the corresponding properties for X that the TS are locally closed subspaces of TX and that the partition of TX in the spaces TS is locally finite. Hence the question arises, whether TX is a stratified space with strata TS and whether π is a projection in the topological sense that means whether X carries the quotient topology of TX. In other words we have to show under which conditions $\mathsf{TR} \subset \overline{\mathsf{TS}}$ for all pieces $\mathsf{R} < \mathsf{S}$ and simultaneously $\pi^{-1}(\overline{\mathsf{Y}}) = \overline{\pi^{-1}(\mathsf{Y})}$ for all subspaces $\mathsf{Y} \subset \mathsf{X}$. It will turn out that this is exactly the case, when X satisfies Whitney's condition (A).

M.J. Pflaum: LNM 1768, pp. 63 - 90, 2001

2.1.2 Theorem *The stratified tangent bundle TX of a stratified space space with smooth structure $(X, \mathcal{C}^\infty_X)$ is a stratified space again and $\pi : TX \to X$ a topological projection, if and only if X satisfies the Whitney condition* (A). *In this case TX inherits from $(X, \mathcal{C}^\infty_X)$ a canonical (weak) smooth structure $\mathcal{C}^\infty_{TX}$ which is determined by the charts $Tx : TU \to \mathbb{R}^{2n}$, where $x : U \to \mathbb{R}^n$ runs through all singular charts of X. The projection $\pi : TX \to X$ then is smooth, and a morphism of stratified space.*

PROOF: Without loss of generality we can assume that X is a subset of $\mathbb{R}^n$, $U = X$ and that X is the identical mapping over U. We will show first that (A) follows from the condition that TX is a stratified space and and π a topological projection. So let $R < S$, $x \in R$ and $(x_k)_{k \in \mathbb{N}}$ a sequence of points of S converging to x such that $\tau = \lim_{k \to \infty} T_{x_k}S$ exists. For the proof of the relation $T_xR \subset \tau$ choose $v \in T_xR$ and let Y be the set of all x_k. By assumption $v \in \pi^{-1}\left(\overline{Y}\right) = \overline{\pi^{-1}(Y)}$, hence there is a sequence of tangent vectors $v_k \in T_{x_k}S$ converging to v. But then $v \in \tau$ must hold, hence Whitney's condition (A) holds as well.

Now let us assume conversely that the condition (A) is satisfied. Then $T_xR \subset \overline{TS}$ has to hold for all $x \in R$, hence $TR \subset \overline{TS}$. Therefore TX becomes a decomposed space. Next choose a subspace Y of X, and $v \in \pi^{-1}\left(\overline{Y}\right)$. Then there exists a sequence of points $x_k \in Y$ with $\lim x_k = x := \pi(v)$. We can suppose that $x \in R$ and $x_k \in S$ for an incident pair of pieces $R < S$; the other case is trivial. By (A) the existence of vectors $v_k \in T_{x_k}S$ with $\lim v_k = v$ follows. Hence $\pi^{-1}\left(\overline{Y}\right) \subset \overline{\pi^{-1}(Y)}$. As the inverse inclusion is trivial, we obtain $\pi^{-1}\left(\overline{Y}\right) = \overline{\pi^{-1}(Y)}$ that means π is a topological projection.

It remains to show that the $Tx : TU \to \mathbb{R}^{2n}$ induce a smooth structure on TX. Let $Tx : TU \to \mathbb{R}^{2n}$ and $T\tilde{x} : T\tilde{U} \to \mathbb{R}^{2\tilde{n}}$ be two singular charts and v a vector $\in TU \cap T\tilde{U} = T(U \cap \tilde{U})$. After transition to smaller neighborhoods U and $\tilde{U}$ of $\pi(v)$ we can find an open neighborhood $O \subset \mathbb{R}^m$ with $m \geq \max(n, \tilde{n})$ of $x(U \cap \tilde{U})$ such that there exists a diffeomorphism $H : O \to \tilde{O} \subset \mathbb{R}^m$ satisfying $x_{|U \cap \tilde{U}} = H \circ \tilde{x}_{|U \cap \tilde{U}}$. But now $T(U \cap \tilde{U})$ is an open neighborhood of v, and $Tx_{|T(U \cap \tilde{U})} = TH \circ T\tilde{x}_{|T(U \cap \tilde{U})}$. Of course, this means that Tx and $T\tilde{x}$ are compatible. By definition of the topology of TX the Tx comprise homeomorphisms which act as diffeomorphisms on every piece TS. As the union of all the TU covers the stratified tangent bundle TX, the Tx form a singular atlas and define a smooth structure on TX. By the just explained construction π becomes a smooth morphism of stratified spaces. This finishes the proof. □

2.1.3 Remark Only in the case, where X is given by a smooth manifold, the stratified tangent bundle is locally trivial. But to emphasize the fact that TX has many geometric properties in common with the tangent bundle of a smooth manifold, we have chosen the notion "stratified tangent bundle" for TX, even if X is not a smooth manifold.

2.1.4 Example Let M be a compact manifold which is supposed to be embedded in a sphere S^n. The cone CM can then be canonically identified with the subspace

$$\left\{ty \in \mathbb{R}^{n+1} \mid t \in [0, 1[, \; y \in M \subset S^n\right\} \subset \mathbb{R}^{n+1}$$

and inherits as such a smooth structure. Obviously CM with this smooth structure becomes an (A)-stratified, even Whitney stratified space. Consequently TCM is a well-defined (A)-stratified space and has the following shape:

$$\mathsf{TCM} = \{(ty, sy + v) \in \mathbb{R}^{2(n+1)} \mid t \in]0,1[,\ y \in M,\ s \in \mathbb{R},\ v \in T_y M\} \cup \{0\}.$$

Note that TCM as a topological space is not homeomorphic, as one might expect, to the cone over $\mathsf{TM} \times \mathbb{R}$. Moreover, this example shows that singular charts of the stratified tangent bundle of a stratified space do in general not map the stratified tangent bundle to locally closed subsets of $\mathbb{R}^n$.

In the special case of the edge and the standard cone let us first identify S^0 resp. S^1 with the subspaces $\{\frac{\sqrt{2}}{2}(\pm 1, 1)\} \subset S^1$ and $\{\frac{\sqrt{2}}{2}(\cos\varphi, \sin\varphi, 1) \mid \varphi \in \mathbb{R}\} \subset S^2$. Then we obtain the stratified tangent bundles of the edge and the standard cone as follows:

$$\begin{aligned}\mathsf{TX}_{\mathrm{Edge}} &= \{(x, |x|, \operatorname{sgn}(x)\, y, \operatorname{sgn}^2(x)\, y) \mid x, y \in \mathbb{R}\} \subset \mathbb{R}^4\\ \mathsf{TX}_{\mathrm{Cone}} &= \{(\tfrac{t\sqrt{2}}{2}(\cos\varphi, \sin\varphi, 1), \operatorname{sgn}(t)(\tfrac{s\sqrt{2}}{2}(\cos\varphi, \sin\varphi, 1) + u(-\sin\varphi, \cos\varphi, 0))) \mid\\ &\qquad t \in [0,1[,\ \varphi \in \mathbb{R},\ s \in \mathbb{R},\ u \in \mathbb{R}\} \subset \mathbb{R}^6,\end{aligned}$$

where $\operatorname{sgn} : \mathbb{R} \to \{\pm 1, 0\}$ is the signum function (which vanishes in the origin).

2.1.5 By a *stratified vector field* over X we understand a section $V : X \to \mathsf{TX}$ such that for every stratum $S \in \mathcal{S}$ the restriction $V_{|S} : S \to \mathsf{TS}$ is a smooth vector field over S. Note that the notion of a stratified vector field does not require that X has a smooth structure. But in case X carries a smooth structure we will speak of a *continuous vector field* on X, if $V : X \to \mathsf{TX}$ is continuous, and, in case X is an (A)-stratified space, of a *vector field of class* $\mathcal{C}^m$ with $m \in \mathbb{N} \cup \{\infty\}$, if $V : X \to \mathsf{TX}$ is a mapping between the (A)-stratified spaces X and TX of class $\mathcal{C}^m$; the space of all such vector fields of class $\mathcal{C}^m$ will be denoted by $\mathfrak{X}^m(X)$. If $m = \infty$, then we speak of a *smooth vector field*. Assigning to every open $U \subset X$ the space $\mathfrak{X}^m(U)$ resp. $\mathfrak{X}^\infty(U)$, one obtains two further sheaves $\mathfrak{X}^m$ and $\mathfrak{X}^\infty$ over X.

2.2 Derivations and vector fields

Given a manifold M the space of derivations $\operatorname{Der}(\mathcal{C}^\infty_x, \mathbb{R})$ of the stalk $\mathcal{C}^\infty_x$ of smooth functions at the footpoint $x \in M$ is canonically isomorphic to the tangent space at x. In case X is stratified, an analogous result does in general not hold, as the following two examples 2.2.1 and 2.2.2 will show.

Besides the derivations on $\mathcal{C}^\infty_x$ with values in $\mathbb{R}$ the derivations from $\mathcal{C}^\infty(X)$ to $\mathcal{C}^\infty(X)$ are of importance as well. For the case of a manifold M the space $\operatorname{Der}(\mathcal{C}^\infty(M), \mathcal{C}^\infty(M))$ is canonically isomorphic to the space $\mathfrak{X}^\infty(M)$ of smooth vector fields over M. In the stratified case on the other hand every smooth vector field on X gives rise to a derivation, as we will show in this section, but the inverse does in general not hold.

2.2.1 Example Consider the half line $X = \mathbb{R}^{\geq 0}$. Then the derivative $D : \mathcal{C}^\infty(X) \to \mathcal{C}^\infty(X)$, $f \mapsto f'$ comprises a derivation, but it cannot be represented as a smooth vector field on X.

2.2.2 Example Let $X_{dc} = \{(x,y,z) \in \mathbb{R}^3 \mid x^2 + y^2 = z^2\}$ be the double cone with the smooth structure induced by $\mathbb{R}^3$. The curve $\gamma :]-1,1[\to X_{dc}$, $t \mapsto (t,0,t)$ then induces on the stalk $\mathcal{C}_0^\infty := \mathcal{C}_{X_{dc},0}^\infty$ the derivation

$$\delta_\gamma : \mathcal{C}_0^\infty \to \mathbb{R}, \quad [f]_0 \mapsto \left.\frac{\partial}{\partial t}\right|_{t=0} (f \circ \gamma)(t).$$

Of course, the tangent space of X_{dc} at the origin is equal to 0, but δ_γ does not vanish, that means the derivations $\mathrm{Der}(\mathcal{C}_0^\infty, \mathbb{R})$ cannot be generated by $T_0 X_{dc}$.

2.2.3 Proposition *Let X be an (A)-stratified space, x a point of X and S the stratum of x. Then every vector $v \in T_x X$ induces a derivation*

$$\partial_v : \mathcal{C}_x^\infty \to \mathbb{R}, \quad [f]_x \mapsto v(f_{|S\cap U}).$$

Conversely, if $\delta : \mathcal{C}_x^\infty \to \mathbb{R}$ denotes a derivation, then there exists a unique tangent vector $v \in T_x X$ with $\delta = \partial_v$, if and only if δ vanishes on the ideal $\mathcal{J}_S = \{[f]_x \in \mathcal{C}_x^\infty \mid f_{|S} = 0\} \subset \mathcal{C}_x^\infty$.

PROOF: Obviously, every vector $v \in T_x X$ generates a derivation $\partial_v : \mathcal{C}_x^\infty \to \mathbb{R}$ by $[f]_x \mapsto v(f_{|S\cap U})$. Moreover, δ_v vanishes by definition on the ideal $\mathcal{J}_S$. Vice versa, if $\delta : \mathcal{C}_x^\infty \to \mathbb{R}$ is a derivation vanishing on $\mathcal{J}_S$, then δ induces a derivation $\overline{\delta} : \mathcal{C}_x^\infty / \mathcal{J}_S \to \mathbb{R}$. As $\mathcal{C}_x^\infty / \mathcal{J}_S$ is canonically isomorphic to the stalk of smooth functions S over the footpoint x, there exists a $v \in T_x S$ with $\overline{\delta} = v$. Hence $\delta = \delta_v$ follows. This finishes the proof. □

2.2.4 Remark One can interpret the results of this section in such a way that the tangent bundle TX of an (A)-stratified space does in general not coincide with the Zariski tangent space $T^Z X$, but that always $TX \subset T^Z X$ holds. Note that only for manifolds (without boundary) $TM = T^Z M$. Nevertheless there are singular stratified spaces with $\mathfrak{X}^\infty(X) = \mathcal{C}^\infty(X, T^Z X) := \mathrm{Der}(\mathcal{C}^\infty(X), \mathcal{C}^\infty(X))$ like for example the standard cone X_{Cone}.

2.2.5 In the last section we introduced for every (A)-stratified space X the notion of a smooth vector field as a smooth section for the projection $\pi : TX \to X$. In the following we give a further and for many purposes very useful characterization of smooth vector fields.

First note that one can define for every (not necessarily continuous or smooth) vector field $V : X \to TX$ and every smooth function $f \in \mathcal{C}^\infty(X)$ a new function $Vf : X \to \mathbb{R}$ by requiring $(Vf)_{|S} = V_{|S} f_{|S}$ for every stratum $S \in \mathcal{S}$. We now show that Vf is an element of $\mathcal{C}^\infty(X)$ in case that V is smooth. As smoothness is a local property we can suppose without loss of generality that X is a closed subset of an open set $O \subset \mathbb{R}^n$ and that X inherits the smooth structure from $\mathbb{R}^n$. By the smoothness of $V : X \to TX \subset \mathbb{R}^{2n}$ one can find functions $v_k \in \mathcal{C}^\infty(O)$, $k = 1, \cdots, n$, such that

$$V(x) = \left(x, \sum_k v_k(x)\, \partial_k \right) \tag{2.2.1}$$

for all $x \in X$. Hence we can extend V to a smooth vector field $\mathbf{V}$ on O. Likewise, there exist functions $\mathbf{f} \in \mathcal{C}^\infty(O)$ with $\mathbf{f}_{|X} = f$. As $\mathbf{Vf}$ is differentiable of any order and as the relation $(\mathbf{Vf})_{|X} = Vf$ is true, the claim follows. Vice versa, if $Vf \in \mathcal{C}^\infty(X)$ is satisfied for every $f \in \mathcal{C}^\infty(X)$, then we can conclude that there exist smooth functions v_k over O such that $v_{k|X} = V\pi_k$ for $k = 1, \cdots, n$. Hereby π_k denotes the projection onto the k-th coordinate. But then Eq. (2.2.1) is true, hence V has to be smooth. Altogether we thus obtain

2.2.6 Proposition *For any $V : X \to TX$ on an (A)-stratified space X the following conditions are equivalent:*

(1) *$V : X \to TX$ is smooth.*

(2) *For every $f \in \mathcal{C}^\infty(X)$ the function Vf is an element of $\mathcal{C}^\infty(X)$.*

(3) *Given a chart domain $U \xrightarrow{x} O \subset \mathbb{R}^n$ with $O \in \mathbb{R}^n$ open and $\overline{x(U)} \cap O = x(U)$ the vector field V is the restriction of a smooth vector field $\mathbf{V} : O \to T\mathbb{R}^n$, more precisely*

$$Tx \circ V_{|U} = \mathbf{V} \circ x. \tag{2.2.2}$$

2.2.7 For any two vector fields $V, W \in \mathfrak{X}^\infty(X)$ we can define their *Lie bracket* $[V, W] \in \mathfrak{X}^\infty(X)$ by

$$[V, W]_{|S} = [V_{|S}, W_{|S}], \qquad S \in \mathcal{S}.$$

By definition $[V, W]$ is tangential to S for every piece S, hence $[V, W]$ is a vector field on X. It remains to show that $[V, W]$ is smooth again. Let $x : U \to O \subset \mathbb{R}^n$ be a chart like in Prop. 2.2.6 (3) and let $\mathbf{V}, \mathbf{W}$ be vector fields on O such that $Tx \circ V_{|U} = \mathbf{V} \circ x$ and $Tx \circ W_{|U} = \mathbf{W} \circ x$. Then we have for every piece S

$$Tx \circ [V, W]_{|S \cap U} = Tx \circ [V_{|S\cap U}, W_{|S\cap U}] = [\mathbf{V}, \mathbf{W}] \circ x_{|S\cap U},$$

hence $Tx \circ [V, W]_{|U} = [\mathbf{V}, \mathbf{W}] \circ x$. As $[\mathbf{V}, \mathbf{W}]$ is a smooth vector field on O, the claim follows.

Now we come back to our original topic, namely the derivations on $\mathcal{C}^\infty(X)$.

2.2.8 Proposition *Let M be a manifold, $X \subset M$ a closed (A)-stratified subspace with the smooth structure induced by M and $\mathcal{J} \subset \mathcal{C}^\infty(M)$ the vanishing ideal of X in M. Denoting by $\mathfrak{X}^\infty_{\mathcal{J}}(M)$ the space of those smooth vector fields V on M which satisfy $V(\mathcal{J}) \subset \mathcal{J}$ we have the following canonical relations:*

$$\mathfrak{X}^\infty(X) \subset \mathfrak{X}^\infty_{\mathcal{J}}(M)/\mathcal{J}\mathfrak{X}^\infty(M) \cong \mathrm{Der}(\mathcal{C}^\infty(X), \mathcal{C}^\infty(X)).$$

PROOF: Obviously $\mathfrak{X}^\infty(X) \subset \mathrm{Der}(\mathcal{C}^\infty(X), \mathcal{C}^\infty(X))$, hence it remains to prove the isomorphy $\mathfrak{X}^\infty_{\mathcal{J}}(M)/\mathcal{J}\mathfrak{X}^\infty(M) \cong \mathrm{Der}(\mathcal{C}^\infty(X), \mathcal{C}^\infty(X))$. By $\mathcal{C}^\infty(X) = \mathcal{C}^\infty(M)/\mathcal{J}$ we obtain a canonical mapping

$$\Pi : \mathfrak{X}^\infty_{\mathcal{J}}(M) \to \mathrm{Der}(\mathcal{C}^\infty(X), \mathcal{C}^\infty(X)), \quad \mathbf{V} \mapsto (f = \mathbf{f} + \mathcal{J} \mapsto \mathbf{Vf} + \mathcal{J}), \quad \mathbf{f} \in \mathcal{C}^\infty(M).$$

First let us show that Π is surjective. So let δ be a derivation on $\mathcal{C}^\infty(X)$. Then choose a locally finite covering $(U_j)_{j\in J}$ of M by chart domains $U_j \xrightarrow{x_j} \mathbb{R}^n$ and a smooth partition of unity $(\varphi_j)_{j\in J}$ subordinate to $(U_j)_{j\in J}$. Moreover, let $\tilde{\varphi}_j \in \mathcal{C}^\infty(M)$ be functions with $\operatorname{supp}\tilde{\varphi}_j \subset U_j$ and $\tilde{\varphi}_j = 1$ over a neighborhood of $\operatorname{supp}\varphi_j$. Then there exists for every index j and every $i = 1, \dots, n$ a mapping $v_{ji} \in \mathcal{C}^\infty(M)$ with $v_{ji} + \mathcal{J} = \delta(\tilde{\varphi}_j x_j^i + \mathcal{J})$, where x_j^i is the i-th component of x_j. Then $V = \sum_j \varphi_j \sum_i v_{ji} \partial_{x_j^i}$ is a smooth vector field on M. We will show that $V \in \mathcal{X}_{\mathcal{J}}^\infty(M)$ and $\Pi(V) = \delta$. Now, δ induces for every $x \in X$ a derivation $\delta^x : \mathcal{C}_x^\infty(M) \to \mathbb{R}$, $[f]_x \mapsto \delta(f + \mathcal{J})(x)$. On the other hand $\sum_i v_{ji}(\partial_{x_j^i} f)(x) = \delta^x([f]_x)$ holds by definition of the functions v_{ji} for all $f \in \mathcal{C}^\infty(M)$ and all j with $x \in \operatorname{supp}\varphi_j$. Consequently, $Vf(x) = \delta^x([f]_x)$. By $\delta^x(\mathcal{J}) = \{0\}$ the relations $V(\mathcal{J}) \subset \mathcal{J}$ and $\Pi(V) = \delta$ follow. Hence Π is surjective. Now, by $\ker\Pi = \mathcal{J}\mathcal{X}^\infty(M)$ the functions $v_{ji} = V(\tilde{\varphi}_j x_j^i)$ have to be elements of $\mathcal{J}$ for every $V \in \ker\Pi$. Altogether, this proves the claim. □

2.3 Differential forms and stratified cotangent bundle

Given a Whitney (A) space X with structure sheaf $\mathcal{C}^\infty = \mathcal{C}_X^\infty$ one can form the $\mathcal{C}^\infty$-module sheaf Ω_X of Kähler differentials over X according to Appendix B and the sheaf of exterior algebras $\Omega_X^\bullet = \bigoplus_{k\in\mathbb{N}} \Omega_X^k$, where $\Omega_X^k = \Lambda^k\Omega_X$. Together with the Kähler derivative $d : \Omega_X^\bullet \to \Omega_X^\bullet$ we thus obtain a complex of sheaves over X. Like in differential geometry we call the elements of the space $\Omega_X^k(X)$ *differential forms*. They provide geometric information about the underlying stratified space, even though not every result about differential forms on manifolds can be carried over to the stratified case. In this section we will derive some of the more fundamental properties of X; cohomological aspects or in other words the deRham cohomology of X will be considered in Chapter 5.

2.3.1 Lemma *If $x : U \to \mathbb{R}^n$ is a singular chart, every differential form $\omega \in \Omega_X^k(U)$ has a representation of the form*

$$\omega = \sum_{1\le i_1,\cdots,i_k\le n} f_{i_1\dots i_k}\, dx^{i_1}\wedge\ldots\wedge dx^{i_k} \qquad f_{i_1\dots i_k} \in \mathcal{C}^\infty(U), \tag{2.3.1}$$

where the functions x^i, $i = 1, \cdots, n$ denote the components of the chart x.

Before we start with the proof let us agree that for the rest of this section we only write Ω^k instead of Ω_X^k as long as this does not lead to any confusion.

PROOF: We prove the claim only for $k = 1$; the other cases are proved analogously. First choose an open neighborhood $O \subset \mathbb{R}^n$ of $x(U)$, such that $x(U)$ is closed in O. Then realize that $\mathcal{C}^\infty(O) \to \Omega^1(U)$, $f \mapsto d\, f_{|U}$ comprises a derivation on $\mathcal{C}^\infty(U)$, hence induces an obviously surjective mapping $\Omega^1_{\mathbb{R}^n}(O) \to \Omega^1(U)$. Now, every form $\alpha \in \Omega^1_{\mathbb{R}^n}(O)$ has a (unique) representation of the form $\alpha = \sum_{i=1}^n f_i\, d\pi_i$, where $f_i \in \mathcal{C}^\infty(O)$ for every i, and where the π_i are the coordinate functions on $\mathbb{R}^n$. By $\pi_{i|U} = x^i$ the claim then follows immediately. □

2.3.2 Proposition *The sheaf Ω^k of k-forms on a Whitney* (A) *space* X *with structure sheaf $\mathcal{C}^\infty$ is fine.*

PROOF: This follows immediately from the fact that $\mathcal{C}^\infty$ is a fine sheaf and Ω^k a $\mathcal{C}^\infty$-module sheaf. □

2.3.3 Besides the stratified tangent bundle we also want to construct a stratified cotangent bundle T^*X for X. In the following we will describe how this can be done. As a set T^*X is defined analogously to the stratified tangent bundle by

$$T^*X = \bigcup_{S \in \mathcal{S}} T^*S,$$

and the projection $\pi = \pi_{T^*X} : T^*X \to X$ is given as the mapping which coincides over every stratum $S \in \mathcal{S}$ with the canonical projection $\pi_{T^*S} : T^*S \to S$. For every singular chart $x : U \to \mathbb{R}^n$ now let $\iota^*_{x(U)}(T^*\mathbb{R}^n) = \{(y, \alpha) \in T^*\mathbb{R}^n \mid y \in x(U)\} \subset T^*\mathbb{R}^n$ be the pullback of $T^*\mathbb{R}^n$ via the embedding $\iota_{x(U)} : x(U) \to \mathbb{R}^n$. Then the mapping x induces a map $T^*x : \iota^*_{x(U)}(T^*\mathbb{R}^n) \to T^*U$ by $(y, \alpha) \mapsto (x^{-1}(y), \alpha \circ T_{x^{-1}(y)}x)$. We supply T^*X with the finest topology such that all T^*x are continuous. Obviously all T^*U are then open. Moreover, we have for every additional singular chart $\tilde{x} : \tilde{U} \to \mathbb{R}^n$ with transition map $H : O \to \tilde{O}$ from x to $\tilde{x}$ over $U_x \subset U \cap \tilde{U}$ a commutative diagram of the following form:

$$\begin{array}{ccc} \iota^*_{x(U_x)}(T^*\mathbb{R}^n) & \xrightarrow{T^*x} & T^*U_x . \\ {\scriptstyle T^*H}\uparrow & \nearrow & \\ \iota^*_{\tilde{x}(U_x)}(T^*\mathbb{R}^n) & & \end{array} \qquad (2.3.2)$$

Hereby $T^*H : T^*\tilde{O} \to T^*O$ is defined by $(y, \alpha) \mapsto (H^{-1}(y), \alpha \circ T_{H^{-1}(y)}H)$. Thus T^*U carries the quotient topology of $\iota^*_{x(U)}(T^*\mathbb{R}^n)$ induced by T^*x.

In the following we will show that T^*X is a decomposed space with pieces T^*S, $S \in \mathcal{S}$. Without loss of generality we can assume that X can be embedded into some $\mathbb{R}^n$ via a chart $x : X \to \mathbb{R}^n$. Now let $T^*R \cap \overline{T^*S} \neq \emptyset$ be two pieces $R, S \in \mathcal{S}$. Because of the continuity of π the relation $R \cap \overline{S} \neq \emptyset$ holds, hence $R \subset \overline{S}$ and R is incident to S. Next choose some $\beta \in T^*R$ and $(y, \alpha) \in T^*\mathbb{R}^n$ with $y = x(\pi(\beta))$ and $T^*x(y, \alpha) = \beta$. Then there exists a sequence $(x_k)_{k\in\mathbb{N}} \subset S$ with $\lim_{k\to\infty} x_k = x := \pi(\beta)$. Consequently, the sequence $(x(x_k), \alpha)_{k\in\mathbb{N}}$ converges in $T^*\mathbb{R}^n$ to (y, α), and $(T^*x(x(x_k), \alpha))_{k\in\mathbb{N}} \subset T^*S$ to β. So finally $T^*R \subset \overline{T^*S}$.

By comparison with the theory of differentiable manifolds the question arises whether the stratified cotangent bundle T^*X is isomorphic to the stratified tangent bundle TX. But this in general not the case as the following examples shows.

2.3.4 Example It has been shown in Example 2.1.4 that the edge $X_{\mathrm{Edge}} = \{(x, |x|) \in \mathbb{R}^2 \mid x \in \mathbb{R}\}$ has the stratified tangent bundle $TX_{\mathrm{Edge}} = \{(x, |x|, \mathrm{sgn}(x)\, y, \mathrm{sgn}^2(x)\, y) \in \mathbb{R}^4 \mid x, y \in \mathbb{R}\}$. On the other hand its stratified cotangent bundle is given by $T^*X_{\mathrm{Edge}} = X_{\mathrm{Edge}} \times \mathbb{R}/\{0\} \times \mathbb{R}$. More generally, if M is a submanifold of the sphere S^n, then the stratified cotangent bundle T^*CM of the cone over M can be identified as a topological space with the cone $C(T^*M \times \mathbb{R})$.

2.3.5 From the stratified cotangent bundle one can construct the alternating powers $\Lambda^k T^*X$ by setting

$$\Lambda^k T^*X = \bigcup_{S\in\mathcal{S}} \Lambda^k T^*S.$$

One supplies $\Lambda^k T^*X$ with the finest topology such that all mappings

$$\begin{aligned}\Lambda^k T^*x : \iota^*_{x(U)}(\Lambda^k T^*\mathbb{R}^n) &\to \Lambda^k T^*X,\\ (y, \alpha_1 \wedge \ldots \wedge \alpha_k) &\mapsto (x^{-1}(y), \alpha_1 \circ T_{x^{-1}(y)}x \wedge \ldots \wedge \alpha_k \circ T_{x^{-1}(y)}x)\end{aligned}$$

are continuous. Moreover, one has a canonical projection $\pi = \pi_{\Lambda^k T^*X} : \Lambda^k T^*X \to X$, and understands by a *section* of $\Lambda^k T^*X$ a mapping $\omega : X \to \Lambda^k T^*X$ such that $\pi \circ \omega = \mathrm{id}_X$.

2.3.6 By the universal property of Kähler differentials one can insert tangent vectors in differential forms. More precisely, one defines the *insertion* of a tangent vector $v \in T_xX$, $x \in X$ (or more generally of a derivation $v \in \mathrm{Der}(\mathcal{C}^\infty_x, \mathbb{R})$) in a differential form $\alpha \in \Omega^1(X)$ by

$$\alpha_x(v) := \langle \alpha, v\rangle := i_v(\alpha_x),$$

where i_v is the morphism associated to the derivation v in the universal property (KÄ) from Appendix B. One now extends the insertion canonically to mappings

$$i : \Omega^k(X) \times T_xX^{\otimes k} \to \mathbb{R}, \quad (\alpha, v_1 \otimes \cdots \otimes v_k) \mapsto i_{v_1\otimes\cdots\otimes v_k}(\alpha) = \alpha_x(v_1 \otimes \cdots \otimes v_k).$$

Every k-form α then provides in a natural way a section $\overline{\alpha} : X \to \Lambda^k T^*X$, $x \mapsto \alpha_x$.

2.3.7 Proposition *For every continuous section $\omega : X \to \Lambda^k T^*X$ there exists exactly one differential form α on X with $\omega = \overline{\alpha}$, if for every singular chart $x : U \to O \subset X$ of X there exists a differential form $\alpha^x \in \Omega^k(O)$ such that*

$$\omega_{|U} = \Lambda^k T^*x \circ \alpha^x \circ x. \tag{2.3.3}$$

In this case there exists a uniquely defined continuous section $d\omega : X \to \Lambda^{k+1}T^*X$ such that

$$(d\omega)_{|U} = (\overline{d\alpha})_{|U} = \Lambda^{k+1}T^*x \circ d\alpha^x \circ x. \tag{2.3.4}$$

PROOF: With the help of Lemma 2.3.1 the claim follows almost immediately by a similar argument like in Proposition 2.2.6; therefore the details are left to the reader. □

2.3.8 For later purposes let us provide some universal constructions on the stratified tangent and cotangent bundle. Let $F : \mathfrak{Vec}^{\mathrm{fin}}_{\mathbb{R}} \to \mathfrak{Vec}^{\mathrm{fin}}_{\mathbb{R}}$ be a (covariant) functor on the category of finite dimensional real vector spaces.

As an example for such a functor we name the k-th tensor product $\bigotimes^k$, the exterior product Λ^k, or the k-times symmetric tensor product Sym^k. Then we define for every (A)-stratified space X stratified spaces FTX and FT^*X by

$$FTX = \bigcup_{S\in\mathcal{S}} FTS \quad \text{and} \quad FT^*X = \bigcup_{S\in\mathcal{S}} FT^*S.$$

Now one checks easily by the methods from Section 2.1 and the ones in this section that FTX is again (A)-stratified and that FT*X is stratified, where the strata are given by FTS resp. FT*S with $S \in \mathcal{S}$. Thus it is clear from now on, what to understand by the spaces $TX^{\otimes^k}$, $\Lambda^k TX$, $\mathrm{Sym}^k T^*X$ and so on.

2.3.9 If $f : X \to Y$ is a smooth map between (A)-stratified spaces then the pullback with f induces a morphism $\Omega^\bullet f = f^* : \Omega^\bullet_Y \to \Omega^\bullet_X$. But the situation is different for the stratified tangent bundle: the smooth curve $\gamma :]1,1[\to X_{dc}$ from Example 2.2.2 which runs through the cusp of the double cone does not induce a smooth mapping between the stratified tangent bundles $]1,1[\times \mathbb{R}$ and TX_{dc}. Insofar it is in general not possible to define for arbitrary smooth maps $f : X \to Y$ a meaningful tangent map. But if f is additionally a morphism of stratified spaces, then one can define a tangent map $Tf : TX \to TY$ by setting $(Tf)_{|S} = T(f_{|S})$ for all strata S of X.

2.4 Metrics and length space structures

The goal of this section is to carry over the notion of a Riemannian metric on a differentiable manifold to stratified spaces. Moreover, we want to examine, under which conditions such a Riemannian metric provides a geodesic distance which reflects the topology of the underlying space.

2.4.1 Definition A continuous section

$$\mu : X \to T^*X \otimes T^*X = \left\{ \alpha_x \otimes \beta_x \middle| \, \alpha_x, \beta_x \in T^*_xX, \; x \in X \right\}$$

over an (A)-stratified space X is called a *Riemannian metric*, if the following conditions are satisfied:

(ME1) For every stratum S the restriction $\mu_{|S} : S \to T^*S \otimes T^*S$ is a Riemannian metric.

(ME2) For every pair V, W of smooth vector fields on X the restriction

$$\langle V, W \rangle_\mu := \mu(V, W) : X \to \mathbb{R}, \quad x \mapsto \langle V_x, W_x \rangle_{\mu_x} := \mu_x(V_x, W_x)$$

is a smooth mapping on X.

We call μ a *smooth Riemannian metric*, if there exists a covering $(U_j)_{j \in J}$ of X by chart domains $U_j \overset{x_j}{\to} O_j \subset \mathbb{R}^n$ and a family of smooth Riemannian metrics μ_j on the open sets $O_j \subset \mathbb{R}^n$ such that for every stratum S and all $j \in J$ the pullback metric $\left(x_{j|S \cap U_j}\right)^* \mu_j$ coincides with the restriction $\mu_{|S \cap U_j}$.

If μ is a smooth Riemannian metric on X, then we often call the pair (X, μ) a *Riemannian Whitney* (A) *space*.

2.4.2 Remark For the case that the underlying stratified space is a manifold-with-boundary Melrose [126] has introduced the notion of a *b-metric*. By definition this notion coincides with the notion of a Riemannian metric defined above for the case of a manifold-with-boundary.

2.4.3 Example The Fubini–Study metric on projective space induces (via its real part) on every projective algebraic variety a smooth Riemannian metric in the above sense.

2.4.4 Example The so-called PL-spaces, where PL stands for piecewise linear, are of significance in the integral geometry of singular spaces. PL-spaces provide beautiful examples for (A)-stratified spaces with a Riemannian metric. The importance of PL-spaces lies in the fact that one can define and study intrinsic curvature measures on PL-spaces, as CHEEGER–MÜLLER–SCHRADER [43] (see also KUPPE [106]). As the underlying stratified space of a PL-space one has given a polyhedron X with an explicit triangulation $h : X \to |K| \subset \mathbb{R}^n$ as singular chart. Hereby K is a simplicial complex (see 1.1.14). Suppose that X is equipped with a continuous section $\mu : X \to T^*X \otimes T^*X$ which is smooth over every stratum. If for every closed simplex $\sigma \in K$ there exists a linear embedding $\iota : \sigma \to \mathbb{R}^m$ such that the composition $h^{-1}(\sigma) \xrightarrow{h} \sigma \xrightarrow{\iota} \mathbb{R}^m$ is an isometry, then the pair (X, μ) is called a *PL-space*. Obviously μ then is a Riemannian metric on X.

2.4.5 Proposition *For every* (A)*-stratified space* X *there exists a smooth Riemannian metric.*

PROOF: First choose a covering $(U_j)_{j \in J}$ of X by coordinate neighborhoods $U_j \xrightarrow{x_j} O_j \subset \mathbb{R}^{n_j}$, and a subordinate smooth partition of unity $(\varphi_j)_{j \in J}$. Denote by $\langle \cdot, \cdot \rangle$ the Euclidean scaler product $\mathbb{R}^{n_j}$. Then define for $x \in X$:

$$\mu_x(v, w) = \sum_j \varphi_j(x) \, \langle Tx_j.v, Tx_j.w \rangle, \qquad v, w \in T_x X.$$

A simple argument now shows that the section $\mu : X \to T^*X \otimes T^*X$, $x \mapsto \mu_x$ comprises a smooth Riemannian metric on X. □

2.4.6 Let X be a Whitney (A) space, and μ a smooth Riemannian metric on X. To every weakly piecewise $\mathcal{C}^1$-curve $\gamma : [t_j^-, t_j^+] \to X$ we can then assign its *geodesic length* $|\gamma|_\mu$ by

$$|\gamma|_\mu := \sum_{j \in J} \int_{I_j} \sqrt{\langle \dot\gamma(t), \dot\gamma(t) \rangle_{\mu(\gamma(t))}} \, dt.$$

Hereby $\bigcup_{j \in J} I_j = I$ is the decomposition of a differentiability set I of γ into its connected components and $\dot\gamma(t) \in TS_j$ the tangent vector of γ at $t \in I_j$. We show that $|\gamma|_\mu$ is finite. To this end choose a compact set $K \subset X$ such that γ has image in K°. By assumption on μ there exists a singular chart $x : U \to O \subset \mathbb{R}^n$ and a smooth metric η over O such that $K \subset U$ and $x^*\eta = \mu_U$. Let $b \in \mathcal{C}^\infty(O; GL(\mathbb{R}^n))$ be the matrix-valued function such that $\langle b_z^{-1}.v, b_z^{-1}.w \rangle_{\eta(z)}$ is equal to the Euclidean scalar product $\langle v, w \rangle$ for all $z \in O$ and $v, w \in \mathbb{R}^n$. Then we have

$$\begin{aligned} |\gamma|_\mu &= \sum_{j \in J} \int_{I_j} \sqrt{\langle Tx(\dot\gamma(t)), Tx(\dot\gamma(t)) \rangle_{\eta(x(\gamma(t)))}} \, dt = \sum_{j \in J} \int_{I_j} \| b_{x(\gamma(t))}.Tx(\dot\gamma(t)) \| \, dt \\ &\le C_K \sum_{j \in J} \int_{I_j} \| Tx(\dot\gamma(t)) \| \, dt = C_K \, |x \circ \gamma| < \infty, \end{aligned} \tag{2.4.2}$$

where $C_K = \sup\{\|b_{x(x)}\| \mid x \in K\}$. Of course, this estimate holds for every weakly piecewise $\mathcal{C}^1$-curve in K. Moreover, the constant C_K depends only on K and μ; a fact which later will turn out to be essential. Finally note that the geodesic length $|\gamma|_\mu$ does not depend on the choice of the singular charts, in contrast to the Euclidean length $|x \circ \gamma|$. Setting $D_K = \sup\{\|b_{x(x)}^{-1}\| \mid x \in K\}$ we also have an estimate inverse to (2.4.2):

$$|x \circ \gamma| \leq D_K\, |\gamma|_\mu. \tag{2.4.3}$$

2.4.7 Theorem *Let $(X, \mathcal{C}^\infty)$ be a connected locally finitely path connected Whitney (A) space and μ a smooth Riemannian metric on X. Then the map*

$$\delta_\mu : X \times X \to \mathbb{R}^{\geq 0}, \quad (x, y) \mapsto \inf\{|\gamma|_\mu \mid \gamma \in \mathcal{C}^{w1}([0,1]; X),\ \gamma(0) = x \text{ and } \gamma(1) = y\}$$

comprises a metric on X. In case X is regular, δ_μ induces the original topology of X.

The metric δ_μ is called the *geodesic distance* of μ.

PROOF: By the estimate (2.4.2) the geodesic distance $\delta_\mu(x, y)$ of two points $x, y \in X$ is well-defined. Obviously, the inequality $\delta_\mu(x, y) \geq 0$ holds, and δ_μ is symmetric in x and y. For the proof that δ_μ comprises a metric on X it remains to show only the triangle inequality and that $\delta_\mu(x, y) = 0$ is satisfied, if and only if $x = y$. Let x, y, z be three points of X and $\varepsilon > 0$. Then there exist weakly piecewise $\mathcal{C}^1$-curves γ and λ from x to y respectively from y to z such that

$$|\gamma|_\mu \leq \delta_\mu(x, y) + \varepsilon \quad \text{and} \quad |\lambda|_\mu \leq \delta_\mu(y, z) + \varepsilon.$$

The composed path $\lambda\gamma$ then has geodesic length $|\gamma|_\mu + |\lambda|_\mu$, hence

$$\delta_\mu(x, z) \leq |\gamma|_\mu + |\lambda|_\mu \leq \delta_\mu(x, y) + \delta_\mu(y, z) + 2\varepsilon$$

follows. As $\varepsilon > 0$ is arbitrary, the triangle inequality follows.

To prove positivity assume $x \neq y$. Then choose a singular chart $x : U \to O \subset \mathbb{R}^n$ with finitely path connected U such that $x, y \in U$. Also choose a compact and finitely path connected neighborhood $K \subset U$ of $\{x, y\}$. Finally choose a finitely path connected open neighborhood $V \subset K$ of x, and $\varepsilon > 0$ such that $B_\varepsilon(x(x)) \cap x(U) \subset x(V)$, where $B_\varepsilon(z)$ is the ball around $z \in \mathbb{R}^n$ of radius $\varepsilon > 0$. Now let η be a Riemannian metric on O inducing η, and let D_K like in the inequality (2.4.3). Then there exists for every weakly piecewise $\mathcal{C}^1$-curve $\gamma : [0,1] \to K \subset X$ from x to y a unique smallest $t > 0$ with $\gamma(t) \in \partial V$. Then the restricted path $\gamma_{|[0,t]}$ lies in $\overline{V}$. By (2.4.3) the path $\gamma_{|[0,t]}$ has geodesic length

$$|\gamma_{|[0,t]}|_\mu \geq \frac{1}{D_K}\, |x \circ \gamma_{|[0,t]}| \geq \frac{1}{D_K}\, |x(\gamma(t)) - x| \geq \frac{1}{D_K}\, \varepsilon.$$

Consequently

$$\delta_\mu(x, y) \geq \frac{1}{D_K}\, \varepsilon > 0,$$

hence all the axioms of metric have been proven for δ_μ. Moreover, it follows by these considerations that the topology generated by δ_μ is finer than the original topology of X.

The assumption that X is regular entails that the topology generated by δ_μ is weaker than the original topology. To check this it suffices to assume that X is a regular stratified subspace of $\mathbb{R}^n$ and that we have given a Riemannian metric η on $\mathbb{R}^n$ such that the pullback of η to X is equal to μ. Let $x \in X$ and K a compact, finitely path connected $\mathfrak{l}$-regular neighborhood of x in X, and C_K like in (2.4.2). Next choose a ball $B_\varepsilon(x)$ with $B_\varepsilon(x) \cap X \subset K$ and a finitely path connected neighborhood $V \subset B_\varepsilon(x) \cap X$ of x in X. Then there exists for $y \in V$ a rectifiable curve λ in V from x to y. As $V \subset B_\varepsilon(x) \cap X$ and $B_\varepsilon(x) \cap X$ is a Whitney (A) space, there exists by Lemma 1.6.10 a weakly piecewise $\mathcal{C}^1$-curve γ in $B_\varepsilon(x) \cap X$ connecting x with y and satisfying $|\gamma| \leq |\lambda|$. Together with (2.4.2) and the definition 1.6.6 of $\mathfrak{l}$-regularity

$$\delta_\mu(x, y) \leq |\gamma|_\mu \leq C_K |\gamma| \leq C_K |\lambda| \leq C\, C_K \|x - y\|^{1/\mathfrak{l}},$$

where $C\, C_K$ depends only on K and η. Hence every neighborhood of x with respect to δ_μ is a neighborhood of x with respect to the topology induced by the Euclidean metric. This proves the theorem. □

2.4.8 Remark Under the assumption that X is a smooth manifold and μ a smooth Riemannian metric on X the geodesic distance $\delta_\mu(x, y)$ is given by the infimum of all lengths $|\gamma|_\mu$, where γ runs through all piecewise continuously differentiable curves from x to y. In other words δ_μ coincides in this case with the geodesic distance as defined in Riemannian geometry.

2.4.9 Corollary *Every regular Riemannian Whitney* (A) *space* (X, μ) *with the geodesic distance* δ_μ *as metric comprises an inner metric space in the sense of the following definition.*

2.4.10 Definition A finitely path connected metric space (X, d) is called an *inner metric space*, if for every pair $x, y \in X$ the distance $d(x, y)$ coincides with the geodesic distance $\delta(x, y)$.

If (X, d) is a complete and locally compact inner metric space, then one calls (X, d) a *length space*.

PROOF OF THE COROLLARY: Let ρ be the geodesic distance with respect to the distance function δ_μ. Then for every two points $x, y \in X$ and every weakly piecewise $\mathcal{C}^1$-curve γ connecting these points γ

$$\rho(x, y) \leq |\gamma|_\mu, \tag{2.4.4}$$

as for every partition $0 = t_0 < \cdots < t_k = 1$ of [0, 1] the relation $\delta_\mu(\gamma(t_i), \gamma(t_{i+1})) \leq |\gamma_{|[t_i,t_{i+1}]}|_\mu$ holds for $0 \leq i < k$. By (2.4.4) the estimate $\rho(x, y) \leq \delta_\mu(x, y)$ follows.

One proves the inverse inequality as follows. Let $\varepsilon > 0$ and $\lambda : [0, 1] \to X$ be a curve from x to y such that with respect to an appropriate partition of [0, 1]

$$\sum_{0 \leq i < k} \delta_\mu(\lambda(t_i), \lambda(t_{i+1})) \leq \rho(x, y) + \frac{\varepsilon}{2}.$$

Then there exists a weakly piecewise $\mathcal{C}^1$-curve $\gamma : [0,1] \to X$ such that $\gamma(t_i) = \lambda(t_i)$ for $i = 0, \cdots, k$ and

$$|\gamma_{|[t_i,t_{i+1}]}|_\mu \leq \delta_\mu(\lambda(t_i), \lambda(t_{i+1})) + \frac{\varepsilon}{2k}.$$

Consequently

$$\delta_\mu(x,y) \leq |\gamma|_\mu \leq \rho(x,y) + \varepsilon.$$

As $\varepsilon > 0$ is arbitrary, the claim follows. □

Before we start to study Riemannian metrics and curves on stratified spaces let us first provide some fundamental properties of metric and inner metric spaces.

2.4.11 Lemma (cf. BUSEMANN [37, I.1.(4)]) *If (X, d) is an inner metric space, then the following relations hold for all $x \in X$ and $\varepsilon > 0$*

$$\overline{B}_\varepsilon(x) = S_\varepsilon(x) \cup B_\varepsilon(x), \quad \textit{where} \quad S_\varepsilon(x) = \{y \in X \mid d(y,x) = \varepsilon\}.$$

PROOF: One only has to show that for every point $y \in S_\varepsilon(x)$ there exists a sequence of points $y_k \in B_\varepsilon(x)$ converging to y. By assumption on X there is a sequence of curves γ_k from x to y such that $|\gamma_k| \to \varepsilon$. Choose for every k a point y_k on γ_k. By $|\gamma_k| \geq d(x, y_k) + d(y_k, y) \geq \varepsilon$ the relation $y_k \to y$ follows, hence the claim follows. □

2.4.12 To every rectifiable curve $\gamma : [t^-, t^+] \to X$ in a metric space one can assign the obviously continuous and monotone function $s : [t^-, t^+] \to [0, |\gamma|]$, $t \mapsto |\gamma_{|[t^-,t]}|$. Then there exists a unique continuous curve $\tilde{\gamma} : [0, |\gamma|] \to X$ with $\gamma = \tilde{\gamma} \circ s$, and the curve $\tilde{\gamma}$ is rectifiable. We call $\tilde{\gamma}$ the *parametrization of γ by arc length.*

Two rectifiable curves are called *equivalent*, if they have the same parametrization by arc length. An equivalence class of rectifiable curves is called a *geometric curve* (cf. BUSEMANN [36, Sec. 5]), a representative of a geometric curve is called a *parametrization.* The lengths, the initial and the end points of all parametrizations of a geometric curve coincide, hence it makes sense to use these notions for geometric curves as well. Often we denote a rectifiable curve and its associated geometric curve by the same symbol.

A sequence of geometric curves in a metric space is called *uniformly convergent* to a geometric curve $\gamma : [t^-, t^+] \to X$, if there exists a sequence of parametrizations $\gamma_k : [t^-, t^+] \to X$ such that the sequence $(\gamma_k)_{k \in \mathbb{N}}$ converges uniformly to the rectifiable curve γ.

Let us add some notation for curves of shortest length. A rectifiable curve $\gamma : [t^-, t^+] \to X$ is called a *segment*, if $|\gamma| = d(\gamma(t^-), \gamma(t^+))$ and a *geodesic*, if for every $s \in [t^-, t^+]$ there exists a $\varepsilon > 0$ such that $|\gamma_{|[s^-,s^+]}| = \delta(\gamma(s^-), \gamma(s^+))$ for all times $s^- < s^+$ in the interval $[t^-, t^+] \cap [s - \varepsilon, s + \varepsilon]$. In other words geodesics are curves locally of shortest length. Moreover, it follows immediately by definition that every segment is a geodesic. Note that the inverse is not true.

2.4.13 Theorem (BUSEMANN [36, 5.16 Thm.]) *Let (X, d) be a length space, and $(\gamma_k)_{k \in \mathbb{N}}$ a sequence of geometric curves in X such that the set of initial points of the*

γ_k as well as the sequence of lengths $|\gamma_k|$ is bounded. Then there exists a rectifiable curve $\gamma : [0, t^+] \to X$ and a subsequence $(\gamma_{k_l})_{l\in\mathbb{N}}$ converging uniformly to γ such that

$$|\gamma| \leq \liminf_{k\to\infty} |\gamma_k|.$$

PROOF: Let us proof the claim under the assumption that every d-bounded closed set is compact. Later it will turn out by the theorem of HOPF–RINOW that this assumption is automatically fulfilled in every length space. But note that for the proof of the theorem of HOPF–RINOW the theorem of BUSEMANN for compact X is needed. Therefore we state the theorem of BUSEMANN already at this point.

We assume that every curve γ_k is parametrized by arc length and first choose a subsequence $(\gamma_{k_l})_{l\in\mathbb{N}}$ of $(\gamma_k)_{k\in\mathbb{N}}$ such that the sequence of lengths $|\gamma_{k_l}|$ converges to $\liminf_{k\to\infty} |\gamma_k|$. By assumption on the curves γ_k the set of all initial points $\gamma_k(0)$ lies in a compact subset of X, hence there exists a point $x \in X$ such that after a possible transition to a subsequence $\lim_{l\to\infty} \gamma_{k_l}(0) = x$.

If $|\gamma_{k_l}| = 0$ from a certain index l_0 on, then the sequence $(\gamma_{k_l})_{l\in\mathbb{N}}$ obviously converges uniformly to the constant curve $\gamma : \{0\} \to X$, $\gamma(0) = x$.

If $|\gamma_{k_l}| > 0$ for all $l \in \mathbb{N}$, we define for every l a new parametrization $\lambda_l : [0, 1] \to X$ of γ_{k_l} by $\lambda_l(t) = \gamma_{k_l}(t\,|\gamma_{k_l}|)$. Let $L > 0$ be an upper bound of the lengths $|\gamma_k|$ and $\rho > 0$ such that all $\gamma_k(0)$ lie in the ball $B_\rho(x)$. Then for every $l \in \mathbb{N}$ and $t \in [0, 1]$

$$d(x, \lambda_l(t)) \leq d(x, \lambda_l(0)) + d(\lambda_l(0), \lambda_l(t)) \leq \rho + L,$$

that means for every t the set $\{\lambda_l(t) \mid l \in \mathbb{N}\} \subset X$ is relatively compact. On the other hand for $t_1, t_2 \in [0, 1]$ the following relation holds:

$$d\big(\lambda_l(t_1), \lambda_l(t_2)\big) = d\big(\gamma_{k_l}(t_1\,|\gamma_{k_l}|), \gamma_{k_l}(t_2\,|\gamma_{k_l}|)\big) = |\gamma_{k_l}|\,|t_1 - t_2| \leq L\,|t_1 - t_2|,$$

that means the sequence $(\lambda_l)_{l\in\mathbb{N}}$ is equicontinuous. By the theorem of ASCOLI–BOURBAKI and after a further transition to a subsequence $(\lambda_l)_{l\in\mathbb{N}}$ converges uniformly to a curve $\gamma : [0, 1] \to X$. For this γ

$$|\gamma| \leq \lim_{l\to\infty} |\lambda_l| = \liminf_{k\to\infty} |\gamma_k|. \tag{2.4.6}$$

Namely, let $0 = t_0 < t_1 < \cdots < t_n = 1$ be a sequence of intermediate points, $\varepsilon > 0$ and l_0 such that $d\big(\gamma(t), \lambda_l(t)\big) < \frac{\varepsilon}{4n}$ for all $t \in [0, 1]$ and $l \geq l_0$. By possibly enlarging l_0 one can achieve that $|\lambda_l| \leq \liminf_{k\to\infty} |\gamma_k| + \frac{\varepsilon}{2}$. Then

$$\sum_{i=0}^{n-1} d\big(\gamma(t_i), \gamma(t_{i+1})\big) \leq \frac{\varepsilon}{2} + \sum_{i=0}^{n-1} d\big(\lambda_{l_0}(t_i), \lambda_{l_0}(t_{i+1})\big) \leq \liminf_{k\to\infty} |\gamma_k| + \varepsilon,$$

hence (2.4.6) and thus the claim follow. □

2.4.14 Corollary (cf. BUSEMANN [37, I.1.(4)]) *If (X, d) is an inner metric space and if the closed ball $\overline{B}_\varepsilon(x)$ around $x \in X$ is compact, then there exists for every $y \in \overline{B}_\varepsilon(x)$ a segment connecting x with y.*

PROOF: First assume $0 < d(x,y) < \varepsilon$. Then there exists a sequence of geometric curves γ_k from x to y with $|\gamma_k| \to d(x,y)$ and $|\gamma_k| < \varepsilon$. Consequently every curve γ_k lies in $B_\varepsilon(x)$. By the compactness of $\overline{B}_\varepsilon(x)$ and the proof of the theorem of BUSEMANN 2.4.13 for compact length spaces the sequence $(\gamma_k)_{k\in\mathbb{N}}$ converges after a possible transition to a subsequence to a curve $\gamma \subset \overline{B}_\varepsilon(x)$ connecting x with y and having length $d(x,y) \le |\gamma| \le \liminf_{k\to\infty} |\gamma_k| = d(x,y)$. Consequently, γ is a segment connecting x and y.

Now let $d(x,y) = \varepsilon$. Then there exists by Lemma 2.4.11 a sequence of points $y_k \in B_\varepsilon(x)$ converging to y. Choose for every y_k a segment from x to y_k. Using the compactness of $\overline{B}_\varepsilon(x)$ and recalling the proof of the theorem of BUSEMANN one checks that after a transition to a subsequence $(\sigma_k)_{k\in\mathbb{N}}$ converges to a segment σ connecting x with y. □

Now we can prove the following fundamental result.

2.4.15 Theorem of Hopf–Rinow (*cf.* BUSEMANN [37, I.1.(8)] and GROMOV [69, Sec. 1.B]) *For a locally compact inner metric space* (X, d) *the following conditions are equivalent:*

(1) (X, d) *is complete, hence a length space.*

(2) *Every closed ball* $\overline{B}_\varepsilon(x) \subset X$ *is compact.*

(3) *If* $\gamma : [t^-, t^+[\to X$ *is a semiopen segment, that means if for every* $t < t^+$ *the restricted path* $\gamma_{|[t^-,t]}$ *is a segment, then* γ *can be extended to a segment* $\overline{\gamma} : [t^-, t^+] \to X$.

If one of these conditions is satisfied, then there exists for every pair of points x, y *a connecting segment that means a segment* $\gamma : [0, d(x,y)] \to X$ *with* $\gamma(0) = x$ *and* $\gamma(d(x,y)) = y$.

2.4.16 Remark HOPF–RINOW were the first who showed in their work [90] dated from 1932 that for any two points of a complete Riemannian manifold there exists a minimizing geodesic. Some time later COHN–VOSSEN [44] succeeded to prove that the main result of HOPF–RINOW is valid without the assumption of differentiability. Thus the "modern" version of the theorem of HOPF–RINOW as formulated here and as one can find it in BUSEMANN [37] or GROMOV [69] goes back essentially to COHN–VOSSEN.

PROOF: (2) ⇒ (1). Let $(x_j)_{j\in\mathbb{N}}$ be a Cauchy-sequence in X and K the closure of the set all points x_j. Then K is closed and d-bounded, hence lies in a compact ball $\overline{B}_\varepsilon(x_0)$. Consequently the Cauchy-sequence $(x_j)_{j\in\mathbb{N}}$ has an accumulation point, hence a limit. Thus X is complete.

Obviously (3) follows from (1). Moreover, under the assumption of (1) the theorem of BUSEMANN entails that for any two points there exists a connecting segment. So it remains to prove the implication (3) ⇒ (2).

We proceed like in [37, I.1.(8)]. For an arbitrary point x consider the set R of all $\rho > 0$ such that $\overline{B}_\rho := \overline{B}_\rho(x)$ is compact. If $\rho \in R$ then every positive $\rho' < \rho$ lies in R

as well. As X is assumed to be locally compact R cannot be empty. We show $\mathsf{R} = \mathbb{R}^{>0}$ by proving that R is an open and closed subset of $\mathbb{R}^{>0}$.

Let us suppose that $\overline{\mathsf{B}}_\rho$ is compact. Then $\overline{\mathsf{B}}_\rho$ can be covered by finitely many balls $\mathsf{B}_{\rho_i}(\mathsf{x}_i)$ such that $\mathsf{x}_i \in \overline{\mathsf{B}}_\rho$ for all i and such that $\overline{\mathsf{B}}_{\rho_i}(\mathsf{x}_i)$ is compact. Then for some $\delta > 0$ the relation $\mathsf{B}_{\rho+\delta}(\mathsf{x}) \subset \bigcup_i \mathsf{B}_{\rho_i}(\mathsf{x}_i)$ holds, consequently $\overline{\mathsf{B}}_{\rho+\delta} \subset \bigcup_i \overline{\mathsf{B}}_{\rho_i}(\mathsf{x}_i)$ is compact. Hence $\rho + \delta$ lies in R which means that R is open.

The proof that R is closed is more difficult. Suppose that $\overline{\mathsf{B}}_{\rho'}$ is compact for every positive $\rho' < \rho$. We will show the claim by proving that every sequence $(\mathsf{x}_k)_{k\in\mathbb{N}}$ of points $\mathsf{x}_k \in \overline{\mathsf{B}}_\rho$ has an accumulation point. We can assume $\mathsf{d}(\mathsf{x}_k, \mathsf{x}) \to \rho$ and $\rho_k := \mathsf{d}(\mathsf{x}, \mathsf{x}_k) < \rho_{k+1} = \mathsf{d}(\mathsf{x}, \mathsf{x}_{k+1})$, because in case $\mathsf{d}(\mathsf{x}, \mathsf{x}_k) = \rho$ one can find by Lemma 2.4.11 a point $\mathsf{x}'_k \in \mathsf{B}_\rho(\mathsf{x})$ with $\mathsf{d}(\mathsf{x}_k, \mathsf{x}'_k) < \frac{1}{k}$ and because an accumulation point of $(\mathsf{x}'_k)_{k\in\mathbb{N}}$ is one of $(\mathsf{x}_k)_{k\in\mathbb{N}}$ as well. By Corollary 2.4.14 there exists for every k a segment γ_k from x to x_k. Choose for every $i \leq k$ a point x^i_k on the curve γ_k with $\mathsf{d}(\mathsf{x}, \mathsf{x}^i_k) = \rho_i$. Let γ^i_k be the restriction of γ_k which connects x with x^i_k. Hence for a subsequence $(k_{1j})_{j\in\mathbb{N}}$ of $(k)_{k\in\mathbb{N}}$ the sequence $\left(\gamma^1_{k_{1j}}\right)_{j\in\mathbb{N}}$ converges by the theorem of BUSEMANN to a segment γ'_1 from x to a point $\mathsf{y}_1 \in \mathsf{B}_\rho(\mathsf{x})$ with $\mathsf{d}(\mathsf{x}, \mathsf{y}_1) = \rho_1$. Furthermore, there exists a subsequence $(k_{2j})_{j\in\mathbb{N}}$ of $(k_{1j})_{j\in\mathbb{N}}$ such that the sequence of segments $\left(\gamma^2_{k_{2j}}\right)_{j\in\mathbb{N}}$ converges to a segment γ'_2 from x to a point $\mathsf{y}_2 \in \mathsf{B}_\rho(\mathsf{x})$ with $\mathsf{d}(\mathsf{x}, \mathsf{y}_2) = \rho_2$. Continuing like this one obtains a sequence of segments $\gamma'_1 \subset \gamma'_2 \subset \cdots$ such that the initial point of γ'_k is x and the final point y_k has distance ρ_k to x.

Now $\bigcup_k \gamma'_k$ is a semiopen segment, hence can be completed by a point y to a segment γ'. Obviously $\mathsf{y}_k \to \mathsf{y}$. The diagonal sequence $\left(\mathsf{x}^{k_{jj}}_{k_{jj}}\right)_{j\in\mathbb{N}}$ converges to y as well, as for $j \geq i$

$$\mathsf{d}\big(\mathsf{y}, \mathsf{x}^{k_{jj}}_{k_{jj}}\big) \leq \mathsf{d}(\mathsf{y}, \mathsf{y}_i) + \mathsf{d}\big(\mathsf{y}_i, \mathsf{x}^i_{k_{jj}}\big) + \mathsf{d}\big(\mathsf{x}^i_{k_{jj}}, \mathsf{x}^{k_{jj}}_{k_{jj}}\big) < \mathsf{d}(\mathsf{y}, \mathsf{y}_i) + \mathsf{d}\big(\mathsf{y}_i, \mathsf{x}^i_{k_{jj}}\big) + \rho - \rho_i.$$

Choose for $\varepsilon > 0$ an index i such that $\mathsf{d}(\mathsf{y}, \mathsf{y}_i) + \rho - \rho_i < \varepsilon/2$. Afterwards choose N large enough such that $\mathsf{d}\big(\mathsf{y}_i, \mathsf{x}^i_{k_{jj}}\big) < \varepsilon/2$ for all $j \geq N$. Therefore y is an accumulation point of $(\mathsf{x}_k)_{k\in\mathbb{N}}$, and the claim is proved. □

Now we come back again to the construction of Riemannian metrics on stratified spaces.

2.4.17 Theorem *Given a connected regular Whitney* (A) *space* X *there exists a smooth Riemannian metric* μ *on* X *such that* (X, δ_μ) *becomes a length space. In that case there exists for every two points* x, y *of* X *a weakly piecewise* $\mathcal{C}^1$*-curve* γ *connecting* x *and* y *and which satisfies* $|\gamma|_\mu = \delta_\mu(\mathsf{x}, \mathsf{y})$.

Moreover, every connected compact regular Riemannian Whitney (A) *space is a length space.*

PROOF: If X is a compact connected regular Whitney (A) space with a smooth Riemannian metric μ, then every closed set is compact, hence by Corollary 2.4.9 (X, δ_μ) has to be a length space.

Now let X be an arbitrary connected regular Whitney (A) space. We will prove in the following the existence of a smooth Riemannian metric μ such that (X, δ_μ) is a length space. To this end choose a compact exhaustion $(\mathsf{K}_j)_{j\in\mathbb{N}}$ of X and then according to Lemma 1.3.17 an inductively embedding atlas with respect to $(\mathsf{K}_j)_{j\in\mathbb{N}}$ consisting of singular charts $x_j : K^\circ_{j+2} \to O_j \subset \mathbb{R}^{n_j}$. As X is connected and locally finitely path connected one can choose the compact sets K_j to be finitely path connected. Fix a point $x \in K_0$ and set $K_{-1} = \emptyset$. After an appropriate transformation we can suppose that $x_0(x) = 0$. Inductively we now construct Riemannian metrics μ_j on $\mathbb{R}^{n_j}$. First let μ_0 be the Euclidean scalar product multiplied with a factor $C > 0$ such that $C\|y\| > 1$ for all $y \in x_0(K^\circ_1 \setminus K_0)$. Assume that one has given $\mu_0, \cdots, \mu_j$ on $\mathbb{R}^{n_0}, \cdots, \mathbb{R}^{n_j}$ such that

$$\left((\iota^{n_{i+1}}_{n_i})^* \mu_{i+1}\right)_{|U_{i+1}} = \mu_{i|U_{i+1}} \tag{2.4.8}$$

for $i = 1, \cdots, j-1$ and appropriate open neighborhoods $U_{i+1} \subset O_i$ of $x_i(K_i)$. Furthermore assume that $d_i(y, 0) \geq i$ for all $y \in x_i(K_i \setminus K^\circ_{i-1})$ and $d_i(y, 0) \geq i+1$ for all $y \in x_i(K^\circ_{i+1} \setminus K_i)$, where d_i is the distance function corresponding to μ_i. We want to extend the sequence of the μ_i and will construct μ_{j+1}. To this end we first define a Riemannian metric η on $\mathbb{R}^{n_{j+1}}$ by

$$\eta_z((v_1, v_2), (w_1, w_2)) = (\mu_j)_{z_1}(v_1, w_1) + \langle v_2, w_2 \rangle, \quad z_1, v_1, w_1 \in \mathbb{R}^{n_j}, z_2, v_2, w_2 \in \mathbb{R}^{n_{j+1}-n_j}.$$

Next let $V_{j+1} \subset O_{j+1}$ be an open neighborhood of $x_{j+1}(K_j)$ such that $V_{j+1} \cap x_{j+1}(K_{j+1})$ is relatively compact in $x_{j+1}(K^\circ_{j+1})$. Finally choose smooth functions $\varphi : \mathbb{R}^{n_{j+1}} \to [0, 1]$ with relatively compact support in V_{j+1} such that $\varphi = 1$ on an open neighborhood $V'_{j+1} \subset V_{j+1}$ of $x_{j+1}(K_j)$ and set $U_{j+1} := V'_{j+1} \cap \mathbb{R}^{n_j}$. As $x_{j+1}(K_j)$ is compact, one can now choose $N \in \mathbb{N}$ such that $N\|y - z\| \geq 2$ for all $z \in x_{j+1}(K_j)$ and $y \in \mathbb{R}^{n_{j+1}} \setminus V_{j+1}$. One checks easily that the Riemannian metric

$$\mu_{j+1} = \varphi\eta + (1 - \varphi)\, N\, \langle \cdot, \cdot \rangle$$

together with the neighborhood U_{j+1} has the desired properties. This completes the induction.

As the Riemannian metrics μ_j are compatible in the above sense, we thus obtain a smooth metric μ on X which on every one of the open sets K°_j is equal to the pullback $(x_j)^*\mu_j$. We show that with this metric (X, δ_μ) is a length space indeed. More precisely we prove that every μ-bounded and closed set $K \subset X$ is compact. Let us assume that this is not the case, that means that there exists a sequence of points $x_j \in K$ having no accumulation point. By transition to a subsequences of $(K_j)_{j\in\mathbb{N}}$ and $(x_j)_{j\in\mathbb{N}}$ we can achieve $x_j \in K_j \setminus K^\circ_{j-1}$, because otherwise all x_j would lie in one of the compact sets K_j and the sequence $(x_j)_{j\in\mathbb{N}}$ would have an accumulation point. Consider the distances $\delta_\mu(x_j, x)$. If $k \geq j$ is large enough, then there exists a rectifiable curve γ_j in K_k from x to x_j with $\delta_\mu(x_j, x) \geq |\gamma_j| - 2^{-j}$. We choose k as small as possible, so there exists a point $y \in \operatorname{im}\gamma_j$ with $y \in K_k \setminus K^\circ_{k-1}$. Let us estimate the geodesic length of γ_j from below:

$$\delta_\mu(x_j, x) \geq |\gamma_j| - 2^{-j} \geq d_k(x_k(y), 0) - 2^{-j} \geq k - 2^{-j} \geq j - 2^{-j}.$$

Hence the sequence $(x_j)_{j\in\mathbb{N}}$ of points of K is not bounded which is in contradiction to the assumption on K. Therefore the sequence must have an accumulation point which by closedness of K lies again in K. This implies the compactness of K.

As we now know that (X, δ_μ) is a length space, the rest of the theorem follows immediately from the theorem of HOPF–RINOW 2.4.15 and Lemma 1.6.10. □

2.5 Differential operators

For every commutative $\mathbb{k}$-algebra $\mathcal{A}$ (with unit) one can define according to GROTHENDIECK [72] the space $\mathcal{D}(\mathcal{A}) \subset \mathrm{Hom}_{\mathbb{k}}(\mathcal{A}, \mathcal{A})$ of differential operators on $\mathcal{A}$ as the set of the in the following sense almost $\mathcal{A}$-linear operators. More precisely, one specifies recursively for every $k \in \mathbb{N}$ a space $\mathcal{D}^k(\mathcal{A})$ and sets afterwards $\mathcal{D}(\mathcal{A}) = \bigcup_{k\in\mathbb{N}} \mathcal{D}^k(\mathcal{A})$. First let $\mathcal{D}^0(\mathcal{A}) = \mathrm{Hom}_{\mathcal{A}}(\mathcal{A}, \mathcal{A}) \cong \mathcal{A}$ be the space of all $\mathcal{A}$-linear endomorphisms of $\mathcal{A}$. Suppose that the set $\mathcal{D}^k(\mathcal{A})$ of *differential operators of order at most* k has been constructed for some natural k. Then define $\mathcal{D}^{k+1}(\mathcal{A})$ by

$$\mathcal{D}^{k+1}(\mathcal{A}) = \{\mathrm{D} \in \mathrm{Hom}_{\mathbb{k}} \mid [\mathrm{D}, a] \in \mathcal{D}^k(\mathcal{A}) \text{ for all } a \in \mathcal{A}\}.$$

Thus, the space $\mathcal{D}^{k+1}(\mathcal{A})$ consists of all endomorphisms of $\mathcal{A}$ which are $\mathcal{A}$-linear up to operators of lower order. One now checks easily by an induction argument that all the spaces $\mathcal{D}^k(\mathcal{A})$ and $\mathcal{D}(\mathcal{A})$ are $\mathbb{k}$-linear. Moreover, the composition $\mathrm{D} \circ \tilde{\mathrm{D}}$ of two differential operators $\mathrm{D} \in \mathcal{D}^k(\mathcal{A})$ and $\tilde{\mathrm{D}} \in \mathcal{D}^l(\mathcal{A})$ is an element of $\mathcal{D}^{k+l}(\mathcal{A})$. Hence $\mathcal{D}(\mathcal{A})$ becomes a (filtered) $\mathbb{k}$-algebra. A first statement about the structure of $\mathcal{D}(\mathcal{A})$ is given by the following.

2.5.1 Lemma *For every commutative* $\mathbb{k}$*-algebra one has*

$$\mathcal{D}^1(\mathcal{A}) = \mathcal{A} + \mathrm{Der}_{\mathbb{k}}(\mathcal{A}, \mathcal{A}).$$

PROOF: Of course $\mathcal{A} + \mathrm{Der}_{\mathbb{k}}(\mathcal{A}, \mathcal{A}) \subset \mathcal{D}^1(\mathcal{A})$. For the proof of the inverse inclusion let $\mathrm{D} \in \mathcal{D}^1(\mathcal{A})$ and set $c = \mathrm{D}(1)$. We will show the inclusion $\tilde{\mathrm{D}} := \mathrm{D} - c \in \mathrm{Der}_{\mathbb{k}}(\mathcal{A}, \mathcal{A})$, which obviously will entail the claim. First note that $\tilde{\mathrm{D}} \in \mathcal{D}^1(\mathcal{A})$. In other words this means that for every $a \in \mathcal{A}$ there exists $c_a \in \mathcal{A}$ with $[\tilde{\mathrm{D}}, a] = c_a$. In particular the relation $\tilde{\mathrm{D}}(a) = \tilde{\mathrm{D}}(a1) = c_a$ is true, hence for all $a, b \in \mathcal{A}$

$$\tilde{\mathrm{D}}(ab) = a\tilde{\mathrm{D}}(b) + b\,c_a = a\tilde{\mathrm{D}}(b) + b\tilde{\mathrm{D}}(a).$$

Therefore $\tilde{\mathrm{D}}$ is a derivation of $\mathcal{A}$. □

Now let M be a manifold. Then it is well-known [46, Thm. 2.3] that the space $\mathcal{D}^k(\mathsf{M}) := \mathcal{D}^k(\mathcal{C}^\infty(\mathsf{M}))$ of differential operators on M of order at most k consists of all linear combinations of endomorphisms $\mathrm{D} \in \mathrm{End}(\mathcal{C}^\infty(\mathsf{M}))$ of the form

$$\mathcal{C}^\infty(\mathsf{M}) \ni f \mapsto \mathrm{D}f = \mathsf{V}_1 \cdots \mathsf{V}_l f \in \mathcal{C}^\infty(\mathsf{M}), \tag{2.5.1}$$

where $l \leq k$ and $\mathsf{V}_1, \cdots, \mathsf{V}_l$ are smooth vector fields on M. Replacing M by a Whitney (A) space X every smooth vector field V on X gives rise to a first order differential operator. Therefore the question arises, whether in analogy to the differentiable case the algebra $\mathcal{D}(\mathsf{X}) := \mathcal{D}(\mathcal{C}^\infty(\mathsf{X}))$ of differential operator is generated by the vector fields on X. But Example 2.2.1 and the above Lemma 2.5.1 show that this need not be the case; in general $\mathcal{D}(\mathsf{X})$ even is not generated by the derivations of $\mathcal{C}^\infty(\mathsf{X})$, as the following example shows.

2.5.2 Example (cf. [46, Exercise 3.8]) Consider Neil's parabola $\mathsf{X}_{\mathrm{Neil}}$ together with its smooth structure as a stratified subspace of $\mathbb{R}^2$ (see Example 1.1.15) and parametrize

X_{Neil} via the embedding $\mathbb{R} \to \mathbb{R}^2$, $t \mapsto (t^2, t^3)$. The image of this embedding is X_{Neil}. Then check that the operator ∂_t does actually not induce a derivation of $\mathcal{C}^\infty(X_{\text{Neil}})$, but that the operators $t\partial_t$ and $t^2\partial_t$ do, and that these two derivations span the $\mathcal{C}^\infty(X_{\text{Neil}})$-module $\text{Der}_{\mathbb{R}}\left(\mathcal{C}^\infty(X_{\text{Neil}}), \mathcal{C}^\infty(X_{\text{Neil}})\right)$. On the other hand $\partial_t^2 - 2t^{-1}\partial_t$, $t\partial_t^2 - \partial_t$ and $\partial_t^3 - 3t^{-1}\partial_t^2 + 3t^{-2}\partial_t$ are differential operators on X_{Neil} which are not generated by $t\partial_t$ and $t^2\partial_t$, hence $\text{Der}_{\mathbb{R}}\left(\mathcal{C}^\infty(X_{\text{Neil}}), \mathcal{C}^\infty(X_{\text{Neil}})\right)$ does not generate the algebra of differential operators on X_{Neil}.

2.5.3 Lemma *Every differential operator* D *on a Whitney* (A) *space* X *is local that means* $\operatorname{supp} Df \subset \operatorname{supp} f$ *for all* $f \in \mathcal{C}^\infty(X)$.

PROOF: The claim is trivial, if D has order 0. So let us assume that the claim holds for all operators lying in $\mathcal{D}^k(X)$. Then let $D \in \mathcal{D}^{k+1}(X)$. Choose for $f \in \mathcal{C}^\infty(X)$ and an open neighborhood U of $\operatorname{supp} f$ a function $\varphi \in \mathcal{C}^\infty(X)$ with $\operatorname{supp} \varphi \subset U$ and $\varphi_{|\operatorname{supp} f} = 1$. As $Df = D(\varphi f) = \varphi\, Df + \tilde{D} f$ for a differential operator $\tilde{D} \in \mathcal{D}^k(X)$, the relation $\operatorname{supp} Df \subset \operatorname{supp} \varphi \subset U$ follows. As U was an arbitrary neighborhood of $\operatorname{supp} f$, the claim now follows. □

2.5.4 Every open subset $U \subset X$ is again (A)-stratified, hence we canonically obtain operator spaces $\mathcal{D}^k(U)$ and $\mathcal{D}(U)$. Next we will show that the spaces $\mathcal{D}^k(U)$ are the sectional space of sheaves $\mathcal{D}^k$ on X. Hereby, locality makes it possible that one can restrict differential operators to smaller sets. To give the restriction morphisms explicitly let $\tilde{U} \subset U \subset X$ be open and $D \in \mathcal{D}(U)$. Then choose for every $x \in \tilde{U}$ a smooth function $\varphi_x \in \mathcal{C}^\infty(X)$ with $\operatorname{supp} \varphi_x \subset\subset \tilde{U}$ and $\varphi_{x|U_x} = 1$ on a neighborhood $U_x \subset \tilde{U}$ of x and define for all $f \in \mathcal{C}^\infty(\tilde{U})$ and $x \in \tilde{U}$ an extension $f_x \in \mathcal{C}^\infty(U)$ by

$$f_x(y) = \begin{cases} \varphi_x(y) f(y) & \text{if } \; y \in \tilde{U}, \\ 0 & \text{if } \; y \in U \setminus \tilde{U}. \end{cases} \tag{2.5.2}$$

The restricted differential operator $D_{|\tilde{U}} = r^U_{\tilde{U}}(D)$ will now be determined uniquely by requiring $D_{|\tilde{U}} f(x) = Df_x(x)$ for all $f \in \mathcal{C}^\infty(\tilde{U})$ and $x \in \tilde{U}$. As D is local, $Df_x(x)$ does not depend on the choice of φ_x, hence $Df_y(y) = Df_x(y)$ holds for all $y \in \tilde{U}$ sufficiently close to x. Therefore $D_{|\tilde{U}} f$ is smooth and $D_{|\tilde{U}}$ is well-defined. Moreover, by definition $r^{\tilde{U}}_{\hat{U}} \circ r^U_{\tilde{U}} = r^U_{\hat{U}}$ follows immediately for all open sets $\hat{U} \subset \tilde{U} \subset U \subset X$, that means $\mathcal{D}$ and $\mathcal{D}^k$ are presheaves on X.

Now it is easy to see that for any open covering $(U_j)_{j \in J}$ of U and a family of differential operators $D_j \in \mathcal{D}^k(U_j)$ with $D_{j|U_j \cap U_i} = D_{i|U_j \cap U_i}$ for all indices $j, i \in J$ there exists a unique differential operator $D \in \mathcal{D}^k(U)$ such that $D_{|U_j} = D_j$ for all j. One just defines for $f \in \mathcal{C}^\infty(U)$ the function $Df \in \mathcal{C}^\infty(U)$ by $(Df)_{|U_j} = D_j(f_{|U_j})$ and verifies immediately that the operator D is well-defined and has the desired properties. Hence the $\mathcal{D}^k$ are sheaves on X indeed. On the other hand, the presheaf $\mathcal{D}$ is in general not a sheaf. The reason lies in the fact that for noncompact X one cannot "glue together" every family $(D_j)_{j \in J}$ of pairwise compatible differential operators to a global one; namely if and only if the order of the set of orders of the D_j is unbounded.

2.5.5 Definition A differential operator $D \in \mathcal{D}(X)$ is called *stratified*, in signs $D \in \mathcal{D}_{\mathcal{S}}(X)$, if for all $S \in \mathcal{S}$ and all $f \in \mathcal{C}^\infty(X)$ the restriction $(Df)_{|S}$ depends only on $f_{|S}$ that means in other words if for $g \in \mathcal{C}^\infty(X)$ with $g_{|S} = 0$ the relation $(Dg)_{|S} = 0$ holds.

2.5.6 Example The smooth vector fields on X are stratified differential operators of first order. By Lemma 2.5.1 and the following proposition the smooth vector fields together with the smooth functions span the stratified differential operators of first order on X.

The following result extends Proposition 2.2.8 from vector fields to the case of differential operators.

2.5.7 Proposition *Let X be an* (A)*-stratified space. If* $x : U \to O \subset \mathbb{R}^n$ *is a singular chart,* $\mathcal{J} \subset \mathcal{C}^\infty(O)$ *the ideal of smooth functions on O vanishing on* $x(U)$ *and* $\mathcal{D}_{\mathcal{J}}(O)$ *the space of differential operators over O mapping the ideal* $\mathcal{J}$ *into itself, then there is a canonical isomorphy:*

$$\mathcal{D}(U) \cong \mathcal{D}_{\mathcal{J}}(O)/\mathcal{J}\mathcal{D}(O).$$

Moreover, a family $(D_S)_{S\in\mathcal{S}}$ *of differential operators* $D_S \in \mathcal{D}^k(S)$ *defines a differential operator* $D \in \mathcal{D}^k(X)$ *with* $(Df)_{|S} = D_S f_{|S}$ *for all* $f \in \mathcal{C}^\infty(X)$ *and* $S \in \mathcal{S}$*, if and only if for every smooth function* $f : X \to \mathbb{R}$ *the function* $X \ni x \mapsto D_{S_x} f_{|S_x}(x)$ *is smooth again, where* S_x *denotes the stratum of* x*. In this case D is determined uniquely and necessarily stratified. Vice versa, any stratified differential operator* $D \in \mathcal{D}^k_{\mathcal{S}}(X)$ *originates in this way from a family* $(D_S)_{S\in\mathcal{S}}$ *of differential operators* $D_S \in \mathcal{D}^k(S)$.

PROOF: As all sheaves involved are fine, we can suppose without loss of generality that $U = X$, that X is closed in O, that O is a ball around the origin of $\mathbb{R}^n$ and that $x = (x^1, \cdots, x^n)$ is given by the identical embedding $X \hookrightarrow \mathbb{R}^n$. Then we consider for $k \in \mathbb{N}$ the canonical morphism

$$\Pi^k : \mathcal{D}^k_{\mathcal{J}}(O) \to \mathcal{D}^k(X), \quad \mathbf{D} \mapsto (f = \mathbf{f} + \mathcal{J} \mapsto \mathbf{Df} + \mathcal{J}), \quad \mathbf{f} \in \mathcal{C}^\infty(O),$$

and first show that Π^k is surjective. To this end let us denote for $\alpha \in \mathbb{N}^n$ and $y \in \mathbb{R}^n$ by $(x-y)^\alpha$ the smooth function $(x^1 - y_1)^{\alpha_1} \cdot \ldots \cdot (x^n - y_n)^{\alpha_n} \in \mathcal{C}^\infty(X)$, and by $(\pi - y)^\alpha \in \mathcal{C}^\infty(O)$ the function $(\pi_1 - y_1)^{\alpha_1} \cdot \ldots \cdot (\pi_n - y_n)^{\alpha_n}$, where $\pi_i : \mathbb{R}^n \to \mathbb{R}$ is the projection onto the i-th coordinate and $y_i = \pi_i(y)$. Then one proves by induction on the order k that for every $D \in \mathcal{D}^k(X)$, $g \in \mathcal{C}^\infty(X)$, every multiindex α with $|\alpha| > k$ and every point $y \in X$ the following relation is true:

$$D(g \cdot (x-y)^\alpha)(y) = 0. \tag{2.5.4}$$

Now fix $D \in \mathcal{D}^k(X)$ and choose for $|\alpha| \le k$ functions $d_\alpha \in \mathcal{C}^\infty(O)$ such that

$$d_\alpha(y) = \frac{1}{\alpha!} D(x-y)^\alpha(y), \qquad y \in X.$$

Setting $\mathbf{D} = \sum_{|\alpha|\le k} d_\alpha \partial^\alpha$, the operator $\mathbf{D}$ lies in $\mathcal{D}^k(O)$. Next let us write down for every $\mathbf{f} \in \mathcal{C}^\infty(O)$ an expansion of the following kind:

$$\mathbf{f} = \sum_{|\alpha|\le k} \mathbf{f}_\alpha (\pi - y)^\alpha + \sum_{|\beta|=k+1} \mathbf{f}_\beta \cdot (\pi - y)^\beta,$$

where $f_\alpha \in \mathbb{R}$ and $f_\beta \in \mathcal{C}^\infty(O)$. Then apply D and calculate with the help of Eq. (2.5.4):

$$Df(y) = \sum_{|\alpha|\leq k} f_\alpha D(\pi-y)^\alpha + \sum_{|\beta|=k+1} D(f_\beta\cdot(\pi-y)^\beta)(y) = \sum_{|\alpha|\leq k} \alpha!\, f_\alpha d_\alpha(y) = D(f+\mathcal{J})(y).$$

But this means $D \in \mathcal{D}^k_{\mathcal{J}}(O)$ and $\Pi^k(D) = D$, hence Π^k is surjective.

If $D = \sum d_\alpha \partial^\alpha \in \mathcal{D}^k_{\mathcal{J}}(O)$ is in the kernel of Π^k, then $d_\alpha(y) = \frac{1}{\alpha!}D(\pi-y)^\alpha(y) = 0$ for all $y \in X$, hence $d_\alpha \in \mathcal{J}$ and therefore $\ker \Pi^k = \mathcal{J}\mathcal{D}^k(O)$. Consequently $\mathcal{D}(U)$ is isomorphic to $\mathcal{D}_{\mathcal{J}}(O)/\mathcal{J}\mathcal{D}(O)$.

Obviously, the family $(D_S)_{S\in\mathcal{S}}$ of differential operators $D_S \in \mathcal{D}^k(S)$ defines a uniquely determined differential operator $D \in \mathcal{D}^k(X)$ with $(Df)_{|S} = D_S f_{|S}$ for $f \in \mathcal{C}^\infty(X)$, if the function $X \ni x \mapsto D_{S_x} f_{|S_x}(x)$ is smooth. So it remains to show that every stratified differential operator is given by such a family. Let $D \in \mathcal{D}^k(X)$ be stratified. For a point $x \in S$ choose a singular chart $\tilde{x} : U \to \mathbb{R}^n$ around x such that $S \cap U = \tilde{x}^{-1}(\mathbb{R}^m \times \{0\})$, where $m = \dim S \leq n$. In other words $x := \tilde{x}_{|S\cap U} : S \cap U \to \mathbb{R}^m$ is a differentiable chart of S. By the above considerations and as D is stratified, we have for all $f \in \mathcal{C}^\infty(X)$

$$(Df)_{|S\cap U} = \sum_{|\alpha|\leq k} d_\alpha\, \partial^\alpha_x f_{|S\cap U} \quad \text{with } d_\alpha(y) = \frac{1}{\alpha!} D(x-y)^\alpha(y) \text{ for } y \in S\cap U.$$

Now, $\sum_{|\alpha|\leq k} d_\alpha\, \partial^\alpha_x$ is a differential operator on $S\cap U$, hence after appropriately gluing together these operators with the help of a smooth partition of unity for S we obtain a differential operator D_S. Altogether, we thus obtain a family $(D_S)_{S\in\mathcal{S}}$ such that $(Df)_{|S} = D_S f_{|S}$ for all $f \in \mathcal{C}^\infty(X)$. □

2.6 Poisson structures

Besides the differential geometric notion of a Riemannian metric one can also transfer the most important objects of symplectic geometry to stratified spaces. With the Marsden–Weinstein reduced spaces, which play an essential role in mathematical physics, we then have a rich and important class of examples for symplectic stratified spaces in our hands.

2.6.1 Definition A smooth section $\Lambda : X \to TX \otimes TX$ over a Whitney (A) space X is called *Poisson bivector*, if the following condition is satisfied:

(PS) For every stratum S the restriction

$$\Lambda_{|S} : S \to TS \otimes TS$$

provides a Poisson structure for S.

A Whitney (A) space X together with a Poisson bivector will be called a *Poisson stratified space*; if every stratum S with the Poisson structure induced by Λ_S is symplectic, then X is called a *symplectic stratified space.*

A Poisson bivector Λ induces a bidifferential operator $\{\cdot,\cdot\}$ on X, the so-called *Poisson bracket*, by setting for every stratum S and all functions $f, g \in \mathcal{C}^\infty(X)$:

$$\{f,g\}_{|S} = \Lambda_{|S} \lrcorner (df_{|S} \otimes dg_{|S}). \tag{2.6.1}$$

The function $\{f,g\}$ then is smooth on X indeed, as Λ can be extended locally in every singular chart to a smooth (antisymmetric) bivector field, hence in every chart $\{f,g\}$ is the restriction of a smooth function on Euclidean space.

2.6.2 Remark The notion of a symplectic stratified space has been introduced by SJAMAAR–LERMAN [162] (see as well SJAMAAR [161], BATES–LERMAN [6] and ORTEGA–RATIU [137]), though in a somewhat different form than in 2.6.1. The difference between our definition and the one of SJAMAAR–LERMAN lies in the fact that in [162] the authors use only a rather weak notion of a smooth structure which in general does not allow the construction of a stratified tangent bundle and therefore also not of continuous or smooth vector fields. More precisely, SJAMAAR–LERMAN understand by a smooth structure on a stratified space a subalgebra $\mathcal{A} \subset \mathcal{C}^0(X)$ such that for every function $f \in \mathcal{A}$ the restrictions $f_{|S}$ on the strata S are smooth. If now $\mathcal{A}$ comes with a bracket $\{\cdot,\cdot\} : \mathcal{A} \times \mathcal{A} \to \mathcal{A}$, then SJAMAAR–LERMAN call the triple $(X, \mathcal{A}, \{\cdot,\cdot\})$ a symplectic stratified space, if the following holds:

1. every stratum is a symplectic manifold,
2. $(\mathcal{A}, \{\cdot,\cdot\})$ is a Poisson algebra, and
3. the embeddings $S \hookrightarrow X$ are Poisson mappings.

One checks immediately that symplectic stratified spaces in our sense are symplectic stratified in the sense of SJAMAAR–LERMAN as well, but not vice versa.

2.6.3 The probably most important class of examples of symplectic stratified spaces is given by reduction of symplectic manifolds with symmetry. In the regular case MARSDEN–WEINSTEIN [119] and independently MEYER [129] have introduced this reduction scheme in mathematical physics; SJAMAAR–LERMAN [162] have extended it to the singular case. In the following we will explain symplectic reduction and will sketch, how one obtains by this construction symplectic stratified spaces. Hereby we will need notions and results from the theory of G-actions and their orbit spaces as they will be provided in Chapter 4. The use of later proven theorems will not lead to any circular arguments, as the results of this paragraph will not be needed anywhere else in this monograph. For possibly necessary supplements from symplectic geometry we refer to ABRAHAM–MARSDEN [1].

Let (M,ω) be a symplectic manifold together with a proper and *symplectic* Lie group action $\Phi : G \times M \to M$. That G acts symplectically hereby means that for all $g \in G$ the map Φ_g is a *canonical transformation* or in other words a *symplectomorphism* of M, i.e.

$$\Phi_g^* \omega = \omega.$$

Suppose further that there exists a G-invariant *moment map* $J : M \to \mathfrak{g}^*$ that means for all $\xi \in \mathfrak{g}$ the relation $dJ^\xi = i_{\xi_M}\omega$ holds, where ξ_M is the fundamental vector field of ξ and J^ξ the smooth function $M \ni x \mapsto \langle J(x), \xi \rangle \in \mathbb{R}$. A quadruple of the form (M, ω, G, J) is called in symplectic geometry a *Hamiltonian* G-*space.* The components of the moment map comprise integrals of the motion for every G-invariant Hamiltonian flow, as for every G-invariant function $h \in \mathcal{C}^\infty(M)^G$ and every $\xi \in \mathfrak{g}$:

$$\{h, J^\xi\} = X_{J^\xi} h = \mathcal{L}_{\xi_M} h = 0,$$

where X_{J^ξ} denotes the Hamiltonian vector field of J^ξ. One can then decrease the degrees of freedom of the Hamiltonian system. Thus one comes to a more simple description of the system and possibly to a complete integration of the system. In mathematical detail this so-called *Marsden–Weinstein reduction* works as follows.

Consider the zero level set $Z = J^{-1}(0)$ of the moment map. Under the assumption that 0 is a regular value of J, Z comprises a submanifold of M. But in many cases this assumption does not hold, and then Z is "only" stratified. The essential observation now is that Z is a G-invariant subspace, hence one can form the orbit space $M_0 = G\backslash Z$, and thus obtains the desired *reduced space.* This space has the following properties.

2.6.4 Theorem (SJAMAAR–LERMAN [162], SJAMAAR [161, 3.1.1. Thm], BATES–LERMAN [6]) *Let* (M, ω) *be a Hamiltonian* G*-space with proper* G*-action and* $J : M \to \mathfrak{g}^*$ *a* G*-equivariant moment map. Then the intersection of the zero level set* Z *with a stratum* $M_{(H)}$ *of orbit type* (H), *where* H *is a closed subgroup of* G *is a manifold. Moreover, the orbit space* $(M_0)_{(H)} = G\backslash(M_{(H)} \cap Z)$ *carries a canonical symplectic structure* $(\omega_0)_{(H)}$ *the pullback of which to* $Z_{(H)} := M_{(H)} \cap Z$ *coincides with the restriction of the symplectic form* ω *to* $Z_{(H)}$. *The decomposition of* $M_0 = G\backslash Z$ *into the manifolds* $(M_0)_{(H)}$ *induces a stratification of* M_0.

PROOF: In [162] the theorem has been proven for the case of a compact Lie group. It has been extended to the case of proper G-actions in [6]. We do not give the prove of the claim at this point but refer the reader to the two cited articles. □

Besides the natural stratification by orbit types the reduced space $M_0 = G\backslash Z$ carries a canonical functional structure given by $\mathcal{C}^\infty(M_0) = \mathcal{C}^\infty(M)^G/\mathcal{J}^G$, where $\mathcal{J}$ is the ideal of smooth functions vanishing on Z. The function algebra $\mathcal{C}^\infty(M_0)$ stems from a smooth structure on M_0. Apparently this follows from the fact that M_0 can locally be embedded in some Euclidean space and that the corresponding local smooth structures coincide with the one given by $\mathcal{C}^\infty(M_0)$ (see PFLAUM [144, Sec. 6] for details). Moreover, as a consequence of [162, Sec. 5] we obtain that M_0 is even Whitney stratified and locally trivial.

2.6.5 Theorem *Every reduced space* M_0 *like in Theorem 2.6.4 comprises a symplectic stratified space in the sense of definition 2.6.1. The Poisson bracket on* $\mathcal{C}^\infty(M_0)$ *satisfies*

$$\{f, g\} \circ \pi = \{\overline{f}, \overline{g}\}_{|Z}, \qquad f, g \in \mathcal{C}^\infty(M_0). \tag{2.6.2}$$

Hereby, $\pi : Z \to M_0$ *denotes the canonical projection, and* $\overline{f}, \overline{g} \in \mathcal{C}^\infty(M)^G$ *are chosen such that* $f \circ \pi = \overline{f}_{|Z}$ *and* $g \circ \pi = \overline{g}_{|Z}$.

PROOF: SJAMAAR–LERMAN have shown in [162] that Eq. 2.6.2 defines a Poisson bracket on $\mathcal{C}^\infty(M_0)$. Moreover, they have shown that for every orbit type (H) and $f, g \in \mathcal{C}^\infty(M_0)$ the relation

$$\{f, g\}_{|(M_0)_{(H)}} = \{f_{|(M_0)_{(H)}}, g_{|(M_0)_{(H)}}\}_{(H)}$$

holds, where $\{\cdot,\cdot\}_{(H)}$ is the Poisson bracket of the symplectic manifold $(M_0)_{(H)}$. Consequently the Poisson bracket on $\mathcal{C}^\infty(M_0)$ is given stratawise by a regular Poisson bivector $\Lambda_{(H)}$, hence every $\Lambda_{(H)}$ comes from a symplectic form. By Proposition 2.2.6 and Eq. 2.6.2 the bivector Λ composed of the $\Lambda_{(H)}$ must be smooth. Thus (M_0, Λ) is a Poisson stratified space and, as all strata are symplectic, even symplectic stratified. □

2.6.6 Remark In the article [137] ORTEGA–RATIU succeeded to generalize singular reduction to Poisson manifolds.

2.6.7 Example (cf. [109, Sec. 1]) Consider the canonical SO(2)-action on $\mathbb{R}^2$ and lift it to the cotangent bundle $M = T^*\mathbb{R}^2 = \mathbb{R}^4$:

$$\begin{pmatrix} x_1 \\ x_2 \\ \xi_1 \\ \xi_2 \end{pmatrix} \mapsto \begin{pmatrix} \cos\varphi & -\sin\varphi & 0 & 0 \\ \sin\varphi & \cos\varphi & 0 & 0 \\ 0 & 0 & \cos\varphi & -\sin\varphi \\ 0 & 0 & \sin\varphi & \cos\varphi \end{pmatrix} \begin{pmatrix} x_1 \\ x_2 \\ \xi_1 \\ \xi_2 \end{pmatrix}$$

One checks by a simple computation that the SO(2)-action is symplectic with respect to the canonical symplectic form $\omega = dx_1 \wedge d\xi_1 + dx_2 \wedge d\xi_2$. A moment map is given by the angular momentum $J(q,p) = x_1\xi_2 - x_2\xi_1$. As 0 is a singular value of J, the reduced space $M_0 = Z/\,SO(2)$ cannot be smooth. One calculates easily that $Z = J^{-1}(0)$ is the union of $Z_{(SO(2))} = \{0\}$ and

$$Z_{(1)} = \{(x_1, x_2, \xi_1, \xi_2) \neq 0 \mid x_1\xi_2 - x_2\xi_1 = 0\}.$$

Consequently M_0 is the disjoint union of the strata $(M_0)_{(SO(2))} = \{0\}$ and $(M_0)_{(1)} \cong \mathbb{R}^2 \setminus \{0\}$, where an appropriate (symplectic) isomorphism is given by

$$\mathbb{R}^2 \setminus \{0\} \ni \begin{pmatrix} x \\ \xi \end{pmatrix} \mapsto SO(2) \cdot \begin{pmatrix} x \\ 0 \\ \xi \\ 0 \end{pmatrix} \in (M_0)_{(1)}. \tag{2.6.4}$$

How does the smooth structure on M_0, which is in fact decisive for the geometric properties of M_0, look like? A homogeneous Hilbert-basis for the SO(2)-invariant polynomials on $T^*\mathbb{R}^2$ is given by the four functions

$$p_1 = x_1^2 + x_2^2 - (\xi_1^2 + \xi_2^2), \quad p_2 = x_1\xi_1 + x_2\xi_2,$$
$$r = x_1^2 + x_2^2 + \xi_1^2 + \xi_2^2, \quad \text{and} \quad J = x_1\xi_2 - x_2\xi_2.$$

As J vanishes on Z, we obtain a proper embedding $\iota : M_0 \to \mathbb{R}^3$, $SO(2)v \mapsto (p_1(v), p_2(v), r(v))$, $v \in \mathbb{R}^4$. Now one proves easily that modulo the vanishing ideal $\mathcal{J}$

of Z the relation $r^2 = p_1^2 + p_2^2$ is satisfied. In other words this means that the image of ι lies in the cone $X_{\text{Cone}} = \{(y_1, y_2, y_3) \in \mathbb{R}^3 \mid y_3^2 = y_1^2 + y_2^2\}$. According to Eq.. 2.6.4 even $\operatorname{im} \iota = X_{\text{Cone}}$ has to hold. By the theorem of SCHWARZ 4.4.3 and the definition of the smooth structure on M_0 the algebra $\mathcal{C}^\infty(M_0)$ coincides canonically with the algebra of smooth functions on $\mathbb{R}^3$ restricted to $X_{\text{Cone}} = \iota(M_0)$. Hence M_0 and X_{Cone} are isomorphic as stratified spaces with smooth structure.

2.6.8 Remark In connection with symplectic stratified spaces the moduli spaces of flat connections on a Riemann surface according to ATIYAH–BOTT [5] are of broad interest. These moduli spaces can be interpreted as Marsden–Weinstein quotients of infinite dimensional symplectic spaces. By GOLDMAN [61] and KARSHON [97] they are symplectic, carry a canonical functional structure (cf. KARSHON [97] and HUEBSCHMANN [93]), which turns out to be a smooth structure in our sense, and finally are stratified (GURUPRASAD–HUEBSCHMANN–JEFFREY–WEINSTEIN [74] and HUEBSCHMANN [91, 92]). Thus moduli of flat connections provide further interesting, nontrivial and important examples for symplectic stratified spaces.

2.6.9 It is possible to coin the notion of a deformation quantization à la BAYEN et al. [7] for symplectic stratified spaces and even for Poisson stratified spaces. So let X be Poisson stratified and Λ the corresponding bivector field. Consider the linear space $\mathcal{A} := \mathcal{C}^\infty(X)[[\lambda]]$ of formal power series in one parameter λ and with coefficients in $\mathcal{C}^\infty(X)$. By a *formal differential deformation quantization* of X or a *star product* on X we then understand nothing else than an associative law of composition $\star : \mathcal{A} \times \mathcal{A} \to \mathcal{A}$ such that the following axioms are satisfied:

(DQ1) The composition $\star$ is $\mathbb{R}[[\lambda]]$-bilinear.

(DQ2) There exist bidifferential operators $b_k : \mathcal{C}^\infty(X) \times \mathcal{C}^\infty(X) \to \mathcal{C}^\infty(X)$ such that for all $f, g \in \mathcal{C}^\infty(X)$

$$f \star g = \sum_{k \in \mathbb{N}} \lambda^k \, b_k(f, g).$$

(DQ3) For all $f, g \in \mathcal{C}^\infty(X)$ the relations

$$f \star g = fg + O(\lambda) \quad \text{and} \quad 1_X \star f = f \star 1_X = f,$$

hold, where 1_X denotes the constant function on X with value 1.

(DQ4) The bidifferential operator b_1 is equal to the Poisson bivector Λ.

By the work of DEWILDE–LECOOMTE [48, 49] and of FEDOSOV [53, 54] one knows that for every symplectic manifold there exists a star product. According to a newer result of KONTSEVICH [102] even every Poisson manifold has a deformation quantization. But concerning symplectic stratified spaces there do not exist up to now general results about the existence of star products on such spaces. Nevertheless it seems rather promising to extend FEDOSOV's geometric scheme for deformation quantization to symplectic stratified spaces which possess only orbifold singularities. (see 4.4.10).

2.6.10 A rather encouraging ansatz to prove even for symplectic stratified spaces that they have a deformation quantization lies in the method of KONTSEVICH, more precisely in his formality theorem which was first stated in [103] and then proved in [102]. Among other things the formality theorem implies that every Poisson manifold has a star product (see as well VORONOV [181]). In the following we will explain the formality theorem and how it could be used for a proof that every symplectic stratified space can be deformation quantized. Hereby we will have to sketch the arguments as a detailed exposition would be beyond the scope of this work.

The essential ingredients for the formality theorem are two differential graded Lie algebras, namely the Lie superalgebra $\mathsf{T}^\bullet(\mathsf{M})$ of antisymmetric polyvector fields on a manifold M and the Lie-superalgebra $\mathsf{D}^\bullet(\mathsf{M})$ of polydifferential Hochschild cochains of the algebra $\mathcal{C}^\infty(\mathsf{M})$. Let me explain these ingredients somewhat more precisely. By a *differential graded Lie algebra* one understands a Lie algebra in the tensor category of complexes of vector spaces, in other words an object of the form

$$\mathfrak{g} = \bigoplus_{k\in\mathbb{Z}} \mathfrak{g}^k, \quad [\cdot,\cdot] : \mathfrak{g}^k \otimes \mathfrak{g}^l \to \mathfrak{g}^{k+l}, \quad \mathsf{d} : \mathfrak{g}^k \to \mathfrak{g}^{k+1}, \quad \mathsf{d}^2 = 0.$$

The Lie superalgebra $\mathsf{T}^\bullet(\mathsf{M})$ has components

$$\mathsf{T}^k(\mathsf{M}) := \mathcal{C}^\infty(\mathsf{M}, \Lambda^{k+1}\mathsf{TM}), \quad k \geq -1,$$

the differential $\mathsf{d} := 0$ and is supplied with the Schouten–Nijenhuis bracket as Lie bracket. On the other hand, the space $\mathsf{D}^k(\mathsf{M})$ of polydifferential Hochschild k-cochains (see Section 6.3 for the definition of Hochschild cochains) consists of all mappings of the form

$$\mathsf{f}_0 \otimes \cdots \otimes \mathsf{f}_k \mapsto \sum_{j\in J} \mathsf{D}_{j0}\mathsf{f}_0 \cdots \mathsf{D}_{jk}\mathsf{f}_k, \qquad \mathsf{f}_0, \cdots, \mathsf{f}_k \in \mathcal{C}^\infty(\mathsf{M}),\ \mathsf{D}_{j0}, \cdots, \mathsf{D}_{jk} \in \mathcal{D}(\mathsf{M}),$$

where J denotes a finite index set. As differential one takes the ordinary differential in the Hochschild complex, the Lie bracket is given by the Gerstenhaber bracket.

There exists a canonical mapping between the just introduced differential graded Lie algebras: $\mathsf{F}_1 : \mathsf{T}^\bullet(\mathsf{M}) \to \mathsf{D}^\bullet(\mathsf{M})$

$$\mathsf{F}_1 : (\mathsf{V}_0 \wedge \cdots \wedge \mathsf{V}_k) \mapsto \left(\mathsf{f}_0 \otimes \cdots \otimes \mathsf{f}_k \mapsto \frac{1}{(k+1)!} \sum_{\sigma\in S_{k+1}} \operatorname{sgn}(\sigma) \prod_{i=0}^{k} \mathsf{V}_{\sigma(i)}\mathsf{f}_i \right),$$

where the V_i are vector fields and the f_i are smooth functions on M. By the topological version of the theorem of HOCHSCHILD–KOSTANT–ROSENBERG [88] F_1 is a quasi isomorphism (see as well Theorem 6.4.5 and PFLAUM [143]), hence we call it the *HKR quasi isomorphism*. Now we can formulate:

2.6.11 Formality Theorem (KONTSEVICH [103, 102]) *There exists an L_∞-morphism* F *from* $\mathsf{T}^\bullet(\mathsf{M})$ *to* $\mathsf{D}^\bullet(\mathsf{M})$ *with first term given by the HKR quasi isomorphism* F_1.

Now the question is what to understand by an L_∞-morphism and what is the connection to deformation theory resp. deformation quantization. First let us explain L_∞-morphisms. Consider a graded vector space $\mathfrak{g}$ and associate to $\mathfrak{g}$ the following graded coalgebra:

$$\mathsf{C}(\mathfrak{g}) := \bigoplus_{k\in\mathbb{N}} \mathsf{Sym}^k(\mathfrak{g}[1]) = \bigoplus_{k\in\mathbb{N}} \Lambda^k(\mathfrak{g})[k],$$

where Sym^k is the functor of the k-times symmetric tensor product. If the coalgebra $\mathsf{C}(\mathfrak{g})$ possesses additionally a differential Q of degree $+1$, then one calls the pair $(\mathsf{C}(\mathfrak{g}),\mathsf{Q})$ a L_∞*-algebra*, and an L_∞*-morphism* is nothing else than a morphism in the category of L_∞-algebras. For the case that $\mathfrak{g}$ is even a differential graded Lie algebra the differential d and the Lie bracket $[\cdot,\cdot]$ of $\mathfrak{g}$ induce automatically a differential Q on $\mathsf{C}(\mathfrak{g})$ with $\mathsf{Q}_1 = \mathsf{d}$, $\mathsf{Q}_2 = [\cdot,\cdot]$ and $\mathsf{Q}_3 = \mathsf{Q}_4 = \cdots = 0$.

Next we come to deformation theory. One assigns to every differential graded Lie algebra $\mathfrak{g}$ the moduli space $\mathcal{M}(\mathfrak{g})$ of solutions of the Maurer–Cartan equations:

$$\mathcal{M}(\mathfrak{g}) := \left\{\xi\in\mathfrak{g}\,\middle|\, d\xi + \tfrac{1}{2}[\xi,\xi] = 0\right\}/\mathsf{G}^0,$$

where G^0 denotes the Lie group corresponding to the Lie algebra $\mathfrak{g}^0$. If now $\mathfrak{m}$ is a commutative algebra without unit, where for our intended applications in deformation quantization we will always have $\mathfrak{m} = \lambda\mathbb{R}[[\lambda]]$, then the functor $\mathrm{Def}_\mathfrak{g}$ of deformations of $\mathfrak{g}$ is given by

$$\mathrm{Def}_\mathfrak{g}(\mathfrak{m}) = \mathcal{M}(\mathfrak{g}\otimes\mathfrak{m}).$$

Hereby one can interpret $\mathfrak{m}$ as the parameter space of the deformation. The fundamental statement now is that every L_∞-morphism F between $\mathsf{C}(\mathfrak{g}_1)$ and $\mathsf{C}(\mathfrak{g}_2)$ such that the first term F_1 is a quasi isomorphism induces an isomorphism between $\mathrm{Def}_{\mathfrak{g}_1}(\mathfrak{m})$ and $\mathrm{Def}_{\mathfrak{g}_2}(\mathfrak{m})$. This is the quintessence which essentially goes back to the approach to deformation theory by SCHLESSINGER–STASHEFF [147].

Now we come back to our original matter of concern, deformation quantization, and remark that (with $\mathfrak{m} = \lambda\,\mathbb{R}[[\lambda]]$) the deformations $\mathrm{Def}_{\mathsf{T}^\bullet(\mathsf{M})}(\mathfrak{m})$ consist of exactly the formal Poisson structures and that the deformations $\mathrm{Def}_{\mathsf{T}^\bullet(\mathsf{M})}(\mathfrak{m})$ are precisely the star products on M. Consequently, an L_∞-morphism like in the formality theorem provides a natural isomorphy between equivalence classes of formal Poisson structures and star products on M.

Let us transfer these considerations to the stratified case and let X be a symplectic stratified space with Poisson bivector Λ. Assume for simplicity that X is a closed subset of an open set $\mathsf{O}\subset\mathbb{R}^n$ and inherits from $\mathbb{R}^n$ a smooth structure. Let $\overline{\Lambda}$ be an antisymmetric bivector field on O such that $\overline{\Lambda}$ coincides over X with Λ. Then the bivector field $\overline{\Lambda}$ induces over O a not necessarily associative "star product" $\overline{\star}$, as for $\overline{\Lambda}$ the Jacobi identity need not hold. At this point we consider the strata S and stress that the canonical embeddings $\iota_\mathsf{S}:\mathsf{S}\to\mathsf{X}$ are Poisson maps. If now the construction of the L_∞-morphism F from $\mathsf{C}(\mathsf{T}^\bullet(\mathsf{M}))$ to $\mathsf{C}(\mathsf{D}^\bullet(\mathsf{M}))$ is in a certain sense natural or functorial with respect to Poisson maps, then for every f in the vanishing ideal $\mathcal{J}$ of X and for every $\mathsf{g}\in\mathcal{C}^\infty(\mathsf{O})$ the relation $\mathsf{f}\overline{\star}\mathsf{g}_{|\mathsf{X}} = \mathsf{g}\overline{\star}\mathsf{f}_{|\mathsf{X}} = 0$ holds. Consequently $\overline{\star}$ can be pushed down to a $\mathbb{R}[[\lambda]]$-bilinear map $\star:\mathcal{A}\times\mathcal{A}\to\mathcal{A}$ with $\mathcal{A} = \mathcal{C}^\infty(\mathsf{X})[[\lambda]] = (\mathcal{C}^\infty(\mathsf{O})/\mathcal{J})[[\lambda]]$. Applying the naturality of the construction again it stratawise becomes clear that the restriction of $\star$ to each of the S gives a star product, hence this must hold also globally on X. Let us summarize:

2.6.12 Result *Under the assumption that the L_∞-morphism in the formality theorem can be constructed in a natural way, then there exists for every symplectic stratified space a star product.*

By KONTSEVICH's proof of the formality theorem it is not immediately clear whether the construction of F is natural in a functorial sense indeed, but this seems to be plausible. Therefore let us formulate the following conjecture, where we do not want to conceal the fact that the notion of "naturality" in the context of the formality conjecture needs further explanation.

2.6.13 Functorial formality conjecture *The graph theoretical construction of the L_∞-morphism according to* KONTSEVICH *is natural in an appropriate sense.*

Chapter 3

Control Theory

3.1 Tubular neighborhoods

In this section we will introduce the notion of a tubular neighborhood and will proof the classical tubular neighborhood theorem. But before we come to this let us provide some useful notation.

3.1.1 Let X be a locally compact Hausdorff space with countable topology, $S \subset X$ a locally closed subset, and T_S an open neighborhood of S in X. Moreover, let $\pi_S : T_S \to S$ be a continuous retraction $(\pi_S)_{|S} = \mathrm{id}_S$, and $\rho_S : T_S \to \mathbb{R}^{\geq 0}$ a continuous function with $\rho_S^{-1}(0) = S$. One finds such a situation for example, if X is given by a metric vector bundle E over a manifold S; then one can choose $T_S = E$, π_S as the projection of E, and $\rho_S : E \to \mathbb{R}$ as the distance function $v \mapsto \|v\|^2 = \eta(v, v)$, where η denotes the scalar product on E.

Assume to be given two (in most cases continuous) functions $\varepsilon, \delta : S \to \overline{\mathbb{R}} := \mathbb{R} \cup \{\pm\infty\}$ with $\varepsilon < \delta$, i.e. $\varepsilon(x) < \delta(x)$ for all $x \in S$. Then we set

$$
\begin{aligned}
[\varepsilon] &= \{(x,t) \in S \times \mathbb{R} \,|\, t = \varepsilon(x)\},\\
[\varepsilon, \delta[&= \{(x,t) \in S \times \mathbb{R} \,|\, \varepsilon(x) \leq t < \delta(x)\},\\
T_S^\varepsilon &= \{x \in T_S \,|\, \rho_S(x) < \varepsilon(\pi_S(x))\},\\
\dot{T}_S^\varepsilon &= T_S^\varepsilon \setminus S,\\
\overline{T}_S^\varepsilon &= \{x \in T_S \,|\, \rho_S(x) \leq \varepsilon(\pi_S(x))\}.
\end{aligned}
$$

Analogously we define $[\varepsilon, \delta]$, $]\varepsilon, \delta[$ and $]\varepsilon, \delta[$. If in some cases we want to stress that the sets T_S^ε, $\dot{T}_S^\varepsilon$ and $\overline{T}_S^\varepsilon$ are subsets of X, then we write $T_{S\subset X}^\varepsilon$, $\dot{T}_{S\subset X}^\varepsilon$ and $\overline{T}_{S\subset X}^\varepsilon$.

3.1.2 Lemma *After possibly shrinking* T_S, π_S *and* ρ_S *the following statements hold:*

(1) *For every open neighborhood W of S in X there exists a continuous function* $\delta : S \to \mathbb{R}^{>0}$ *with* $S \subset T_S^\delta \subset W$.

(2) *There exists a continuous function* $\varepsilon : S \to \mathbb{R}^{>0}$ *such that the restricted map* $(\pi_S, \rho_S) : T_S^\varepsilon \to [0, \varepsilon[$ *is proper and surjective.*

M.J. Pflaum: LNM 1768, pp. 91 - 149, 2001

PROOF: Choose for every $x \in S$ an open neighborhood V_x in X with compact closure $\overline{V_x} \subset T_S$. We claim that the sets $W_{x,n} := V_x \cap \pi_S^{-1}(U_{x,n}) \cap \rho_S^{-1}[0, \delta_{x,n}[$ form a basis of neighborhoods of x in X, if $U_{x,n} \subset S$ runs through a basis of neighborhoods of x in S and $(\delta_{x,n})_{n\in\mathbb{N}}$ is a monotone decreasing sequence of positive numbers with $\lim_{n\to\infty} \delta_{x,n} = 0$. Suppose the claim is not true. Then there exists an open neighborhood V of x in X and for every $n \in \mathbb{N}$ a point $y_n \in W_{x,n} \setminus V$. As the y_n lie in the compact set $\overline{V_x} \subset T_S$ we can achieve after transition to a subsequence that $(y_n)_{n\in\mathbb{N}}$ converges to some $y \in T_S$. By definition of the sets $W_{x,n}$ the relations $\pi_S(y) = \lim_{n\to\infty} \pi_S(y_n) = x$ and $\rho_S(y) = \lim_{n\to\infty} \rho_S(y_n) = 0$ then would hold, hence $y = x$. But this contradicts the fact that in the neighborhood V of $y = x$ there is no element of the converging sequence $(y_n)_{n\in\mathbb{N}}$. Thus the $W_{x,n}$ form a basis of neighborhoods of x.

Now let us choose a locally finite covering $(V_n)_{n\in\mathbb{N}}$ of S by sets $V_n \subset T_S$ open in X such that $V_n \subset V_{x_n}$ for appropriate $x_n \in S$. Then define $\tilde{T}_S = \bigcup_{n\in\mathbb{N}} V_n$ and note that $\tilde{T}_S$ is an open neighborhood of S in X. We denote the restrictions of π_S and ρ_S to $\tilde{T}_S$ by $\tilde{\pi}_S$ and $\tilde{\rho}_S$. By the results proven above the sets $\tilde{W}_{x,n} := \tilde{\pi}_S^{-1}(U_{x,n}) \cap \tilde{\rho}_S^{-1}[0, \delta_{x,n}[$ form a basis of neighborhoods of x in X. We show that the triple $(\tilde{T}_S, \tilde{\pi}_S, \tilde{\rho}_S)$ satisfies (1). To this end choose for every $x \in S$ an open set $U_{x,0} \subset S$ and a number $\delta_{x,0} = \delta_{x,1} > 0$ such that $\tilde{W}_{x,0} \subset W$. As S is paracompact, we can find a family $(x_m)_{m\in J}$, $J \subset \mathbb{N}$, of points of S as well as open neighborhoods $U_m = U_{x_m,1}$ of x_m in S such that $(U_m)_{m\in J}$ is a locally finite open covering of S and $U_m \subset U_{x_m,0}$. If one now sets $\tilde{W} = \bigcup_{m\in J} \tilde{W}_{x_m,1}$, then $\tilde{W}$ is an open neighborhood of S and $\tilde{W} \subset W$. As S is paracompact, there exists a continuous function $\delta : R \to \mathbb{R}^{>0}$ with $\delta(x) < \delta_{x_m,1}$ for all $x \in S$ and $m \in J$ with $x \in U_m$. But this means that $\tilde{T}_S^\delta \subset \tilde{W} \subset W$, hence (1) follows.

As X is locally compact, there exists a locally finite covering $(V_n)_{n\in\mathbb{N}}$ of S by in X open sets V_n with compact closure $\overline{V_n} \subset \tilde{T}_S$. Let V be the neighborhood $V = \bigcup_{n\in\mathbb{N}} \overline{V_n}$ of S and $\varepsilon : S \to \mathbb{R}^{>0}$ a continuous function with $\tilde{T}_S^\varepsilon \subset V$. Furthermore let

$$\rho(x) = \inf \left\{ t \in \mathbb{R}^{\geq 0} \,\middle|\, \nexists y \in \tilde{\pi}_S^{-1}(x) \text{ with } \tilde{\rho}_S(y) = t \right\}.$$

Then the mapping $\rho : R \to \mathbb{R}^{\geq 0}$ is lower semicontinuous and vanishes nowhere by (1). Hence we can choose ε such that $\varepsilon < \rho$. For the proof of (2) for this ε it remains to show that for every compact $K \subset S$ and $\delta > 0$ with $\delta < \varepsilon(x)$ for all $x \in K$ the set $C_{K,\delta} = \tilde{\pi}_S^{-1}(K) \cap \tilde{\rho}_S^{-1}([0, \delta])$ is compact. But this follows from the fact that $C_{K,\delta}$ is closed in $\tilde{T}_S$ and contained in the compact set $\bigcup_{V_n \cap K \neq \emptyset} \overline{V_n} \subset \tilde{T}_S$. Therefore the restriction $(\tilde{\pi}_S, \tilde{\rho}_S) : \tilde{T}_S^\varepsilon \to [0, \varepsilon[$ is proper, and by the choice of $\varepsilon < \rho$ even surjective. This proves the claim. □

Now we have all prerequisites to introduce the notion of a tubular neighborhood.

3.1.3 Definition Let $m \in \mathbb{N}^{>0} \cup \{\infty\}$ and S a $\mathcal{C}^m$-submanifold of M. A *tubular neighborhood* of S in M of *class* $\mathcal{C}^m$ then is a triple $T = (E, \varepsilon, \varphi)$, where $\pi_E : E \to S$ is a $\mathcal{C}^m$-vector bundle over S with scalar product η, $\varepsilon : S \to \mathbb{R}^{>0}$ a $\mathcal{C}^m$-map (or even a lower semicontinuous function), and φ a $\mathcal{C}^m$-diffeomorphism from $T_{S\subset E}^\varepsilon = \{v \in E \mid \rho_E(v) := \|v\|^2 = \eta(v, v) < \varepsilon(\pi(v))\}$ to an open neighborhood T of S, the so-called *total space* of T, such that the diagram

$$\begin{array}{ccc} T^{\varepsilon}_{S\subset E} & & \\ \uparrow & \overset{\varphi}{\searrow} & \\ S & \longrightarrow & M \end{array} \tag{3.1.1}$$

commutes. Hereby S is embedded canonically into $T^{\varepsilon}_{S\subset E} \subset E$ as its zero section. Sometimes we also call the total space T a tubular neighborhood even though this is not formally correct.

If $f : M \to N$ is a $\mathcal{C}^m$-mapping, then one calls a tubular neighborhood T of $S \subset M$ of class $\mathcal{C}^m$ *compatible* with f, if $f \circ \pi = f_{|T}$.

Every tubular neighborhood $T = (E, \varepsilon, \varphi)$ induces via the projection $\pi_E : E \to S$ a continuous retraction $\pi_S : T_S \to S$, called *projection*, by $\pi_S = \pi_E \circ \varphi^{-1}$. Moreover, one obtains the so-called *tubular function* $\rho_S : T_S \to \mathbb{R}$ of T by $\rho_S = \rho_E \circ \varphi^{-1}$, where ρ_E is the distance function on E defined above. Obviously we then have $S = \rho_S^{-1}(0)$. Additionally, the tubular function is submersive exactly over the $T \setminus S$.

3.1.4 Example Consider the natural embedding of $\mathbb{R}^m$ into the Euclidean space $\mathbb{R}^n$, $n = m + k$ via the first m coordinates. Let E be the trivial vector bundle over $\mathbb{R}^m$ with typical fiber $\mathbb{R}^k$ and the Euclidean scaler product as fiber metric. Moreover, let $\varepsilon = 1$ and $\varphi : T^{\varepsilon}_{\mathbb{R}^m} \to \mathbb{R}^n$ the restriction of the identical mapping $\mathrm{id}_{\mathbb{R}^n}$ to

$$T^n_m := T^{\varepsilon}_{\mathbb{R}^m} = \{x = (y, v) \in \mathbb{R}^m \times \mathbb{R}^k \mid \|v\| < 1\}.$$

Then the triple $(\mathbb{R}^n, \varepsilon, \varphi)$ comprises a tubular neighborhood of $\mathbb{R}^m$ in $\mathbb{R}^n$ of class $\mathcal{C}^\infty$. We call it the *standard tubular neighborhood* T^n_m. Its projection $\pi^n_m := \pi_{\mathbb{R}^m}$ is given by the orthogonal projection $(x_1, \cdots, x_n) \mapsto (x_1, \cdots, x_m, 0 \cdots, 0)$ from $\mathbb{R}^n$ to $\mathbb{R}^m$. The tubular function $\rho^n_m := \rho_{\mathbb{R}^m}$ is given by the square of the function measuring the Euclidean distance of a point $x \in \mathbb{R}^n$ to the subspace $\mathbb{R}^m$, hence by $(x_1, \cdots, x_n) \mapsto x^2_{m+1} + \cdots + x^2_n$.

3.1.5 Example If $h : (M', S') \to (M, S)$ is a diffeomorphism mapping S onto S' and $T = (E, \varepsilon, \varphi)$ a tubular neighborhood of S in M, then $h^*T = (h^*E, \varepsilon \circ h_{|S'}, h^{-1} \circ \varphi)$ is a tubular neighborhood of S' in M'. Analogously one defines h_*T' for every tubular neighborhood T' of S' in M'.

3.1.6 Classical tubular neighborhood theorem *Let $m \in \mathbb{N}^{>0} \cup \{\infty\}$, M a Riemannian manifold of class $\mathcal{C}^{m+2}$ with Riemannian metric μ of class $m+1$, and $S \subset M$ a $\mathcal{C}^{m+2}$-submanifold. Furthermore denote by N the subbundle of $T_{|R}M$ orthogonal with respect to μ to TS in $T_{|R}M$. Then there exists a $\mathcal{C}^m$-function $\varepsilon : S \to \mathbb{R}^{>0}$, such that over $T^{\varepsilon}_{S\subset N}$ the restriction $\varphi := \exp_{|T^{\varepsilon}_{S\subset N}} : T^{\varepsilon}_{S\subset N} \to M$ of the exponential function is well-defined and such that the triple $(N, \varepsilon, \varphi)$ comprises a tubular neighborhood of S in M of class $\mathcal{C}^m$.*

A tubular neighborhood $(N, \varepsilon, \varphi)$ like in the classical tubular neighborhood theorem will be called *induced* by μ. We denote it often in the form $(N, \varepsilon, \exp_{|N})$.

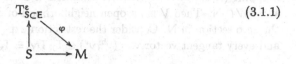

PROOF: Let W an open neighborhood of the zero section of TM, such that the mapping $(\pi, \exp) : W \to M \times M$ comprises a $\mathcal{C}^m$-diffeomorphism onto its image. Set $V := W \cap N$. Then V is an open neighborhood of S in N, where S is identified with the zero section in N. Consider the restriction $\exp_{|V} : V \to M$. Now, for every $x \in M$ and every tangent vector $v = (v^h, v^v) \in T_{0_x}TM \cong T_xM \oplus T_xM$ the relation

$$T_{0_x} \exp v = v^h + v^v$$

is true, hence $\exp_{|V}$ must be submersive after possibly shrinking V to a somewhat smaller neighborhood of the zero section. Hereby one can choose V such that $\exp^{-1}(S) \cap V = S$. By $\dim M = \dim V$ the map $\exp_{|V}$ is not only submersive, but also a local diffeomorphism which coincides over S with the identical map. By the following lemma there exists an open neighborhood $\tilde{T} \subset V$ of S such that $\exp_{|\tilde{T}}$ is a diffeomorphism onto its image. By Lemma 3.1.2 there exists a continuous and even a $\mathcal{C}^m$-function $\varepsilon : S \to \mathbb{R}^{>0}$ such that $T^{\varepsilon}_{S \subset N} \subset \tilde{T}$. Consequently $\varphi = \exp_{|T^{\varepsilon}_{S \subset N}}$ comprises a diffeomorphism onto its image, hence the last component of a tubular neighborhood of S. □

3.1.7 Lemma *Let M, N be manifolds and and $S \subset N$ a submanifold possibly with boundary. If under these assumptions $f : N \to M$ comprises a $\mathcal{C}^1$-function such that the restriction $f_{|S}$ is an embedding and for every point $x \in S$ the tangent map T_xf is bijective then there exists an open neighborhood T of S in N such that f maps T diffeomorphically to an open subset of M.*

PROOF: We proceed like in LANG [108, IV.5] and use an argument given by GODEMENT [60] on page 150.

For simplicity we identify the image of S under f again with S. Now let $(N_j)_{j \in J}$ be a locally finite covering of S by in N open subsets N_j such that the restrictions $f_j := f_{|N_j} : N_j \to M_j$ are diffeomorphisms onto open sets $M_j \subset M$, and such that every N_j has nonempty intersection with S. Afterwards choose a covering $(Y_j)_{j \in J}$ of S subordinate to $(M_j)_{j \in J}$ by in M open subsets Y_j such that the relations $Y_j \cap S \neq \emptyset$ and $\overline{Y_j} \subset M_j$ hold. Let $\overline{f}_j : M_j \to N_j$ be the function inverse to f_j and $Y \subset M$ the set of all points $y \in \bigcup_j Y_j$ such that $\overline{f}_j(y) = \overline{f}_i(y)$, if y lies in the intersection $\overline{Y_j} \cap \overline{Y_i}$. Obviously $S \subset Y$. We show that Y is even a neighborhood of S. To this end choose an arbitrary point $y \in S$. Then there exist $Y_{j_1}, \cdots, Y_{j_k}$ such that y lies exactly in the closed hulls $\overline{Y_{j_l}}$, $l = 1, \cdots, k$. One can now find an open neighborhood $Y_y \subset \bigcup_j Y_j$ of y such that $Y_y \subset M_{j_l}$ for $l = 1, \cdots, N$ and $Y_y \cap \overline{Y_j} = \emptyset$ for $j \neq j_1, \cdots, j_N$. Obviously $\overline{f}_{j_k}(y) = \overline{f}_{j_l}(y)$. Hence there exists an open neighborhood V_y of y in N such that

$$V_y \subset f^{-1}(Y_y) \cap \bigcap_{1 \leq l \leq k} N_{j_l}.$$

Now set $\tilde{Y}_y = f(V_y) \cap Y_y$. Then $\tilde{Y}_y$ is an open neighborhood of y in $\bigcup_j Y_j$ and for all $\tilde{y} \in \tilde{Y}_y \cap Y_{j_k} \cap Y_{j_l}$ the relation $f_{j_k}(\tilde{y}) = f_{j_l}(\tilde{y})$ holds. Consequently Y is a neighborhood of S. Moreover, the function $\overline{f} : Y \to N$ which assigns to every point $y \in Y \cap Y_j$ the value $\overline{f}_j(y)$ is well-defined. By definition $\overline{f}$ is left inverse to the restriction of f to

$T := f^{-1}(Y)$, hence $f_{|T}$ is injective. On the other hand T is a neighborhood of S in N. Thus the claim follows. □

3.1.8 Let $T = (E, \varepsilon, \varphi)$ be a tubular neighborhood of S in M, where we suppose that E is equipped with a metric connection. Then the following propositions hold; the corresponding proofs follow immediately from the definition of a tubular neighborhood and the corresponding properties of the metric vector bundle E.

(1) Identify E with the vertical bundle of TE over $S \subset TE$. Then the restricted mapping

$$T_{|E}\varphi : E \to N, \quad v_x \mapsto T_{0_x}\varphi.v_x, \quad x \in S$$

comprises an isomorphism of vector bundles from E to a vector bundle $N := T\varphi(E)$ which is normal to $TS \to S$ in $T_{|S}M$ that means $T_{|S}M = TS \oplus N$.

(2) The tubular neighborhood T generates for every point $x \in T$ a canonical decomposition of the tangent space T_xM into the *vertical subspace* $V_xM \cong T_v\varphi(E_{\pi(x)})$, $v = \varphi^{-1}(x)$ and the *horizontal subspace* H_xM which is canonically isomorphic to $T_{\pi(x)}S$. Hereby H_xM is the image of the horizontal space of T_vE under $T_v\varphi$; the canonical isomorphy H_xM with $T_{\pi(x)}S$ is given by parallel transport of $T_{\pi(x)}S$ along the curve $\varphi(tv)$, $t \in [0, 1]$.

(3) Let θ be an arbitrary Riemannian metric on T, η the scalar product on E, and for every $x \in T$ let $P_x : T_xM \to T_xM$ be the projection onto H_xM along V_xM. Finally let $Q_x = \mathrm{id}_{T_xM} - P_x$. Then there exists after possibly shrinking T

$$\mu_x(v, w) = \eta_{\pi_S(x)}(T_x\varphi^{-1}.Q_x.v, T_x\varphi^{-1}.Q_x.w) + \theta(P_x.v, P_x.w), \quad x \in T,\ v, w \in T_xM$$

a Riemannian metric μ on T such that $\varphi = \exp \circ T_{|E}\varphi$. Hereby exp is the exponential function of the Levi–Civita connection of μ. Moreover, the vector bundle N is μ-orthogonal to TS that means in other words

$$N_x = \{v \in T_xM \mid \mu(v, w) = 0 \text{ for all } w \in T_xS\}, \quad x \in S.$$

(4) The Hessian of ρ_S has rank $\mathrm{codim}_M S = \dim M - \dim S$ over S.

The statement (3) can be interpreted as saying that the tubular neighborhoods constructed according to the classical tubular neighborhood theorem are universal. This means that every tubular neighborhood T can be written in the form $(N, \varepsilon, \exp_{|N})$, where N is the normal bundle and exp the exponential function exp with respect to an appropriate Riemannian metric defined on a neighborhood of S.

3.2 Cut point distance and maximal tubular neighborhoods

The classical tubular neighborhood theorem provides an important existence result for tubular neighborhoods, but we do not know, how "large" such a tubular neighborhood can be, whether there are maximal tubular neighborhoods and so on. It is the goal of this section to give an answer to such questions. But first we have to define an order relation on the set of tubular neighborhoods, so that we can speak of "larger" or "maximal" tubular neighborhoods. Consider two tubular neighborhoods $T_1 = (N, \varepsilon_1, \exp_{|N})$ and $T_2 = (N, \varepsilon_2, \exp_{|N})$ of S, both induced by a Riemannian metric μ on M. Then we can compare their total spaces T_1 and T_2 with respect to the set theoretic inclusion. If $T_1 \subset T_2$ or equivalently if $\varepsilon_1 \leq \varepsilon_2$, then we write $T_1 \leq T_2$. The relation $\leq$ obviously comprises an order relation on the set of all tubular neighborhoods of S induced by μ. With the help of several quite nontrivial geometric tools we will succeed in this section to construct a maximal element with respect to the order relation $\leq$ in case (M, μ) is a complete Riemannian manifold.

3.2.1 Denote by t_ν^+ for $\nu \in T_x M$ and $x \in M$ the *escape time* of exp in direction ν, that means the supremum of all $t \in \mathbb{R}^{\geq 0}$ such that $\exp t\nu$ is defined. Furthermore, denote by SN the sphere bundle of N. Then for every $\nu \in S_x N$ with $x \in S$ the *focal point distance* in direction ν is given by

$$e_f(\nu) := e_{f,S}(\nu) := \sup\left\{t \in [0, t_\nu^+[\;\middle|\; \ker(T_{s\nu} \exp_{|N}) = 0 \text{ for all } s \in [0, t]\right\},$$

and the *cut point distance* in direction ν by

$$e_c(\nu) := e_{c,S}(\nu) := \sup\left\{t \in [0, t_\nu^+[\;\middle|\; \delta_\mu(\exp_{|N}(t\nu), S) = t\right\}.$$

Hereby δ_μ means the geodesic distance with respect to μ. In case S consists of only one point x, then we define

$$r_i(x) := \inf\left\{e_{c,\{x\}}(\nu) \;\middle|\; \nu \in S_x M\right\}$$

and call $r_i(x)$ the *injectivity radius* of exp at x. Altogether we thus obtain three functions $e_f : SN \to \overline{\mathbb{R}}^{\geq 0} := \mathbb{R}^{\geq 0} \cup \{\infty\}$, $e_c : SN \to \overline{\mathbb{R}}^{\geq 0}$ and $r_i : M \to \overline{\mathbb{R}}^{\geq 0}$.

3.2.2 Proposition *For every $\nu \in S_x N$ with $x \in S$ one has*

$$0 < e_{c,S}(\nu) \leq e_{f,S}(\nu) \quad \textit{and} \quad 0 < e_{c,S}(\nu) \leq e_{f,\{x\}}(\nu).$$

Moreover, the relation $r_i(y) > 0$ holds for every $y \in M$.

PROOF: We only sketch the proof; for more details we refer the reader to KLINGENBERG [101, Thm. 1.12.13 & Thm. 2.5.15] or BISHOP–CRITTENDEN [16, Chap. 11, Thm. 5, Cor. 2].

As the exponential function $\exp_x$ is a local diffeomorphism mapping a sufficiently small ball around the origin of $T_x M$ to a strong convex neighborhood of x, the relations $e_c(\nu) > 0$, $e_f(\nu) > 0$ and $r_i(y) > 0$ have to be true. So it remains to show $e_{c,S}(\nu) \leq e_{f,S}(\nu)$ and $e_{c,S}(\nu) \leq e_{f,\{x\}}(\nu)$. We will prove only the second inequality; the proof of the first one can be carried out analogously with the help of S-Jacobi-fields [16, Chap. 11]. It suffices to consider the nontrivial case $t_f := e_{f,\{x\}}(\nu) < t_\nu^+$. Let t_1 be a real number with $t_f < t_1 < t_\nu^+$ and $\gamma : I = [0, t_1] \to M$ the geodesic $t \mapsto \exp t\nu$. In the following considerations we want to construct a piecewise continuously differentiable variation $F :]\varepsilon, \varepsilon[\times I \to M$ of the geodesic γ such that for sufficiently small $|s|$ the variation curves $\gamma_s = F(s, \cdot)$ connect the points x and $\gamma(t_1)$ and have a smaller length than γ. As $T\exp_x$ has a nonvanishing kernel in the point $t_f v$, there exists a nontrivial Jacobi-field Y along γ with $\mu(Y(t), \dot\gamma(t)) = 0$ and $Y(0) = Y(t_f) = 0$. As Y is nontrivial, there exists a point t_0 between 0 and t_f with $Y(t_0) \neq 0$. Let Z be a differentiable vector field along γ with $Z(t) = 0$ for $0 \leq t \leq t_0$, $Z(t_f) = -\nabla Y(t_f) \neq 0$ and $Z(t_1) = 0$. Then we set for some $\eta \geq 0$ which will be determined later

$$X_\eta := Y^*(t) + \eta Z(t) \quad \text{where } Y^*(t) = \begin{cases} Y(t) & \text{if } 0 \leq t \leq t_f, \\ 0 & \text{if } t_f \leq t \leq t_1. \end{cases}$$

The vector field X_η generates a piecewise continuously differentiable variation $F :]\varepsilon, \varepsilon[\times I \to M$ of γ, in other words there exists an F with $X_\eta(t) = \left.\frac{\partial F(s,t)}{\partial s}\right|_0$. Now we consider for each path $\gamma_s = F(s, \cdot)$ the energy integral

$$E(s) = E(\gamma_s) = \frac{1}{2}\int_I \langle \dot\gamma_s(t), \dot\gamma_s(t) \rangle_{\mu(\gamma(t))}\, dt$$

and calculate the second variation of E:

$$\begin{aligned} D^2E(\gamma)(X_\eta, X_\eta) &= \int_I \langle \nabla X_\eta, \nabla X_\eta \rangle_\mu(t)\, dt - \int_I \langle R_{\dot\gamma} X_\eta, X_\eta \rangle_\mu(t)\, dt \\ &= D^2E(\gamma)(Y^*, Y^*) + 2\eta D^2E(\gamma)(Y^*, Z) + \eta^2 D^2E(\gamma)(Z, Z). \end{aligned} \tag{3.2.2}$$

Via partial integration one observes that the first term on the right side vanishes and that the second one is given by

$$2\eta \left\langle \lim_{t \nearrow t_f} \nabla Y^*(t) - \lim_{t \searrow t_f} \nabla Y^*(t), Z(t_f) \right\rangle_{\mu(\gamma(t_f))} = -2\eta\, |\nabla Y(t_f)|^2.$$

Hence, for η sufficiently small we thus have $D^2E(\gamma)(X_\eta, X_\eta) < 0$. Using the fact that the length of γ is given by $|\gamma|_\mu^2 = 2E(\gamma)\, |I|$ we obtain for sufficiently small s

$$|\gamma_s|_\mu^2 \leq 2E(\gamma_s)\, |I| < 2E(\gamma)\, |I| = |\gamma|_\mu^2,$$

that means γ_s has a length smaller than $|\gamma|$. Hence, as $X_\eta(0) = X_\eta(t_1) = 0$ holds as well, there must exist a path shorter than γ connecting $x = \gamma(0)$ with $\gamma(t_1)$. Thus $\gamma(t_1)$ has shorter distance to S than t_1. But this means $e_{c,S}(v) \leq e_{f,\{x\}}(v)$ which proves the claim. □

3.2.3 Proposition *The functions e_f, e_c and r_i are lower semicontinuous. If (M, μ) is a complete Riemannian manifold, then e_f, e_c and r_i are even continuous.*

PROOF: First recall that by the well-known theorems on the existence and uniqueness of solutions of differential equations the escape time t^+ has to be a lower semicontinuous function on SN (see e.g. [2, Lem. 10.5]). Consequently $W^{\max} := \{tv \in TM \mid v \in SM \text{ and } 0 \leq t < t_v^+\} \subset TM$ is an open neighborhood of the zero section of TM, and $(\pi, \exp) : W^{\max} \to M \times M$ is well-defined. For x and y sufficiently close let us denote in the following by $[x, y]$ the segment $\{\exp(tw) \mid w = \exp_x^{-1}(y) \text{ and } t \in [0, 1]\}$

connecting x and y. After these agreements on the notation we can now start with the proof.

Let us suppose that e_f is not lower semicontinuous in the point v. Then there exists some $\varepsilon > 0$ and a sequence $(v_k)_{k\in\mathbb{N}}$ of unit vectors $v_k \in N$ converging to v such that $t = \lim_{k\to\infty} t_k$ with $t_k := e_f(v_k)$ exists and such that $t < e_f(v)$. Let $w_k \in T_{t_k v_k} N$ be a unit vector in the kernel of $T_{t_k v_k} \exp_{|N}$. After transition to an appropriate subsequence $(w_k)_{k\in\mathbb{N}}$ converges to a nonvanishing vector $w \in T_{tv}N$. By continuity of the exponential function w must be in the kernel of $T_{tv}\exp_{|N}$ in contradiction to $e_f(v) \le t < e_f(v)$. Hence e_f is lower semicontinuous in v.

Next we prove by contradiction that e_c is lower semicontinuous in v. So let us suppose that there exists a sequence of v_k converging to v such that $t_\infty = \lim_{k\to\infty} e_c(v_k) < e_c(v)$. We denote by x the footpoint of v and abbreviate: $t_k := e_c(v_k)$. Then we embed M into its complete hull $\overline{M}$ with respect to the geodesic distance; obviously $\overline{M}$ then is a length space (though in general not a Riemannian manifold). Under these prerequisites there exists by the theorem of BUSEMANN 2.4.13 resp. by the theorem of HOPF–RINOW 2.4.15 for every k a positive $\delta_k < \frac{1}{2^k}$ and a rectifiable path $\gamma_k : [0, s_k] \to \overline{M}$ parametrized by arc length and having the following properties:

(1) $\gamma_k(0) \in \overline{S}$ and $\gamma_k(1) = \exp((t_k + \delta_k)v_k)$, where $\overline{S}$ denotes the closure of S in $\overline{M}$.

(2) The curve γ_k minimizes the distance from $\gamma_k(1)$ to S that means

$$\delta_\mu(\gamma_k(1), S) = |\gamma_k|_\mu.$$

(3) The length $s_k = |\gamma_k|_\mu$ is smaller than $t_k + \delta_k$.

The curves γ_k then fulfill $\lim_{k\to\infty} \gamma_k(s_k) = \exp(t_\infty v)$. After transition to subsequences $(s_k)_{k\in\mathbb{N}}$ then converges by the theorem of BUSEMANN 2.4.13 to some $s_\infty \le t_\infty$ and $(\gamma_k)_{k\in\mathbb{N}}$ uniformly to a rectifiable path $\gamma : [0, s_\infty] \to \overline{M}$ with $\gamma(0) \in \overline{S}$ and $\gamma(s_\infty) = \exp(t_\infty v)$.

We now consider the mapping $(\pi, \exp) : W^{\max} \to M \times M$. As $e_c(v) \le e_{f,\{x\}}(v)$, the map $(\pi, \exp)$ is of maximal rank over the segment $[0, t_\infty]\, v \subset TM$ and injective. Therefore by Lemma 3.1.7 $(\pi, \exp)$ maps a relatively compact open neighborhood $W \subset W^{\max}$ of $[0, t_\infty]\, v$ diffeomorphically onto an open set in $M \times M$. After possibly shrinking W one can suppose that $(\pi, \exp)(W)$ has the form $B \times U$, where B is a ball around x and U an open neighborhood of the segment $[x, \exp(t_\infty v)]$. Furthermore we can assume that U is connected and that $\exp(tv)$ for $\frac{1}{2}(e_c + t_\infty) \le t < t_v^+$ does not lie in $\overline{U}$. As the restricted exponential map $n := \exp_{|N} : N \cap W^{\max} \to M$ is of maximal rank over $[0, t_\infty]\, v$ and injective, n maps by Lemma 3.1.7 an open neighborhood W' of $[0, t_\infty]\, v$ in $N \cap W^{\max}$ diffeomorphically onto an open neighborhood U' of the segment $[x, \exp(t_\infty v)]$. After shrinking W' and W appropriately one can achieve $U = U'$.

Suppose for a moment that the image $\operatorname{im}\gamma$ does not lie completely in U. Under this assumption let s_c be the minimum of all $s' \le s_\infty$ such that $\gamma([s', s_\infty]) \subset \exp([0, e_c(v)[\, v) \cap U$. As γ does not completely lie in U, we have $s_c > 0$. Let further t_c be a real number between 0 and $e_c(v)$ such that $\gamma(s_c) = \exp(t_c v)$. In case we had $t_c > t_\infty$, then we would have $s_c < s_\infty$, hence by $s_\infty \le t_\infty$ the restricted path $\gamma_{|[0,s_c]}$ would have a shorter length than the segment connecting x and $\exp(t_c v)$.

This would imply $t_c > e_c(v)$ which is impossible. Consequently $t_c \leq t_\infty$ and $\gamma(s_c)$ lies in U. Moreover, there exists a strongly convex ball $B \subset U$ around $\gamma(s_c)$ such that γ intersects the boundary of B in a point not lying on $\exp([0, e_c(v)[\, v) \cap U$. Let $0 < s_{\partial B} < s_c$ such that $\gamma(s_{\partial B}) \in \partial B \setminus \exp([0, e_c(v)[\, v)$. Let further $\tilde{\gamma} : [0, \tilde{s}] \to M$ with $\tilde{s} = s_{\partial B} + \delta_\mu(\gamma(s_{\partial B}), \gamma(s_c))$ be the curve composed by $\gamma_{|[0,s_{\partial B}]}$ and the segment $[\gamma(s_{\partial B}), \gamma(s_c)]$. By $t_\infty < e_c(v)$ and the definition of $e_c(v)$ the restricted path $\gamma_{|[s_c,s_\infty]}$ has length $\geq t_\infty - t_c$, consequently the length $\tilde{s}$ of $\tilde{\gamma}$ is smaller or equal to t_c. By construction the vectors $\dot{\tilde{\gamma}}(\tilde{s})$ and $T_{s_c v} \exp_x(v)$ are not collinear. Hence there exist points $\tilde{\gamma}(s')$ and $\exp(t'v)$ in B (where $s_{\partial B} < s' < s_c$ and $t_c < t' < e_c(v)$) such that the segment $[\tilde{\gamma}(s'), \exp(t'v)]$ is shorter than the path composed of $\tilde{\gamma}_{|[s',\tilde{s}]}$ and $[\tilde{\gamma}(\tilde{s}), \exp(t'v)]$. Next consider the curve ν composed by the curves $\tilde{\gamma}_{|[0,s']}$ and $[\tilde{\gamma}(s'), \exp(t'v)]$. Then

$$|\nu|_\mu < |\tilde{\gamma}|_\mu + |[\tilde{\gamma}(\tilde{s}), \exp(t'v)]|_\mu \leq |[x, \exp(t'v)]|_\mu = t'.$$

As $\nu(0) = \gamma(0) \in \overline{S}$, the point $\exp(t'v)$ has shorter distance to S than t'. This entails the contradiction $e_c(v) \leq t' < e_c(v)$. Hence $\operatorname{im}\gamma \subset U$.

Next we set $W_0 = W$ and let $(W_n)_{n\in\mathbb{N}}$ run through a basis of neighborhoods of the segment $[0, t_\infty]v \subset TM$ such that $W_{n+1} \subset W_n$ for all n. Obviously one can choose the W_n such that $(\pi, \exp)(W_n) = B_n \times U_n$, where the B_n form a basis of neighborhoods consisting of open balls around x and the U_n a basis of neighborhoods of the segment $[x, \exp(t_\infty v)]$. By the above argument $\operatorname{im}\gamma$ must lie in each of the neighborhoods U_n. As the sequence $(\gamma_k)_{k\in\mathbb{N}}$, regarded as a sequence of geometric curves, converges by the theorem of BUSEMANN 2.4.13 uniformly to γ, there exists for every n some $k_n \in \mathbb{N}$ such that $\operatorname{im}\gamma_k \subset U_n$ for all $k \geq k_n$. Hence there exists for every $k \geq k_0$ some $w_k \in S_{y_k}M$ with $\gamma_k(t) = \exp(tw_k)$ for $0 \leq t \leq s_k$. Now recall that γ_k minimizes the distance from $\gamma_k(1)$ to S. Thus $w_k \in S_{y_k}N$ follows. As $y_k := \gamma_k(0)$ is an element of S for sufficiently large k and $\operatorname{im}\gamma_k \subset U_n$, the sequence $(y_k)_{k\in\mathbb{N}}$ of footpoints of the w_k converges to x. After the choice of an appropriate subsequence we can therefore assume that the sequence $(w_k)_{k\in\mathbb{N}}$ converges to some vector $w \in S_xN$. At this point recall that

$$\exp((t_k + \delta_k)v_k) = \gamma_k(s_k) = \exp(s_k w_k).$$

After passing to the limit $k \to \infty$ the relation $\exp(t_\infty v) = \exp(s_\infty w)$ follows. We now consider the following two cases.

1. CASE $w \neq v$. Then one can connect the points x and $\exp(t_\infty v)$ in two different ways by geodesics, hence the geodesic $(\exp[0, t_v^+[\, v)$ does not anymore minimize the distance to x beyond t_∞, which is impossible.

2. CASE $w = v$. Then by $t_\infty < e_c(v)$ the relation $t_\infty = s_\infty$ follows. Hence for sufficiently large k the vectors $s_k w_k$ and $(t_k + \delta_k)v_k$ both lie in W'. By $s_k < t_k + \delta_k$ the vectors $s_k w_k$ and $(t_k + \delta_k)v_k$ have to be different. On the other hand $\eta(s_k w_k) = \eta((t_k + \delta_k)v_k)$ holds, which contradicts the fact that $\eta_{|W'}$ is injective.

Altogether $t_\infty < e_c(v)$ does not hold, so e_c is lower semicontinuous. The lower semicontinuity of r_i follows immediately by the one of e_c.

The proof of the continuity of e_f, e_c and r_i for the case that M is geodesically complete will not be performed here, as one can find the corresponding proofs in the literature [16, 101]. Moreover, the continuity results will not be needed further in this work. □

Now we can define a special open neighborhood of the zero section of N which later will yield the desired maximal tubular neighborhood.

3.2.4 Lemma *Set*

$$T^{\max}_{SCN} := \{tv \in N \mid x \in S,\ \ v \in S_xN \text{ and } 0 \leq t < e_c(v)\}.$$

Then $\exp_{|T^{\max}_{SCN}}$ *is an open embedding.*

PROOF: We show first that $\exp_{|T^{\max}_{SCN}}$ is injective. Suppose this were not the case. Then there exist two distinct points $x, y \in S$, two vectors $v \in S_xN$ and $w \in S_yN$ as well as a real number $t < \min\{e_c(v), e_c(w)\}$ with

$$\exp tv = \exp tw.$$

As $t < e_c(v)$, there exists $\delta > 0$ with $t + \delta < e_c(v)$. Let $z = \exp(t+\delta)v$. If v and w were collinear, then $\mu(v, w) = -1$ would hold as otherwise $x = y$. Therefore collinearity would entail $\delta_\mu(z, S) \leq \delta_\mu(z, y) = t - \delta$, which contradicts $t + \delta < e_c(v)$. Hence v and w are not collinear. Therefore the following path connecting y and z

$$\gamma : [0, t+\delta] \to M, \quad s \mapsto \begin{cases} \exp sw & \text{if } s \leq t, \\ \exp sv & \text{if } s \geq t, \end{cases}$$

is not differentiable in t. Hence $\delta_\mu(y, z) < |\gamma|_\mu = t + \delta$ follows which contradicts $t + \delta < e_c(v)$. Therefore $\exp_{|T^{\max}_{SCN}}$ is injective.

By definition of e_f and Proposition 3.2.2 the map $\exp_{|T^{\max}_{SCN}}$ has to be immersive. By reasons of dimension the image $\exp(T^{\max}_{SCN})$ is open in M. Putting all this together we obtain the claim. □

3.2.5 Proposition *Let* (M, μ) *be a complete Riemannian manifold and* S *a closed submanifold. Then* $T^{\max}_{SCN}$ *is maximal with respect to* $\subset$ *among all open neighborhoods* T *of the zero section of* N *such that* $\exp_{|T} : T \to M$ *is an open embedding and such that for every* $v \in T$ *the segment* $[0, 1]v$ *lies in* T.

From the Proposition we obtain the following main result of this section:

3.2.6 Corollary *Define* $\varepsilon^{\max} : S \to \overline{\mathbb{R}}^{\geq 0}$ *by*

$$\varepsilon^{\max}(x) = \inf\{e_c(v) \mid v \in S_xN\}.$$

Then $T^{\max} = (N, \varepsilon^{\max}, \exp_{|N})$ *is maximal among the tubular neighborhoods of* S *in* M *induced by* μ.

PROOF: Let us suppose $T^{\max}_{SCN}$ were not maximal in the claimed sense. Then there exists an open neighborhood T of the zero section of N with $T^{\max}_{SCN} \subsetneq T$ such that the conditions for T in the claim are satisfied. Under these assumptions there exist $v \in SN$ and $\delta > 0$ with $(e_c(v) + \delta)v \in T$. Let $\gamma : [0, s] \to M$ be a geodesic minimizing the distance from $y = \exp((e_c(v) + \delta)v)$ to S. In other words this means that $\delta_\mu(y, S) = |\gamma|_\mu = s$. By the theorem of HOPF–RINOW 2.4.15 such a curve γ exists indeed, and $\gamma(0)$ has to lie in S, as S is closed. Then γ is normal to S, hence there exists $w \in S_{\gamma(0)}N$ with $\gamma(t) = \exp(tw)$ for $0 \leq t \leq s$. As γ is distance minimizing, the relation $e_c(w) \geq s$ is true. Therefore the set $[0, s[w$ lies in $T^{\max}_{SCM}$, hence in T. If $w \neq v$, then the sets $(e_c(v) + \delta)v$ and $[0, s[w$ would have disjoint neighborhoods U_v and U_w in T. But by $y \in \exp(U_v) \cap \overline{\exp(U_w)}$ the map $\exp_{|T}$ could not be an open embedding anymore. Therefore we have $w = v$. So $s \leq e_c(v)$ follows, hence $y \in \overline{\exp(T^{\max}_{SCN})}$. On the other hand $(e_c(v) + \delta)v \in \overline{T \setminus T^{\max}_{SCN}}$, which by the fact that $\exp_{|T}$ is open entails the contradiction $y \notin \overline{\exp(T^{\max}_{SCN})}$. This proves that $T^{\max}_{SCN}$ is maximal as claimed. □

3.2.7 Remark In case (M, μ) is not complete or S is not closed in M the tubular neighborhood $T^{\max}_{SCM}$ need not be maximal anymore in the sense of the last proposition. Nevertheless we can define in this case $\varepsilon^{\max}$ and $T^{\max}$ like in the Corollary. For our purposes we do not need any tubular neighborhoods larger than $T^{\max}$. Therefore we slightly abuse the language and call $T^{\max}$ in every case the *maximal* tubular neighborhood of S in M induced by μ.

3.2.8 Remark Finally let us remark that even in case M is complete and S is closed there might exist other maximal tubular neighborhoods of S induced by μ besides $T^{\max}$. Therefore the claim appearing occasionally in the mathematical literature that $T^{\max}$ is the largest among the tubular neighborhoods of S induced by μ is wrong.

3.3 Curvature moderate submanifolds

In the following we will introduce a notion which describes how a stratum or a submanifold curves within the ambient stratified space respectively ambient manifold when approaching the boundary of the stratum or submanifold. Take for instance the standard cone. Then it is intuitively clear that the behavior of the curvature of the top stratum near the cusp does not change much. More generally consider a real or complex algebraic variety with its natural Whitney stratification. Then the curvature is - speaking again intuitively - bounded by a rational function, so cannot grow "too" fast while approaching a lower stratum. But the situation is different when considering the slow or fast spiral. Here the curvature of the top stratum grows exponentially with the distance to the origin. The notions introduced in this section will help to separate the first two cases, which in the following we will regard as curvature moderate, from the latter cases.

3.3.1 To simplify notation let us agree for this section that $m \in \mathbb{N}^{>0} \cup \{\infty\}$ and that S is alway a submanifold of $\mathbb{R}^n$ or of a manifold M. Moreover denote by T always an open neighborhood of S in $\mathbb{R}^n$ or M such that S is closed in T. In most cases T will be given by the total space of a tubular neighborhood of S.

We consider first a submanifold $S \subset \mathbb{R}^n$. According to the classical tubular neighborhood theorem and Section 3.2 the Euclidean scalar product induces a maximal tubular neighborhood $T^{\max}_S$ of S in $\mathbb{R}^n$; the projection corresponding to $T^{\max}_S$ will be denoted by π_S or shortly by π. For every point $x \in T^{\max}_S$ the tangent space $T_x\mathbb{R}^n \cong \mathbb{R}^n$

possesses a unique orthogonal decomposition $T_x\mathbb{R}^n = T_xS \oplus \ker T_x\pi$. Hereby T_xS originates from $T_{\pi(x)}S$ by parallel transport along the line connecting $\pi(x)$ and x. Now, we denote by $P_{S,x} : T_x\mathbb{R}^n = \mathbb{R}^n \to T_xS$ the corresponding orthogonal projection and write, if any misunderstandings are not possible, simply P_x instead of $P_{S,x}$. Hence $P_S : T_S^{max} \to \mathrm{End}(\mathbb{R}^n)$ becomes a projection valued section in the sense of 1.4.14.

3.3.2 Definition Let $m < \infty$, $S \subset \mathbb{R}^n$ be a submanifold of class $\mathcal{C}^m$ and $P : T \to \mathrm{End}(\mathbb{R}^n)$ a projection valued section of class $\mathcal{C}^m$, that means for all $x \in T$ let P_x be an endomorphism of $\mathbb{R}^n$ with $P_x^2 = P_x$. Then the pair (P, S) or, if any misunderstandings are not possible, simply P is called *curvature moderate* of *order* m, if the following holds:

(CM1) For every point $x \in \mathbb{R}^n$ the set germ of S at x has *finitely many connected components* that means there exists a neighborhood $V \subset \mathbb{R}^n$ such that for every ball $B \subset V$ around x the intersection $S \cap B$ has only finitely many connected components.

(CM2) T is a regularly situated neighborhood of S.

(CM3) For every point of ∂S there exists a neighborhood $V \subset \mathbb{R}^n$ as well as constants $c \in \mathbb{N}$ and $C > 0$ such that for all $\alpha \in \mathbb{N}^n$ with $|\alpha| \leq m$ the following estimate is satisfied for the partial derivatives of P in dependence on the Euclidean distance $d(x, \partial S)$:

$$\|\partial^\alpha P_x\| \leq C\left(1 + \frac{1}{d(x,\partial S)^c}\right), \quad x \in V \cap T. \tag{3.3.1}$$

If P is curvature moderate of every order, then we say say that P is *curvature moderate* of *order* ∞ or briefly that P is *curvature moderate*. In other words P is curvature moderate if and only if the components of P are Whitney functions on T tempered relative ∂S of class $\mathcal{C}^\infty$.

A $\mathcal{C}^m$-submanifold $S \subset \mathbb{R}^n$, where now we allow $m \in \mathbb{N}^{>0} \cup \{\infty\}$, is called *curvature moderate* (of *order* m), if there exists a regularly situated open neighborhood $T_S \subset T_S^{max}$ such that the corresponding projection valued mapping $P_S : T_S \to \mathrm{End}(\mathbb{R}^n)$ is curvature moderate (of order m).

One checks by direct calculation that for every $\mathcal{C}^m$-diffeomorphism $H : O \to \tilde{O} \subset \mathbb{R}^n$ with $O \subset \mathbb{R}^n$ open and $\partial S \cap O \neq \emptyset$ the manifold $S \cap O$ is curvature moderate up to order m, if and only if this holds for $H(S \cap O) \subset \mathbb{R}^n$ as well. Hence it makes sense to define a stratum S of a stratified space X with smooth structure as *curvature moderate* (of *order* m), if there exists a covering of X by singular charts $x : U \to \mathbb{R}^n$ of class $\mathcal{C}^m$ such that $x(S \cap U)$ is curvature moderate (of order m). In particular it is now clear what to understand by a curvature moderate submanifold S of a manifold M. The stratum S resp. the stratified space X is called *curvature moderate*, if S resp. if every stratum of X is curvature moderate.

3.3.3 Example Subanalytic sets with their coarsest Whitney stratification are curvature moderate. In particular all algebraic varieties are curvature moderate. This result can be proved only with some larger technical expense. It can be derived by unpublished work of PARUSIŃSKI [140] or with the help of Newton–Puiseux-expansions as they have been used in MOSTOWSKI [130].

3.3.4 Example The slow and fast spiral are not curvature moderate.

In the following we will consider several situations where one can naturally find projection valued sections.

3.3.5 Example Let $f : \mathbb{R}^n \to \mathsf{N}$ be a $\mathcal{C}^{m+1}$-mapping between manifolds and $f_{|S} : S \to \mathsf{N}$ submersive. Then there exists an open neighborhood T of S in $\mathbb{R}^n$ such that the restricted map $f_{|T} : T \to \mathsf{N}$ is a submersion. Hence the mapping $P^f : T \to \mathrm{End}(\mathbb{R}^n)$ which assigns to every $x \in T$ the Euclidean projection onto the kernel $\ker T_x f$ is of class $\mathcal{C}^m$. We call the pair (f, S) or, if any misunderstandings are not possible, only f *curvature moderate*, if the projection valued section P^f is curvature moderate; finally we call f *strongly curvature moderate* over S, if additionally S and after possibly shrinking T even the mapping $P^f_S : T \to \mathrm{End}(\mathbb{R}^n)$ which assigns to every $x \in T$ the projection onto the kernel of $P^f_x - P_{S,x}$ is curvature moderate.

In the more general case that $f : M \to \mathsf{N}$ is defined over a manifold M and $f_{|S} : S \to \mathsf{N}$ is submersive we call f *(strongly) curvature moderate*, if there exists a covering of ∂S by differentiable charts $x : U \to \mathbb{R}^n$ of M such that every one of the mappings $f \circ x^{-1} : x(U) \to \mathsf{N}$ is (strongly) curvature moderate.

In all of the last definitions we should have mentioned explicitly the order m; by reasons of linguistic aesthetics we have abstained from this. By the same reason we will often not mention in the following the order m, if the context makes clear which order is meant.

The following lemma provides some further means how to generate curvature moderate projection valued sections.

3.3.6 Lemma *Let $S \subset \mathbb{R}^n$ be a submanifold which is curvature moderate of order m and $a : T \to \mathrm{End}(\mathbb{R}^n)$ a $\mathcal{C}^m$-mapping defined over a regularly situated neighborhood of S such that the components of a are tempered relative ∂S of class $\mathcal{C}^m$. Additionally let us assume the following:*

(1) *The rank* $\mathrm{rk}\, a_x$, $x \in T$ *is constant over* T.

(2) *Each of the operators* $a_x \in \mathrm{End}(\mathbb{R}^n)$ *is normal.*

(3) *Denote by λ_x the eigenvalue of a_x having smallest nonvanishing absolute value. Then there exists for every point of ∂S a neighborhood V in $\mathbb{R}^n$ as well as a constant $c \in \mathbb{N}$ and a $C > 0$ such that the following estimate holds:*

$$|\lambda_x| \geq C\, d(x, \partial S)^c, \quad x \in V \cap T.$$

Define for every $x \in T$ the endomorphism $P^a_x = P_{\ker a_x}$ as the projection onto $\ker a_x$ along $\mathrm{im}\, a_x$. Then $P^a : T \to \mathrm{End}(\mathbb{R}^n)$ is a projection valued section and curvature moderate of order m.

PROOF: To ease the formulation of this and further proofs we say that a continuous function $g : A \to \mathbb{R}$ with $A \subset \mathbb{R}^n$ behaves *bounded away from* Z, where $Z \cap A = \emptyset$, if there exists a $d \in \mathbb{N}$ such that

$$\sup_{x \in A} |g(x)|\, d(x, Z)^d < \infty.$$

As we can switch without loss of generality from the mapping a to $aa^* : T \to \mathrm{End}(\mathbb{R}^n)$, $x \mapsto a_x a_x^*$ (where a_x^* denotes the adjoint operator) and as $a_x a_x^*$ has the same image and the same kernel like a_x, one can assume that a_x is for every $x \in T$ selfadjoint and positive semidefinite. In particular $\lambda_x > 0$ then follows. Now let γ_x be the smooth closed curve around the origin of $\mathbb{C}$ with $\gamma_x(t) = \frac{1}{2}\lambda_x e^{2\pi i t}$, $t \in [0,1]$. Then it is well-known that

$$P_{\ker a_x} = \frac{1}{2\pi i} \int_{\gamma_x} \frac{1}{z - a_x}\, dz,$$

hence

$$\partial_i P_{\ker a_x} = \frac{1}{2\pi i} \int_{\gamma_x} \frac{1}{z - a_x} D a_x \frac{1}{z - a_x} e_i\, dz.$$

Now choose for a point of ∂S a neighborhood V in $\mathbb{R}^n$ according to the assumption. Then we have for $x \in V \cap T$ the following estimate

$$\begin{aligned} \|\partial_i P_{\ker a_x}\| &\le \sup_{t \in [0,1]} \frac{1}{2\pi} \frac{\|\dot\gamma_x(t)\|}{\|\gamma_x(t) - a_x\|^2} \|DA_x\| \\ &\le \frac{4}{\lambda_x} \|DA_x\| \le \frac{4}{C}\left(1 + \frac{1}{d(x, \partial S)^c}\right) \|DA_x\|. \end{aligned}$$

After possibly shrinking V we can achieve that $\overline{V}$ is compact and that $V \cap \partial S$ is relatively compact in ∂S. Hence by the temperedness of the components of a the map $(\partial_i P_{\ker a})_{|V \cap T}$ restricted to $V \cap T$ must be bounded away from ∂S. Analogously one shows that the higher derivatives $\partial^\alpha P_{\ker a}$ restricted to $V \cap T$ are bounded away from ∂S. Altogether one thus shows that the components of $P_{\ker a}$ are tempered relative ∂S of class $\mathcal{C}^m$. This proves the claim. □

3.3.7 Proposition *A submanifold* $S \subset \mathbb{R}^n$ *of class* $\mathcal{C}^{m+1}$ *is curvature moderate of order* m, *if there exists a regularly situated open neighborhood* $T_S \subset T_S^{\max}$ *such that the components of the Euclidean projection* $\pi_S : T_S \to S$ *are tempered relative* ∂S *of class* $\mathcal{C}^{m+1}$.

For the proof we need the lemma below which comprises a generalization of the result [59, Lem. II.1].

3.3.8 Lemma *Let* μ *be a Riemannian metric defined on an open neighborhood* S *and* T^μ *a tubular neighborhood of* S *induced by* μ. *Then one can represent the tangent map of the projection* π^μ *of* T^μ *in the form*

$$T_x \pi^\mu = F_x \circ P_x, \qquad x \in T^\mu, \tag{3.3.2}$$

where $P : T^\mu \to \operatorname{End}(\mathbb{R}^n)$ *is the projection valued section* $x \mapsto T_{\pi^\mu(x)}\pi^\mu$ *and* F_x *the automorphism of* $T_{\pi(x)}S$ *given by*

$$F_x = \Big(\mathrm{id} + \sum_{i=1}^{n} n_i\, P_{\pi^\mu(x)} \circ T_{\pi^\mu(x)} N_i\Big)^{-1}.$$

Hereby $N_i : S \to \mathbb{R}^n$ *denotes the vector field* $N_i(\pi^\mu(x)) = (\mathrm{id} - P_{\pi^\mu(x)})e_i$ *which is* μ*-orthogonal to* S*,* n_i *denotes the smooth function* $x \mapsto \Big\langle \exp^{-1}_{\pi^\mu(x)}(x), e_i\Big\rangle$ *and* exp *the exponential function with respect to* μ*.*

PROOF: After abbreviating π^μ by π we expand x with the help of the exponential function exp in the following way:

$$x = \exp_{\pi(x)}\left(\sum_{i=1}^{n} \Big\langle \exp^{-1}_{\pi(x)}(x), e_i\Big\rangle N_i(\pi(x))\right).$$

This is possible indeed, as $\exp^{-1}_{\pi(x)}(x)$ is μ-orthogonal to S at the footpoint $\pi(x)$. Replace in this equation x by $\gamma(t) = tv + x$, $v \in \mathbb{R}^n$, and differentiate with respect to t at the point 0, then one obtains

$$v = T_x\pi.v + \sum_{i=1}^{n} n_i(x) TN_i.T_x\pi.v + \Big\langle T(\exp^{-1}_{\pi(\cdot)} x)v + T(\exp^{-1}_{\pi(x)}(\cdot))v, N_i(\pi(x))\Big\rangle N_i(\pi(x)).$$

Now let the operator $P_{\pi(x)}$ act on this equation. Then one obtains by $P_{\pi(x)}.N_i(\pi(x)) = 0$ and $P_{\pi(x)}.T_x\pi.v = T_x\pi.v$ the relation

$$P_{\pi(x)}.v = G_x.T_x\pi.v \quad \text{with} \quad G_x = \mathrm{id} + \sum_{i=1}^{n} n_i(x)\, P_{\pi(x)} \circ T_{\pi(x)} N_i.$$

Hence G_x is an automorphism of $T_{\pi(x)}S$, and $F_x = G_x^{-1}$ is well-defined. Thus the claim follows. □

PROOF OF THE PROPOSITION: Now we denote by π the projection π_S and abbreviate $T := T_S$. Note that in the Euclidean case, which is the case we consider at the moment, the exponential function is given by $\exp_x v = x + v$. Its inverse is given by $\exp_x^{-1} y = y - x$, where $x, y \in T_S$, $v \in \mathbb{R}^n$. Hence F_x has the form

$$F_x = G_x^{-1} \quad \text{with} \quad G_x = \mathrm{id} + \sum_{i=0}^{n} \langle \pi(x) - x, e_i\rangle\, P_{\pi(x)} \circ T_{\pi(x)} N_i. \tag{3.3.3}$$

Let us suppose first that S is curvature moderate that means the components of P lie in $\mathcal{M}^m(\partial S; T)$. Choose for every point of ∂S a compact neighborhood $V \subset \mathbb{R}^n$ according to Definition 3.3.2. Thus it remains to show that all restricted partial derivatives $(\partial^\alpha T\pi)_{|V \cap T}$ with $|\alpha| \le m$ are bounded away from ∂S over $V \cap T$. Obviously the product of functions bounded away from ∂S is again bounded away from ∂S. Now, the components of the function $P_{|S}$ are bounded, hence bounded away from ∂S. Therefore, this holds for $T\pi_{|S} = P_{|S}$ as well. From Equation (3.3.3) one derives easily that T can be restricted to a regularly situated neighborhood of S such that $\|G_x\| \geq \frac{1}{2}$ and $\|F_x\| \leq 2$ hold for all $x \in T_S$. Together with (3.3.2) this implies that the first derivative $(T\pi)_{|V\cap T}$ is bounded away from ∂S. But this entails that $(Dn_i)_{|V\cap T}$ is bounded away from ∂S as well. As S is curvature moderate, the same holds for $DP_{|V\cap T}$ and $DT_{\pi(\cdot)}N_i$, hence $DG_{|V\cap T}$ is bounded away from ∂S. By $F_x = G_x^{-1}$ and $\|G_x\| \geq \frac{1}{2}$ Equation (3.3.2) then entails that all partial derivatives $(\partial^\alpha T\pi)_{|V\cap T}$ with $|\alpha| = 1$ have this property as well. By induction one moves to higher derivatives $(\partial^\alpha T\pi)_{|V\cap T}$ with $|\alpha| > 1$ and shows by an analogous argument that these are bounded away from ∂S over $V \cap T$. Hence the components of π lie in $\mathcal{M}^{m+1}(\partial S; T)$.

Now let the components of $\pi \in \mathcal{C}^{m+1}(T)$ be tempered relative ∂S of class $\mathcal{C}^{m+1}$. By the representation $P_x = T_{\pi(x)}\pi$, $x \in T_S$ and a repeated use of the chain and LEIBNIZ rule one proves that the projection valued mapping P has to be curvature moderate of order m. This proves the proposition. □

The notion of a curvature moderate submanifold of $\mathbb{R}^n$ has been defined with respect to tubular neighborhoods induced by the Euclidean metric. But it makes sense and will later prove to be necessary for the extension theory of smooth functions that one has available even a notion of curvature moderate Riemannian metrics and of curvature moderate tubular neighborhoods.

3.3.9 Definition Let $S \subset \mathbb{R}^n$ and μ a Riemannian metric of class $\mathcal{C}^m$ defined on a regularly situated neighborhood T of S. We call the pair (μ, S), or if any misunderstandings are not possible, only μ *curvature moderate* of *order* m, if the following conditions are satisfied:

(CM4) The components $\mu_{ij} = \mu(e_i, e_j) : T \to \mathbb{R}$ of μ with respect to the canonical basis of $\mathbb{R}^n$ are tempered relative ∂S of class $\mathcal{C}^m$.

(CM5) For every point of ∂S there exists a neighborhood $V \subset \mathbb{R}^n$ as well as a constant $c \in \mathbb{N}$ and a $C > 0$ such that the following estimate is true:

$$\frac{d(x, \partial S)^c}{C} \|v\| \leq \|v\|_{\mu_x} \leq C\left(1 + \frac{1}{d(x, \partial S)^c}\right) \|v\|, \qquad x \in V \cap T, \ v \in \mathbb{R}^n.$$

Hereby $\|\cdot\|$ denotes the Euclidean norm on $\mathbb{R}^n$ and $\|\cdot\|_{\mu_x}$ the norm induced by μ_x.

In the more general case that S is a submanifold of M and μ a Riemannian metric on a neighborhood T of S in M the metric μ is called *curvature moderate* of *order* m, if there exists a covering of ∂S by differentiable charts $x : U \to \mathbb{R}^n$ of M such that $x_*(\mu_{|T\cap U})$ is curvature moderate of order m.

3.3.10 Proposition *Let* $m \in \mathbb{N}^{>0}$, $S \subset \mathbb{R}^n$ *be curvature moderate of order* $m + 2$ *and* T *a regularly situated open neighborhood of* S. *Then the following statements hold for any connection* ∇ *on the tangent bundle of* T.

(1) *If* μ *is a Riemannian metric on* T *and curvature moderate of order* $m+1$ *and if* ∇ *is the Levi–Civita connection of* μ, *then the Christoffel symbols* Γ_{ij}^k *of* ∇ *(formed with respect to the canonical basis of* $\mathbb{R}^n$*) comprise functions which are tempered relative* ∂S *of class* $\mathcal{C}^m$.

(2) *In case the Christoffel symbols Γ_{ij}^k of ∇ are tempered relative ∂S of class $\mathcal{C}^m$, the mappings*

$$V \ni (x, y) \mapsto \langle \exp_y^{-1}(x), e_i \rangle \in \mathbb{R}$$

and

$$W \ni v \mapsto \langle \exp v, e_i \rangle \in \mathbb{R}$$

comprise functions which are tempered relative ∂S of class $\mathcal{C}^m$. Hereby exp *is the exponential map with respect to ∇, $V \subset T \times T \subset \mathbb{R}^{2n}$ an appropriate regularly situated neighborhood of S, W is a regularly situated neighborhood of S in $TT \subset \mathbb{R}^{2n}$, and S will be canonically identified with the diagonal of $(S \times S) \cap V$ resp. the zero section of $TS \cap W$.*

PROOF: For the proof of (1) we first regard $\mu : T \to \mathrm{End}(\mathbb{R}^n)$ as a matrix valued function of class $\mathcal{C}^{m+1}$ such that μ_x is for every $x \in T$ selfadjoint and positive definite. Then the function $\overline{\mu} : T \to \mathrm{End}(\mathbb{R}^n)$ with $\mu_x.\overline{\mu}_x = \mathrm{id}_{\mathbb{R}^n}$ is well-defined and smooth. It satisfies

$$\partial_i \overline{\mu}_x = -\overline{\mu}_x \, (\partial_i \mu_x) \, \overline{\mu}_x, \qquad x \in T.$$

The component functions on T belonging to μ are tempered relative ∂S of class $\mathcal{C}^{m+1}$. Moreover, by assumption we have for every sufficiently small ball B around a point of ∂S

$$\|\overline{\mu}_x\| = \sup_{\|v\|=1} \|\overline{\mu}_x.v\| \le C \left(1 + \frac{1}{d(x, \partial S)^c} \right), \qquad x \in B \cap T,$$

where $c \in \mathbb{N}$ and $C > 0$ are appropriate. Hence the restriction $(\partial_i \overline{\mu})_{|B \cap T}$ has to be bounded away from ∂S. Analogously one shows that even the higher partial derivatives $(\partial^\alpha \overline{\mu})_{|B \cap T}$ with $|\alpha| \le m + 2$ are bounded away from ∂S. Hence $\overline{\mu}$ is tempered relative ∂S of class $\mathcal{C}^{m+1}$. As it is well-known the Christoffel symbols of the Levi–Civita connection have the form

$$\Gamma_{ij}^k = \frac{1}{2} \sum_l \overline{\mu}^{kl} (\partial_i \mu_{lj} + \partial_j \mu_{li} - \partial_l \mu_{ij}).$$

Thus by the considerations above the Christoffel symbols have to be tempered relative ∂S of class $\mathcal{C}^m$. This proves (1).

Now we come to the proof that the exponetial map is tempered. Denote by t_v^+ for $v \in TT$ the escape time of the exponential map like in 3.2.1, and let $W^{\max} = \{v \in TT | \, t_v^+ > 1\}$ be the maximal domain of exp. With the help of Theorem 1.7.11 we show first that $W^{\max}$ is a regularly situated neighborhood of S in $\mathbb{R}^{2n}$. Now for every vector $v \in W \subset TT \cong T \times \mathbb{R}^n$ the curve $\gamma_v(t) = \exp_x(tv)$, $t \in [0, 1]$ satisfies the initial value problem

$$\ddot{\gamma}_v^k(t) + \sum_{ij} \Gamma_{ij}^k(\gamma_v(t)) \, \dot{\gamma}_v^i(t) \, \dot{\gamma}_v^j(t) = 0, \quad \gamma_v(0) = x, \quad \dot{\gamma}_v(0) = v.$$

By assumption on the Christoffel symbols and by Theorem 1.7.11 there exists a regularly situated tubular neighborhood $\tilde{T} \subset \mathbb{R} \times \mathbb{R}^{2n}$ of $\{0\} \times S \times \{0\}$, such that the mapping $\gamma_{|\tilde{T}} : \tilde{T} \to \mathbb{R}^n$, $(t, v) \mapsto \gamma_v(t)$ is tempered relative $\{0\} \times \partial S \times \{0\}$ of class $\mathcal{C}^m$. Hence there exist positive continuous mappings $\delta, \varepsilon, \varepsilon'$ on S such that

$$\delta(x) \geq D\, d(x, \partial S)^d, \quad \varepsilon(x) \geq C\, d(x, \partial S)^c, \quad \text{and} \quad \varepsilon'(x) \geq C'\, d(x, \partial S)^{c'}$$

for all $x \in S$ and appropriate constants $c, c', d \in \mathbb{N}$ and $C, C', D > 0$ and such that

$$B^1_{\delta(x)}(0) \times B^n_{\varepsilon(x)}(x) \times B^n_{\varepsilon'(x)}(0) \subset \tilde{T}, \quad x \in S.$$

Define

$$W := \bigcup_{x \in S} \left(B^n_{\varepsilon(x)}(x) \cap \pi_S^{-1}(x) \right) \times B^n_{\frac{1}{2}\delta \cdot \varepsilon'(x)}(0).$$

Then for every pair $(t, v_x) \in [-1, 1] \times W$ the point $\left(\delta(x)t, \frac{1}{\delta(x)} v_x\right)$ lies in $\tilde{T}$, hence $\gamma_{|[-1,1] \times W}$ and $\exp_{|W}$ are tempered relative $\{0\} \times \partial S \times \{0\}$ resp. $\partial S \times \{0\}$ of class $\mathcal{C}^m$.

A similar argument shows that an appropriate restriction of $\exp^{-1}$ is tempered, hence (2) follows. $\qquad\qquad\qquad\qquad\qquad\qquad\qquad\qquad\qquad\qquad\qquad\qquad\qquad\qquad\qquad\square$

3.3.11 Corollary *Let S and T be like in the proposition and let μ be a Riemannian metric which is curvature moderate of class $\mathcal{C}^{m+1}$ on T. Then the maximal tubular neighborhood $T^{\mu,\max}$ induced by μ is regularly situated to S that means $T^{\mu,\max} := \exp(T^{\max}_{S \subset N^\mu})$ with*

$$T^{\max}_{S \subset N^\mu} = \left\{ v \in N^\mu_x \,\middle|\, x \in S,\ \exp v \in T,\ \delta_\mu(x, S) = \|v\|_{\mu_x} \right\}$$

is a regularly situated neighborhood of S.

Next we will examine, under which assumptions projection valued sections associated to a Riemannian metric are tempered or in other words under which assumptions these projection valued sections are curvature moderate. To this end we study in the propositions below local orthogonal systems with respect to μ.

3.3.12 Proposition *Let $m \in \mathbb{N}^{>0}$, $S \subset \mathbb{R}^n$ be a submanifold which is curvature moderate of order $m + 2$, T a regularly situated open neighborhood of S such that the Euclidean projection $P_S : T \to \operatorname{End}(\mathbb{R}^n)$ is defined over T and such that P_S is curvature moderate. Let μ be a Riemannian metric on T which is curvature moderate of order $m + 1$. Furthermore let $B = B_\varepsilon(z) \subset \mathbb{R}^n$ be an open ball of radius $\varepsilon < 1$ around a point of ∂S such that $K = \overline{B_{2\varepsilon}}(z) \cap \partial S$ is compact. Then there exist two countable families $(\phi_j)_{j \in J}$ and $(f_{jl})_{j \in J, 1 \leq l \leq n}$ of smooth functions ϕ_j resp. smooth vector fields f_{jl} on $\mathbb{R}^n \setminus K$ such that the following holds:*

(1) *The maps ϕ_j have compact support and comprise a smooth partition of unity over $B \cap T$. Moreover, the number $N(x)$ of indices $j \in J$ with $x \in \operatorname{supp}\phi_j$ is bounded by 4^n.*

(2) *There exist constants $c_m \in \mathbb{N}$ and $C_m > 0$ such that for all $j \in J$ the following estimates are satisfied:*

$$\|\partial^\alpha \phi_j(x)\| \leq C_m \left(1 + \frac{1}{d(x, \partial S)^{c_m|\alpha|}} \right), \quad x \in B \cap T,\ |\alpha| \leq m,\ \alpha \in \mathbb{N}^n.$$

(3) *For every j the n-tuple $(f_{j1}, \cdots, f_{jn})$ is an orthogonal frame with respect to μ over the set $\operatorname{supp} \phi_k \cap B \cap T$.*

(4) *There exist constants $c'_m \in \mathbb{N}$ and $C'_m > 0$ such that for all $j \in J$ and l, $1 \le l \le n$, the following estimate is satisfied:*

$$\|\partial^\alpha f_{jl}(x)\| \le C'_m \left(1 + \frac{1}{d(x, \partial S)^{c'_m}}\right), \qquad x \in \operatorname{supp} \phi_j \cap B \cap T, \ |\alpha| \le m.$$

(5) *Let $T^\mu = (E, \varepsilon, \varphi)$ be the tubular neighborhood of S in M induced by μ, let $\pi^\mu : T^\mu \to S$ be the corresponding projection and $P^\mu : T^\mu \to \operatorname{End}(\mathbb{R}^n)$ the projection valued mapping which assigns to every $x \in T$ the orthogonal projection with respect to μ onto the horizontal space of the tubular neighborhood at the point x. Then P^μ has over $B \cap T$ the following representation:*

$$P^\mu_x v = \sum_{j \in J} \sum_{l=1}^{\dim S} \phi_j(x) \frac{\mu_x(v, f_{jl}(x))}{\mu_x(f_{jl}(x), f_{jl}(x))} f_{jl}(x) \qquad x \in B \cap T, \ v \in \mathbb{R}^n. \tag{3.3.9}$$

Moreover, P^μ is curvature moderate of order m.

PROOF: As $P := P_S$ is curvature moderate, there exist $d_m \in \mathbb{N}$ and $D_m > 0$ such that all partial derivatives $\partial^\alpha P e_l$ at most of order m satisfy the estimate

$$\sup_{x \in B \cap T} \|\partial^\alpha P e_l(x)\| \, d(x, \partial S)^{d_m} < \frac{1}{\sqrt{n}} D_m, \qquad x \in T. \tag{3.3.10}$$

Moreover, by assumption on μ there exist constants $c \in \mathbb{N}$ and $C > 0$ such that for all $v \in \mathbb{R}^n$ and $x \in \overline{B_{2\varepsilon}} \cap T$

$$\frac{d(x, \partial S)^c}{C} \|v\| \le \|v\|_{\mu_x} \le \frac{C}{d(x, \partial S)^c} \|v\|.$$

Now choose for $\delta_m = C'_m \cdot \max\{1, C^2\}$ and $c_m = c'_m + c$ according to Lemma 1.7.9 a locally finite smooth partition of unity $(\phi_j)_{j \in J}$ of $\mathbb{R}^n \setminus K$ such that

$$d(\operatorname{supp} \phi_j, K)^{c_m} \ge 2n\delta_m \operatorname{diam}(\operatorname{supp} \phi_j), \tag{3.3.11}$$

$$|\partial^\alpha \phi_j(x)| \le C_m \left(1 + \tfrac{1}{d(x,K)^{c_m |\alpha|}}\right), \qquad x \in \mathbb{R}^n \setminus K, \ |\alpha| \le m, \tag{3.3.12}$$

where $C_m > 0$ depends only on c_m, δ_m, n and m. By construction of the ϕ_j in 1.7.9 one can assume that all the supports $\operatorname{supp} \phi_j$ are convex. We now restrict the index sets J to exactly those indices j such that $\operatorname{supp} \phi_j$ has nonvanishing intersection with $T \cap B$ and choose $x_j \in \operatorname{supp} \phi_j \cap T \cap B$. Next choose linear mappings $O_j \in GL(\mathbb{R}^n)$ such that $(O_j e_1, \cdots, O_j e_n)$ is an orthogonal basis of $\mathbb{R}^n$ with respect to the scalar product μ_{x_j}, $(O_j e_{\dim S+1}, \cdots, O_j e_n)$ spans the vertical space $T_{\varphi^{-1}(x)} \varphi(E_{\pi^\mu(x)})$ and such that $\|O_j e_1\| = \cdots = \|O_j e_n\| = 1$ holds. For x sufficiently close to x_j the following recursively fixed vectors $f_{jl}(x)$, $l = 1, \cdots, n$ are then well-defined:

$$f_{j1}(x) = P_x O_j e_1,$$

$$f_{j(l+1)}(x) = P_x O_j e_{l+1} - \sum_{k=1}^{l} \frac{\mu_x(P_x O_j e_{l+1}, f_{jk}(x))}{\mu_x(f_{jk}(x), f_{jk}(x))} f_{jk}(x).$$

Now we will show that the $f_{jk}(x)$ are well-defined for $x \in \operatorname{supp}\phi_j \cap B \cap T$ and they satisfy the estimates

$$1 - \frac{l}{2n} \le \|f_{jl}(x)\| \le 1 + \frac{l-1}{2n}, \qquad x \in \operatorname{supp}\phi_j \cap B \cap T. \tag{3.3.13}$$

To this end we estimate the norm $\|P_x v\|$ for $x \in \operatorname{supp}\phi_j \cap B \cap T$ and $v \in \mathbb{R}^n$ with the help of (3.3.11) and (3.3.10):

$$\begin{aligned} \|P_x v\| &\ge \Big| \|P_{x_j} v\| - \|P_x v - P_{x_j} v\| \Big| \ge \|P_{x_j} v\| - \sup_{y \in [x_j, x]} \|T_y P v\| \, \|x - x_j\| \\ &\ge \|P_{x_j} v\| - \frac{1}{2nC_m} \sup_{y \in [x_j, x]} \|T_y P v\| \, d(x, \partial S)^{c'_m} \ge \|P_{x_j} v\| - \frac{1}{2n}\|v\|, \end{aligned} \tag{3.3.14}$$

where it has been used that the segment $[x_j, x]$ lies in $\operatorname{supp}\phi_j \cap B \cap T$. This entails $1 - \frac{1}{2n} \le \|f_{j1}(x)\| \le 1$. Let us suppose that (3.3.13) holds for all $l \le l_0$. Then we show with the help of (3.3.11), (3.3.10) and (3.3.14) that (3.3.13) holds for $l = l_0 + 1$ as well:

$$\begin{aligned} \|f_{j(l_0+1)}(x)\| &\ge \|P_x O_j e_{l_0+1}\| - \sum_{k=1}^{l_0} \left\| \frac{\mu_x(P_x O_j e_{l_0+1}, f_{jk}(x))}{\mu_x(f_{jk}(x), f_{jk}(x))} f_{jk}(x) \right\| \\ &\ge 1 - \frac{1}{2n} - \sum_{k=1}^{l_0} \frac{\|P_x O_j e_{l_0+1}\|_{\mu_x}}{\|f_{jk}(x)\|_{\mu_x}} \|f_{jk}(x)\| \\ &\ge 1 - \frac{1}{2n} - \sum_{k=1}^{l_0} C^2 \sup_{y \in [x_j, x]} \|T_y P O_j e_{l_0+1}\| \, \|x - x_j\| \, \frac{1}{d(x, \partial S)^c} \\ &\ge 1 - \frac{1}{2n} - \sum_{k=1}^{l_0} \frac{1}{2nC_m} \sup_{y \in [x_j, x]} \|T_y P O_j e_{l_0+1}\| \, d(x, \partial S)^{c_m} \\ &\ge 1 - \frac{l_0 + 1}{2n}, \end{aligned}$$

$$\begin{aligned} \|f_{j(l_0+1)}(x)\| &\le \|P_x O_j e_{l_0+1}\| + \sum_{k=1}^{l_0} \left\| \frac{\mu_x(P_x O_j e_{l_0+1}, f_{jk}(x))}{\mu_x(f_{jk}(x), f_{jk}(x))} f_{jk}(x) \right\| \\ &\le 1 + \sum_{k=1}^{l_0} C^2 \sup_{y \in [x_j, x]} \|T_y P O_j e_{l_0+1}\| \, \|x - x_j\| \, \frac{1}{d(x, \partial S)^c} \\ &\le 1 + \frac{l_0}{2n}. \end{aligned}$$

Inductively we thus obtain (3.3.14) for all l, hence $f_{j1}(x), \cdots, f_{jn}(x)$ comprises for every $x \in \operatorname{supp}\phi_j \cap B \cap T$ a μ_x-orthogonal basis of $\mathbb{R}^n$. Now it is not difficult to check that the thus defined f_{jl} can be extended to smooth vector fields over $\mathbb{R}^n \setminus K$ such that every f_{jl} has compact support and such that $\|f_{jl}(x)\| \le 2$ holds for every $x \in \mathbb{R}^n \setminus K$.

We already know by the fact that S is curvature moderate that the components of the functions $PO_j e_l$ lie in $\mathcal{M}^m(\partial S; T)$, hence there exist constants $c_{m1} \in \mathbb{N}$ and $C_{m1} > 0$ such that for all α with $|\alpha| \le m$

$$\sup_{x \in \operatorname{supp}\phi_j \cap B \cap T} \|\partial^\alpha f_{j1}(x)\| \, d(x, \partial S)^{c_{m1}} < C_{m1}. \tag{3.3.15}$$

We will show that such an estimate holds for the other f_{jl} as well. Recall that for every fixed l there exists a polynomial function $r_l : \mathbb{R}^{n(l+n)+l-1} \to \mathbb{R}^n$ such that

$$f_{jl}(x) =$$
$$= r_l\left(P_x O_j e_l, f_{j1}(x), \cdots, f_{j(l-1)}(x), \frac{1}{\mu(f_{j1}, f_{j1})}, \cdots, \frac{1}{\mu(f_{j(l-1)}, f_{j(l-1)})}, \mu_{11}, \cdots, \mu_{nn}\right).$$

As f_{jl} satisfies the estimate (3.3.13) with constants independent of the j and as μ is curvature moderate over S, induction by l shows the existence of constants $c_{ml} \in \mathbb{N}$ and $C_{ml} > 0$ such that

$$\sup_{x \in \operatorname{supp} \phi_j \cap B \cap T} \|\partial^\alpha f_{jl}(x)\| \, d(x, \partial S)^{c_{ml}} < C_{ml}. \tag{3.3.16}$$

By the the orthogonalization scheme of Gram–Schmidt and the definition of f_{jl} the right hand side of (3.3.9) is the μ-orthogonal projection P^μ_x onto the horizontal space of the tubular neighborhood T^μ indeed. Moreover, (3.3.12) and the results proved so far entail that the component functions of P^μ induce Whitney functions on T which are tempered relative ∂S of class $\mathcal{C}^m$. Altogether the claim follows. □

3.3.13 Definition Let $S \subset \mathbb{R}^n$ be a $\mathcal{C}^m$-submanifold, $T = (E, \varepsilon, \varphi)$ a tubular neighborhood of S, $\pi : T \to S$ the corresponding projection and $P : T \to \operatorname{End}(\mathbb{R}^n)$ the continuous mapping which assigns to every $x \in T$ the projection onto the horizontal space of the tubular neighborhood along the vertical space. Moreover, let $Q : T \to \operatorname{End}(\mathbb{R}^n)$ be the projection valued section with $Q_x = \mathrm{id}_{\mathbb{R}^n} - P_x$ for every $x \in T$. Then the tubular neighborhood T is called *curvature moderate* of *order* m, if the following axioms are satisfied:

(CM6) The projection valued section $P : T \to \operatorname{End}(\mathbb{R}^n)$ is curvature moderate of order m.

(CM7) Denote by η the scalar product on E and by μ the Riemannian metric on T given by

$$\mu_x(v, w) = \eta_{\pi(x)}(T_x\varphi^{-1}.Q_x.v, T_x\varphi^{-1}.Q_x.w) + \langle P_x.v, P_x.w\rangle, \quad x \in T, v, w \in \mathbb{R}^n.$$

Then μ is curvature moderate of order m.

In case that S is a submanifold of a manifold M, we say that the tubular neighborhood T is *curvature moderate*, if there exists a covering of ∂S by differentiable charts $x : U \to \mathbb{R}^n$ of M of the form that the tubular neighborhoods $x_*(T_{|S\cap U})$ of $x(S \cap U)$ are curvature moderate.

A curvature moderate tubular neighborhood T of S has the following additional property:

(CM8) The function $\rho : T \to \mathbb{R}$ is tempered relative ∂S of class $\mathcal{C}^m$.

3.3.14 Proposition *Let $m \in \mathbb{N}^{>0}$, $S \subset M$ be a submanifold which is curvature moderate of order $m+3$, let T be a regularly situated open neighborhood of S and μ a Riemannian metric on T which is curvature moderate of order $m+3$. Then the maximal tubular neighborhood induced by μ according to the classical tubular neighborhood theorem is curvature moderate of order m. Moreover, in case $M = \mathbb{R}^n$ the following holds:*

(CM9) *The components of the projection π^μ of T^μ are tempered relative ∂S of class $\mathcal{C}^m$.*

PROOF: As the statement is a local one we can assume without loss of generality that $M = \mathbb{R}^n$. Under this assumption let T be the tubular neighborhood of S with respect to the Euclidean scalar product, $\pi : T \to S$ the corresponding projection onto S and $P : T \to \mathrm{End}(\mathbb{R}^n)$ the projection valued section according to 3.3.1. After possibly shrinking T the projection $\pi^\mu : T \to S$ of T^μ and the mapping $P^\mu : T \to \mathrm{End}(\mathbb{R}^n)$ which assigns to every $x \in T$ the projection onto the horizontal space of T^μ along the vertical space are well-defined over T. By Corollary 3.3.11 we can achieve hereby that $\complement T$ and S remain regularly situated.

By (5) of Proposition 3.3.12 we already know that the components of the vector valued functions $P^\mu e_i$ are tempered relative ∂S of class $\mathcal{C}^{m+1}$, hence (CM6) is satisfied.

By Lemma 3.3.8 and the argument given in the proof of Proposition 3.3.7 it is immediately clear that the projection π^μ induces Whitney functions over T which are tempered relative ∂S. This proves (CM9).

Next recall Proposition 3.3.10 and check that the components of $\exp^{-1}$ comprise relative ∂S tempered functions of class $\mathcal{C}^{m+1}$. Now use the already proven axiom (CM6) and the assumption that μ is curvature moderate and check that the Riemannian metric μ' on T given by

$$\mu'_x(v, w) = \mu_{\pi^\mu(x)}(T_x \exp^{-1}_{\pi^\mu(x)} .Q^\mu_x.v, T_x \exp^{-1}_{\pi^\mu(x)} .Q_x\mu.w) + \langle P^\mu_x.v, P^\mu_x.w\rangle, \quad x \in T,$$

has to be curvature moderate of order m. This finishes the proof. □

3.4 Geometric implications of the Whitney conditions

In this section we will introduce some geometric results about tubular neighborhoods of submanifolds satisfying the Whitney conditions. Hereby we will often need a pair of disjoint submanifolds of the manifold M. To simplify notation (R, S) will always mean in this section such a pair of disjoint submanifolds.

3.4.1 Proposition (cf. MATHER [122, Lem 7.3]) *Let T_R be a tubular neighborhood of R in M. If the pair (R, S) satisfies the Whitney condition (A), then there exists a smooth function $\delta : R \to \mathbb{R}^{>0}$ such that*

$$(\pi_R)_{|S \cap T^\delta_R} : S \cap T^\delta_R \longrightarrow R$$

is a submersion. If even Whitney (B) *is satisfied, then one can choose* δ *such that*

$$(\pi_R, \rho_R)_{|S\cap T_R^\delta} : S \cap T_R^\delta \longrightarrow R \times \mathbb{R}$$

is submersive.

PROOF: First let us abbreviate: $\pi := \pi_R$ and $\rho := \rho_R$. By Lemma 3.1.2 it suffices to show that for every $x \in R \cap \overline{S}$ there exists a neighborhood U of X in M such that $\pi_{|S\cap U}$ resp. $(\pi, \rho)_{|S\cap U}$ is submersive. But this is a local statement, hence it suffices to prove the claim for the case that $M = \mathbb{R}^n$, $R = \mathbb{R}^m$ with $n = m + k$ and that T_R is the standard tubular neighborhood T^n_m.

We first consider the (A) case. Suppose the claim does not hold. Then there exists a sequence of elements y_j of $S \cap T^n_m$ converging to x and a sequence of unit vectors v_j of TR with $v_j \in \left(T_{y_j}\pi(T_{y_j}S)\right)^\perp$. After transition to an appropriate subsequence $(v_j)_{j\in\mathbb{N}}$ converges to a unit vector $v \in T_xR$ and $T_{y_j}S \to \tau$. By Whitney (A) the relation $T_xR \subset \tau$ holds, hence $v \in T\pi(\tau)^\perp \subset (T_xR)^\perp$ follows. This contradicts the fact that v is a unit vector of T_xR.

Now we come to the (B) case. Suppose there exists a sequence $(y_j)_{j\in\mathbb{N}}$ of elements of $S \cap T^n_m$ converging to x such that $\left(T_{y_j}\rho\right)_{|T_{y_j}S} = 0$. Then the secant $\ell_j = \overline{y_j\,\pi(y_j)}$ is orthogonal to $\mathbb{R}^m$ and $\ker T_{y_j}\rho$. After transition to an appropriate subsequence of $(y_j)_{j\in\mathbb{N}}$ the sequence of spaces $T_{y_j}S$ converges to a subspace τ and the sequence of secants ℓ_j to a line ℓ, which is orthogonal to $\mathbb{R}^m$ as well. By Whitney (B) $\ell \subset \tau$ must hold, that means for all j larger than an appropriate j_0 there exist unit vectors v_j in $T_{y_j}S$ with nonvanishing orthogonal projection onto ℓ and nonvanishing projection onto ℓ_j. As $\ell_j \subset \left(\ker T_{y_j}\rho\right)^\perp$, the relation $T_{y_j}\rho(v_j) \neq 0$ is satisfied for all v_j with $j > j_0$. This is a contradiction to the assumption. Hence there exists a neighborhood $U \subset T^n_m$ of x such that $\rho_{|S\cap U} : S \cap U \to \mathbb{R}$ is submersive. Furthermore, the kernel bundle $E = \ker\left(T\rho_{|U\cap S}\right)$ is well-defined over $U \cap S$ and possesses the fiber dimension $\dim S - 1$. Suppose now there exists a sequence $(y_j)_{j\in\mathbb{N}} \subset S \cap U$ converging to x such that $\dim T_{y_j}\pi(E_{y_j}) < m$. Define τ and ℓ like above. Let $K_j = \ker(T_{y_j}\pi)_{|E_{y_j}}$. After transition to a subsequence of $(y_j)_{j\in\mathbb{N}}$ we can assume that the sequence of vector spaces K_j converges to a vector space K which by assumption must have dimension $> \dim S - 1 - m$. By Whitney's condition (B) the relation $K + \ell + T_xR \subset \tau$ holds. By definition the vector spaces K, ℓ and T_xR intersect pairwise only in the origin, hence the contradiction $\dim K \leq \dim \tau - \dim \ell - \dim T_xR = \dim S - 1 - m$ follows. Therefore, after shrinking the neighborhood U of x the relation $\dim T_y\pi(E_y) = m$ has to be satisfied for all $y \in S \cap U$. This entails our second claim. □

3.4.2 Remark In a certain sense the inverse of the preceding proposition holds as well. The inverse statement has been considered more precisely by TROTMAN in his article [172] about geometric versions of Whitney-regularity. It is shown in [172] that the pair (R, S) satisfies Whitney's condition (A) at $x \in R$ if and only if for every chart $x : U \to \mathbb{R}^n$ around x of class $\mathcal{C}^1$ such that $R \cap U$ is mapped to an open set of $\mathbb{R}^m \subset \mathbb{R}^n$ the projection

$$\pi^x : T_R \cap S \to R, \quad x \mapsto \left(x^{-1} \circ \pi^n_m \circ x\right)(x), \qquad T_R := x^{-1}(T^m_m),$$

is submersive. On the other hand condition (B) is true, if and only if for every such chart $x : U \to \mathbb{R}^n$ around x of class $\mathcal{C}^1$ the mapping

$$(\pi^x, \rho^x) : T_R \cap S \to R \times \mathbb{R}^{>0}, \quad x \mapsto (\pi^x(x), \rho^x(x)), \qquad \rho^x(x) := (\rho^n_m \circ x)(x),$$

is submersive. Interestingly enough it does not suffice in either cases to consider only charts of class $\mathcal{C}^2$. Finally, the result by TROTMAN implies immediately that the conditions (A) and (B) are both $\mathcal{C}^1$-invariant.

Independent proofs of the result of TROTMAN have been given by HAJTO [76] and PERKAL [142].

3.4.3 Corollary (BEKKA [8]) *If the pair* (R, S) *satisfies Whitney* (B) *at* $x \in R$, *then Bekka's condition* (C) *is satisfied at* x.

PROOF: Without loss of generality we can suppose $M = \mathbb{R}^n$. Moreover, let π and ρ like in the proof of the proposition. By the argument given in the proposition there exists a neighborhood U of x in $\mathbb{R}^n$ such that $(\pi, \rho)_{|S\cap U} : S \cap U \to R \times \mathbb{R}$ is submersive. Now let $(y_k)_{k\in\mathbb{N}}$ be a sequence in $S \cap U$ converging to x such that the sequence $(\ker T_{y_k}(\rho_{|S\cap U}))_{k\in\mathbb{N}}$ converges to a subspace $\lambda \subset \mathbb{R}^n$. After transition to a subsequence one can achieve that the sequence $(\ell_k)_{k\in\mathbb{N}}$ of connecting lines $\ell_k = \overline{y_k \pi(y_k)}$ converges to line ℓ perpendicular to $T_x R$, and that $(T_{y_k} S)_{k\in\mathbb{N}}$ converges to a subspace $\tau \subset \mathbb{R}^n$. By Whitney (B) and (A) $T_x R \oplus \ell \subset \tau$ holds. On the other hand $\lambda \oplus \ell = \tau$ is true as well. The second sum hereby follows from Whitney (B) and the fact that according to the proof of the proposition the sum of the projection of ℓ_k to $T_{y_k} S$ and the subspace $\ker T_{y_k}(\rho_{|S\cap U})$ is equal to $T_{y_k} S$. But as the line ℓ is perpendicular to both subspaces $T_x R$ and λ, the relation $T_x R \subset \lambda$ must hold. This proves the claim. □

3.4.4 Corollary (BEKKA [8]) *If the pair* (R, S) *satisfies the conditions* (A)+(δ) *at* $x \in R$, *then there exists a neighborhood* U *of* x *such that the mappings*

$$(\pi_R)_{|S\cap U} : S \cap U \longrightarrow R \quad \text{and} \quad (\rho_R)_{|S\cap U} : S \cap U \longrightarrow \mathbb{R}$$

are submersive. Moreover, (R, S) *then satisfies Bekka's condition* (C) *at* x.

PROOF: By 3.4.1 we already know that there exists U such that $(\pi)_{|S\cap U}$ is submersive. By (δ) one can achieve after shrinking U that $\ker(T_y \rho_{|S})$ has codimension 1 in $T_y S$ for every $y \in U \cap S$, as $P_{S,y}(y - \pi(y))$ is perpendicular to $\ker(T_y \rho_{|S}) = T_y S \cap \ker(T_y \rho)$. But this implies that $\rho_{|S}$ is submersive. For the proof of (C) let $(y_k)_{k\in\mathbb{N}}$ be a sequence of points of $S \cap U$ converging to x such that the sequence of kernels converges to a subspace $\tau \subset \mathbb{R}^n$. After transition to a subsequence the sequence of vector spaces $\operatorname{span}\{P_{S,y_k}(y_k - \pi(y_k))\}$ converges to a line $\ell \subset \mathbb{R}^n$. By Whitney (A) we have $T_x R \subset \tau \oplus \ell = \lim_{k\to\infty} T_{y_k} S$. On the other hand ℓ is perpendicular to τ and $T_x R$, hence $T_x R \subset \tau$ must be true. This proves the claim. □

3.4.5 Lemma *Let* $R, S \subset \mathbb{R}^n$ *with* $R \subset \partial S$, *let* π_R *be the projection and* ρ_R *the tubular function of the Euclidean tubular neighborhood of* R, *and let finally* T_S *be the Euclidean tubular neighborhood of* S. *Denote by* $P^\pi : T_R \to \operatorname{End}(\mathbb{R}^n)$ *the projection* *valued section onto the kernel bundle of* $T\pi_R$, *and by* $P_S : T_S \to \mathrm{End}(\mathbb{R}^n)$ *the projection onto the horizontal bundle of* T_S. *By* $Q^\pi : T_R \to \mathrm{End}(\mathbb{R}^n)$ *and* $Q_S : T_S \to \mathrm{End}(\mathbb{R}^n)$ *we understand the projection valued sections given by* $Q^\pi_x = \mathrm{id}_{\mathbb{R}^n} - P^\pi_x$ *and* $Q_{S,x} = \mathrm{id}_{\mathbb{R}^n} - P_{S,x}$. *Then for every* $x \in R$ *the following four statements are equivalent:*

(1) *The pair* (R, S) *satisfies Whitney (A) at* x.

(2) *Given* $\varepsilon > 0$ *there exists a neighborhood* U *of* x *such that* $d_{\mathrm{Gr}}(T_xR, T_yS) < \varepsilon$ *for all* $y \in S \cap U$.

(3) *Given* $\varepsilon > 0$ *there exists a neighborhood* $U \subset T_R$ *of* x, *such that* $\|Q^\pi_y Q_{S,y} w\| \leq \varepsilon \|Q_{S,y} w\|$ *for all* $y \in S \cap U$ *and* $w \in \mathbb{R}^n$.

(4) *Given* $\varepsilon > 0$ *there exists a neighborhood* $U \subset T_R$ *of* x, *such that* $\|Q_{S,y} Q^\pi_y w\| \leq \varepsilon \|Q^\pi_y w\|$ *for all* $y \in S \cap U$ *and* $w \in \mathbb{R}^n$.

PROOF: The equivalence of (1) and (2) follows immediately from the definition of Whitney (A), the definition of the vector space distance d_{Gr} in Appendix A.1 and Proposition A.1.1 (2).

Now let $(y_k)_{k\in\mathbb{N}}$ be a sequence of points of S converging to x and $(T_{y_k}S)_{k\in\mathbb{N}}$ be convergent to $\tau \subset \mathbb{R}^n$. As Q^π_x is the orthogonal projection onto T_xR, and Q_{S,y_k} the one onto $(T_{y_k}S)^\perp$, (3) implies that $\tau^\perp \subset (T_xR)^\perp$, hence $T_xR \subset \tau$. Therefore Whitney (A) follows. Property (4) entails immediately $T_xR \subset \tau$, hence Whitney (A) follows again.

Next let us suppose that (1) holds but not (3). Then there exists $\varepsilon > 0$, a sequence $(y_k)_{k\in\mathbb{N}}$ of points of S with limit x and a sequence of unit vectors $v_k \in (T_{y_k}S)^\perp$ with $\|Q^\pi_{y_k} v_k\| \geq \varepsilon$. By transition to subsequences one can achieve that $(v_k)_{k\in\mathbb{N}}$ converges to a unit vector $v \in \mathbb{R}^n$. As the projections $Q^\pi_{y_k}$ converge to the orthogonal projection onto T_xR, the vector v has nonvanishing projection to T_xR, which contradicts Whitney (A). Analogously one proves (1)⇒(4). □

3.4.6 Proposition *Let* $X \subset M$ *be an* (A)+(δ)*-stratified closed subspace of* M *consisting of two strata* $S = X^\circ$ *and* ∂S. *Let further* μ *be a Riemannian metric on* M *and for every stratum* R *of* X *let* T_R *be the maximal tubular neighborhood of* R *in* M *induced by* μ, π_R *the projection and* ρ_R *the tubular function. If* X *is curvature moderate of order* m, *then after appropriately restricting* $T_{\partial S}$ *the submersion* $(\pi_{\partial S}, \rho_{\partial S}) : T_S \cap T_{\partial S} \to \partial S \times \mathbb{R}^{>0}$ *is even strongly curvature moderate of order* m.

PROOF: As the claim is essentially a local one, we can assume without loss of generality that M is an open subset of $\mathbb{R}^n$. Moreover, we can suppose that μ is given by the Euclidean scalar product $\langle \cdot, \cdot \rangle$; the case of arbitrary curvature moderate μ is proved analogously, only somewhat more technical. Finally we abbreviate $\rho := \rho_{\partial S}$, $\pi := \pi_{\partial S}$ and set $R := \partial S$.

Denote by $P_S : T_S \to \mathrm{End}(\mathbb{R}^n)$ the projection valued section onto the horizontal bundle of T_S along the vertical bundle. Further projection valued sections $P^\rho : T_R \setminus R \to \mathrm{End}(\mathbb{R}^n)$ and $P^\pi : T_R \to \mathrm{End}(\mathbb{R}^n)$ are given by the orthogonal projection onto the kernel bundle of $T(\rho_{|T_R \setminus R})$ resp. onto the kernel bundle of $T\pi$. More explicitly P^ρ_x has the form

$$P^\rho_y w = w - \langle w, y - \pi(y) \rangle \frac{y - \pi(y)}{\|y - \pi(y)\|^2}, \qquad y \in T_R \setminus R, \ w \in \mathbb{R}^n. \qquad (3.4.1)$$

As the claim has to be proved only locally around a point of R we can suppose (after shrinking M and thus shrinking R) that P^π has the form

$$P^\pi_y w = w - \sum_{i=1}^{\dim R} \langle w, f_i(\pi(y)) \rangle f_i(\pi(y)), \qquad y \in T_R, \ w \in \mathbb{R}^n, \qquad (3.4.2)$$

where the $f_i : R \to \mathbb{R}^n$ denote vector fields spanning an orthonormal frame of R around x. The two projection valued sections P^ρ and P^π are obviously curvature moderate. Moreover P^ρ_y and P^π_y commute, hence $P^\rho P^\pi$ comprises the projection onto the kernel bundle of $T(\pi, \rho)$ over $T_R \setminus R$. Furthermore, as P^ρ and P^π are curvature moderate, this holds for $P^\rho P^\pi$ as well. For the proof of the claim we thus only have to show that the projection valued section onto the kernel bundle of $P_S - P^\rho P^\pi$ is curvature moderate. To this end we will apply Lemma 3.3.6. As $P_S - P^\rho P^\pi$ is selfadjoint, it suffices by Lemma 3.3.6 to show that for $x \in R$ there exists a neighborhood U and a constant $\lambda > 0$ such that for every $y \in S \cap U$ and every eigenvector $v \in \left(\ker(P_{S,y} - P^\rho_y P^\pi_y) \right)^\perp$ the corresponding eigenvalue λ_v has absolute value $|\lambda_v| \geq \lambda$. We will show this in the following. Let us calculate:

$$\left(\ker(P_{S,y} - P^\rho_y P^\pi_y) \right)^\perp \subset \left(\operatorname{im} P_{S,y} \cap \operatorname{im} P^\rho_y P^\pi_y \right)^\perp = \ker P_{S,y} + \ker P^\rho_y + \ker P^\pi_y,$$

where the vector space sum is not direct in general; in the course of the following considerations it will turn out that instead of the relation $\subset$ even equality holds. We now estimate the norms $\|(P_{S,y} - P^\rho_y P^\pi_y)w\|$ for a vector $w \in \ker P_{S,y} + \ker P^\rho_y + \ker P^\pi_y$ from below. First let $w \in \ker P^\rho_y$. Then

$$w = \frac{\langle w, y - \pi(y) \rangle}{\|y - \pi(y)\|^2}(y - \pi(y)),$$

hence by (δ) for appropriate U and $\delta > 0$

$$\|(P_{S,y} - P^\rho_y P^\pi_y)w\| = \|P_{S,y}w\| = \frac{\|P_{S,y}(y - \pi(y))\|}{\|y - \pi(y)\|}\|w\| \geq \delta \|w\|. \qquad (3.4.3)$$

Next let $w \in \ker P_{S,y}$. We shrink U such that according to Lemma 3.4.5 (3) for a $\varepsilon > 0$ to be determined later the relation $\|Q^\pi_y v\| \leq \varepsilon \|v\|$ holds for all $y \in S \cap U$ and $v \in \ker P_{S,y}$. Then one calculates with Eq. (3.4.1) and relation $P^\pi_y(y - \pi(y)) = y - \pi(y)$ that

$$\begin{aligned} \|(P_{S,y} - P^\rho_y P^\pi_y)w\| = \|P^\rho_y P^\pi_y w\| &\geq \|P^\pi_y w\| - \frac{|\langle w, y - \pi(y) \rangle|}{\|y - \pi(y)\|} \\ &= \|P^\pi_y w\| - \frac{|\langle w, Q_{S,y}(y - \pi(y)) \rangle|}{\|y - \pi(y)\|} \\ &\geq \sqrt{\|w\|^2 - \|Q^\pi_y w\|^2} - \frac{\|Q_{S,y}(y - \pi(y))\|}{\|y - \pi(y)\|}\|w\| \\ &\geq \left(\sqrt{1 - \varepsilon^2} - \sqrt{1 - \delta^2} \right)\|w\| = \delta^\pi \|w\|^2. \end{aligned} \qquad (3.4.4)$$

At this point we determine $\varepsilon > 0$ such that $\delta^\pi := \sqrt{1-\varepsilon^2} - \sqrt{1-\delta^2} > 0$. Finally let $w \in \ker P^\pi_y$. By Lemma 3.4.5 (4) we can shrink U such that for the orthonormal frame $(f_1, \cdots, f_{\dim R})$ of R around x the estimate

$$\|Q_{S,y} f_i(\pi(y))\| \leq \frac{1}{2n}, \qquad y \in S \cap U,$$

holds. On the other hand by Eq. (3.4.2) $w = \sum_{i=1}^{\dim R} \langle w, f_i(\pi(y))\rangle\, f_i(\pi(y))$, hence

$$\begin{aligned} \|(P_{S,y} - P^\rho_y P^\pi_y) w\| = \|P_{S,y} w\| &\geq \|w\| - \sum_{i=1}^{\dim R} |\langle w, f_i(\pi(y))\rangle|\, \|Q_{S,y} f_i(\pi(y))\| \\ &\geq \|w\| \left(1 - \sum_{i=1}^{\dim R} \|Q_{S,y} f_i(\pi(y))\|\right) \geq \frac{1}{2} \|w\|. \end{aligned} \tag{3.4.5}$$

We now set $\lambda := \min\{\delta, \delta^\pi, \frac{1}{2}\}$. Our considerations so far now imply that the eigenvalue for the eigenvector $v \in \left(\ker(P_{S,y} - P^\rho_y P^\pi_y)\right)^\perp$ must have absolute value $|\lambda_v| \geq \lambda > 0$. This finishes the poof. □

The preceding proposition suggests the following definition.

3.4.7 Definition An (A)-stratified space X is called *strongly curvature moderate* of order m, if every stratum is curvature moderate of order m, and if for every pair $R < S$ of strata (after possibly shrinking the tubular neighborhoods) the submersion $(\pi_R, \rho_R)_{|S \cap T_R} : S \cap T_R \to R \times \mathbb{R}^{>0}$ is strongly curvature moderate of order m.

3.4.8 Example Subanalytic sets with their coarsest Whitney stratification are strongly curvature moderate of any order.

3.5 Existence and uniqueness theorems

J. MATHER has generalized the classical tubular neighborhood theorem in his notes [122]. In particular he proved far reaching theorems about the existence and uniqueness of tubular neighborhoods. In this section we will explain and proof the results of J. MATHER. Moreover, we will supplement these results by "curvature moderate versions". But before we will come to this let us provide some necessary terminology, which has been used in [122] as well.

3.5.1 As prerequisites assume to be given the same objects like in the preceding section: $m \in \mathbb{N} \cup \{\infty\}$, S is a submanifold of M; moreover let T, T_0, T_1 and so on always be tubular neighborhoods of S in M.

For every subset $U \subset S$ we denote by $T_{|U}$ the *restriction* of the tubular neighborhood T to U that means the triple $(E_{|U}, \varepsilon_{|U}, \varphi_{|E_{|U} \cap T^\varepsilon_{SCE}})$.

If $T = (E, \varepsilon, \varphi)$ and $\tilde{T} = (\tilde{E}, \tilde{\varepsilon}, \tilde{\varphi})$ are two tubular neighborhoods of S in M, then we understand by a *morphism* of tubular neighborhoods from T to $\tilde{T}$ (of *class* $\mathcal{C}^m$)

the pair (ψ, δ), where $\psi : E \rightarrow \tilde{E}$ is an isometric morphism of vector bundles (of class $\mathcal{C}^m$), $\delta : S \rightarrow \mathbb{R}^{>0}$ a $\mathcal{C}^m$-function such that $\delta \leq \min(\varepsilon, \tilde{\varepsilon})$ and

$$\tilde{\varphi} \circ \psi_{|T^\delta_{S \subset E}} = \varphi_{|T^\delta_{S \subset E}}.$$

In particular this implies the following two relations to hold:

$$\pi_{R|\varphi(T^\delta_{S\subset E})} = \tilde{\pi}_{R|\varphi(T^\delta_{S\subset E})},$$
$$\rho_{R|\varphi(T^\delta_{S\subset E})} = \tilde{\rho}_{R|\varphi(T^\delta_{S\subset E})}.$$

As the dimension of the fiber of E is equal to $\operatorname{codim}_M S$, the pair (ψ^{-1}, δ) is well-defined as well and comprises a morphism of tubular neighborhoods from $\tilde{T}$ to T. Therefore we say in this situation that T and $\tilde{T}$ are *isomorphic*, and denote it by the following symbols:

$$(\psi, \delta) : T \rightarrow \tilde{T}, \quad \psi : T \rightarrow \tilde{T}, \quad T \sim_{(\psi,\delta)} \tilde{T}, \quad \text{or briefly} \quad T \sim \tilde{T}.$$

By a $\mathcal{C}^m$-*isotopy* from M to N one understands a $\mathcal{C}^m$-homotopy $H : M \times [0,1] \rightarrow N$ such that all mappings $H_t : M \rightarrow N$, $t \in [0,1]$ are embeddings of class $\mathcal{C}^m$.

Let $S \subset M$ be a submanifold, $Z \subset \partial S$ be locally closed and T an open neighborhood of S such that S is closed in T. If $h : T \rightarrow M$ denotes an embedding which leaves S invariant, that means if $h_{|S} = \mathrm{id}_S$, then we say that h is *tempered relative* Z of *class* $\mathcal{C}^m$, if the following holds: there exists a covering of ∂S by sets V open in M together with $\mathcal{C}^m$-charts $x : U \rightarrow \mathbb{R}^n$ of M such that $V \cap T \subset h^{-1}(U)$ and such that the mappings $V \cap T \ni x \mapsto x(h(x))$ comprise functions which are tempered relative $Z \cap V$ of class $\mathcal{C}^m$. If $H : T \times [0,1] \rightarrow M$ is a $\mathcal{C}^m$- isotopy leaving S invariant, that means if $H_t(x) = x$ holds for all $x \in S$, then we say that H is *tempered relative* Z of *class* $\mathcal{C}^m$, if every embedding $H_t : T \rightarrow M$ is tempered relative Z of class $\mathcal{C}^m$.

Given an embedding $h : T \rightarrow M$ the *support* of h is the closure of the set of all points $x \in M$ with $h(x) \neq x$. Analogously we define the *support* of an isotopy $H : M \times [0,1] \rightarrow M$ as the closure of the set of all points $x \in M$ with $H(x,t) \neq x$ for some $t \in [0,1]$.

If finally $f : M \rightarrow N$ is a $\mathcal{C}^m$-mapping, then a $\mathcal{C}^m$-mapping $h : M \rightarrow M$ resp. a $\mathcal{C}^m$-homotopy $H : M \times [0,1] \rightarrow M$ is called *compatible* with f, if $f \circ h = f$ resp. $f \circ H_t = f$ holds for all $t \in [0,1]$.

Now we have all ingredients needed for the two main theorems about tubular neighborhoods.

3.5.2 Uniqueness of tubular neighborhoods *Let $m \in \mathbb{N}^{>0}$, $S \subset M$ be a submanifold of M of class $\mathcal{C}^{m+2}$ and $f : M \rightarrow N$ a $\mathcal{C}^{m+2}$-mapping such that the restriction $f_{|S} : S \rightarrow N$ is submersive. Further let T_0 and T_1 be two tubular neighborhoods of S in M of class $\mathcal{C}^m$ which are compatible with f and let $\tilde{\psi} : T_{0|U} \rightarrow T_{1|U}$ be an isomorphism of tubular neighborhoods of class $\mathcal{C}^m$ over the open subset $U \subset S$. Finally let $A, Z \subset S$ with $A \subset U$ be two relatively S closed subsets and V an open neighborhood of Z in M. Then the following statements hold:*

(TU1) *There exists an open neighborhood* $\mathsf{T} \subset \mathsf{M}$ *of* S *with* S *closed in* T *and a* $\mathcal{C}^m$*-isotopy* $\mathsf{H} : \mathsf{T} \times [0,1] \to \mathsf{M}$ *which leaves* S *invariant, is compatible with* f *and has support in* V *such that the tubular neighborhoods* $\mathsf{h}^* (\mathsf{T}_{0|A\cup Z})$ *and* $\mathsf{T}_{1|A\cup Z}$ *are isomorphic. Hereby* h *is the embedding* H_1. *The isomorphism* $\psi : \mathsf{h}_* (\mathsf{T}_{0|A\cup Z}) \to \mathsf{T}_{1|A\cup Z}$ *can be constructed such that* $\psi_{|A} = \tilde{\psi}_{|A}$ *is satisfied.* (*cf.* MATHER [122, Prop. 6.1])

(TU2) *Under the assumption that* V *is a regularly situated neighborhood of* Z, *that* S *and the two tubular neighborhoods* T_i, $i = 0, 1$ *are curvature moderate of order* m *and finally that* f *is curvature moderate of order* $m + 1$ *the isotopy* H *can be constructed such that* H *is tempered relative* $\partial \mathsf{S}$ *of class* $\mathcal{C}^m$.

PROOF: We will prove the claim in two steps. In the first one we show the properties (TU1) and (TU2) for the local case that means under the assumption $\mathsf{M} = \mathbb{R}^n$. In the second step we will reduce the general case to the local one.

1. STEP Let $\mathsf{M} = \mathbb{R}^n$ and $\mathsf{S} \subset \mathbb{R}^n$ be a submanifold. Further let π_i, $i = 0, 1$ be the projection of T_i and π the projection of the tubular neighborhood $\mathsf{T} = (\mathsf{E}, \varepsilon, \varphi)$ of S induced by the Euclidean scalar product. Denote by P_i and P the projection valued section associated to the tubular neighborhoods T_i and T. Vector bundle isomorphisms from E_0 resp. E_1 to E are now given by

$$\xi_{i,x} : \mathsf{E}_{i,x} \to \mathsf{E}_x, \quad v \mapsto \mathsf{P}_x . T\varphi_i . v, \qquad x \in \mathsf{S}.$$

Let $\xi : \mathsf{E}_0 \to \mathsf{E}_1$ be the isomorphism of vector bundles of class $\mathcal{C}^{m-1}$ constructed by composition of ξ_0 and ξ_1^{-1}. Over U the isomorphism ξ coincides by assumption with $\tilde{\psi}$: $\xi_{|\mathsf{U}} = \tilde{\psi}$. By changing ξ outside a neighborhood $\mathsf{U}' \subset \mathsf{U}$ of A appropriately we can achieve that ξ is even a $\mathcal{C}^m$-vector bundle isomorphism. By the polar decomposition A.2.1 there exists for every $x \in \mathsf{S}$ a unique positive definite operator $\zeta_x : \mathsf{E}_{1,x} \to \mathsf{E}_{1,x}$ such that $\psi_x := \zeta_x^{-1} \circ \xi_x : \mathsf{E}_{0,x} \to \mathsf{E}_{1,x}$ is unitary, hence $\psi : \mathsf{E}_0 \to \mathsf{E}_1$ is an isomorphism of vector bundles with scalar product. Furthermore the bundle map $(1-t)\xi + t\psi$: $\mathsf{E}_0 \to \mathsf{E}_1$ is well-defined for $t \in [0,1]$ and comprises an isomorphism of vector bundles which coincides over U with $\tilde{\psi}$. But this implies the existence of an open neighborhood T of S (in which S is closed) such that for every $t \in [0,1]$

$$\mathsf{G}_t : \mathsf{T} \to \mathbb{R}^n, \quad x \mapsto \left(\varphi_1 \circ ((1-t)\xi + t\psi) \circ \varphi_0^{-1}\right)(x)$$

is well-defined and comprises a diffeomorphism onto its image. Obviously then

$$\mathsf{G}_{t|\mathsf{S}} = \mathrm{id}_\mathsf{S} \quad \text{and} \quad \mathsf{G}_{t|\mathsf{U}'} = \mathrm{id}_{\mathsf{U}'}$$

holds, where now U' is chosen to be of the form $\mathsf{U}' := \pi_0^{-1}(A) \cap \mathsf{T}_0^\delta$ and $\delta : \mathsf{S} \to \mathbb{R}^{>0}$ is an appropriate $\mathcal{C}^m$-function with $\mathsf{T}_0^\delta \subset \mathsf{T}$. As the tubular neighborhoods T_i are assumed to be compatible with f, we have

$$\mathsf{f}(\mathsf{G}_t(x)) = \mathsf{f}(x), \qquad x \in \mathsf{T}. \tag{3.5.1}$$

After possibly shrinking T we can achieve that G_0 maps the neighborhood T diffeomorphically onto an open neighborhood $\tilde{\mathsf{T}}$ of S. Then the mapping

$$\mathsf{G}_{t,0} : \tilde{\mathsf{T}} \to \mathbb{R}^n, \; x \mapsto \mathsf{G}_t(\mathsf{G}_0^{-1}(x))$$

is well-defined and a diffeomorphism onto its image. By construction $G_{t,0}$ has the following shape:

$$G_{t,0}(x) = \big(\varphi_1 \circ ((1-t)\mathrm{id} + t\zeta^{-1}) \circ \varphi_1^{-1}\big)(x), \qquad x \in \tilde{T}.$$

As ζ_x is positive definite, the derivative $DG_{t,0}(x)$ is for every $x \in S$ diagonalizable with only positive eigenvalues. For later purposes we will keep this fact in mind.

Next consider the projection valued $\mathcal{C}^{m+1}$-mapping $P^f : \tilde{T} \to \mathrm{End}(\mathbb{R}^n)$ which we obtain according to Example 3.3.5 out of f. Set $Q^f_x = \mathrm{id}_{\mathbb{R}^n} - P^f_x$, $x \in \tilde{T}$. Then

$$\nabla_v w = P^f(D(P^f w).v) + Q^f(D(Q^f w).v), \qquad v, w \in \mathcal{X}^{m+1}(\tilde{T}) \tag{3.5.2}$$

defines a connection ∇ on T. Let exp be the corresponding exponential function. By definition of ∇ the map exp is of class $\mathcal{C}^m$ and

$$f(\exp tv) = f(x) \tag{3.5.3}$$

must hold for all $t \in [0,1]$ and sufficiently small $v \in \ker T_x f = P^f_x \mathbb{R}^n$, $x \in \tilde{T}$.

After possibly shrinking T we can suppose that $\exp_x^{-1}(y)$ is defined for all $x \in S$ and $y \in T \cap f^{-1}(f(x))$ and that the geodesic $\gamma(t) = \exp(t \exp_x^{-1}(y))$ lies in $\tilde{T}$. Now choose a smooth function $\kappa : \tilde{T} \to [0,1]$, with the following properties. For all x of a sufficiently small open neighborhood of S in $\tilde{T}$ let $\kappa(x) = \kappa(\pi_1(x))$. The support of κ is contained in V, and over an open neighborhood $V' \subset V$ of Z the equality $\kappa_{|V'} = 1$ holds. Then define:

$$F_{t,0}(x) = \exp\big(\kappa(x) \exp_x^{-1}(G_{t,0}(x))\big), \qquad x \in \tilde{T},\ t \in [0,1]. \tag{3.5.4}$$

From now on we have to consider both cases of the claim separately.

ad (TU1): In this case we show first that after appropriately shrinking T and $\tilde{T}$ the mappings $F_{t,0} : \tilde{T} \to \mathbb{R}^n$ comprise diffeomorphisms onto their image. To this end it suffices by Lemma 3.1.7 to prove that for every $x \in S$ the derivatives $DF_{t,0}(x)$ are bijective. As $F_{t,0|S} = \mathrm{id}_S$ the derivative $DF_{t,0}(x)$ acts identically on tangent vectors $v \in T_x S$. If on the other hand $v \in \ker T_x \pi_1$, then

$$DF_{t,0}(x).v = (1 - \kappa(x))\, v + \kappa(x) DG_{t,0}(x).v$$

As mentioned above, the operator $DG_{t,0}(x)$ has only positive eigenvalues, hence the first derivative is invertible. After further shrinking T finally

$$H : T \times [0,1] \to M,\ (x,t) \mapsto (F_{t,0} \circ G_0)(x) \tag{3.5.6}$$

is well-defined and comprises a $\mathcal{C}^m$-isotopy from T to M which is compatible with f. Hereby, the compatibility of H with f follows from (3.5.1) and (3.5.3). The support of κ lies in V, hence $\operatorname{supp} H \subset V$ has to be true. As κ can be chosen such that $\kappa_{|V'} = 1$, the relation $H_t(x) = G_t(x)$ is true for $x \in V'$ after possible shrinking the neighborhood V' of Z, hence

$$\varphi_1^{-1} \circ h\,(x) = \varphi_1^{-1} \circ G_1\,(x) = \psi \circ \varphi_0^{-1}\,(x), \qquad x \in V'.$$

On the other hand, we have $G_t(x) = x$ over the set U' defined above, hence

$$H_t(x) = x \quad \text{and} \quad \varphi_1^{-1} \circ h(x) = \varphi_1^{-1}(x) = \psi \circ \varphi_0^{-1}(x), \qquad x \in U'.$$

Altogether, we thus obtain that $\psi_{|A\cup Z} : h_*(T_0)_{|A\cup Z} \to T_{1|A\cup Z}$ is an isomorphism of tubular neighborhoods. By construction $\psi_{|A} = \tilde{\psi}$, which is the last part of (TU1) for the case $M = \mathbb{R}^n$.

ad (TU2): We consider the just constructed isotopy H and will show that the mappings $G_{t,s}$, κ and $\exp$ involved in the construction are tempered relative ∂S of class $\mathcal{C}^m$. By the representation (3.5.6) of the isotopy H the claim that H is tempered relative ∂S of class $\mathcal{C}^m$ then follows immediately.

The Christoffel symbols $\Gamma_{ij}^k = \langle e_k, \nabla_{e_i} e_j \rangle$ are tempered relative ∂S of class $\mathcal{C}^m$, because P^f is curvature moderate of order $m+1$ and because the vector fields $\nabla_{e_i} e_j$ are defined by Equation (3.5.2). By Proposition 3.3.10 one can achieve after shrinking T that the functions $\exp$ and $\exp^{-1}$ are tempered relative ∂S of class $\mathcal{C}^m$. Furthermore it is easy to check using Lemma 1.7.10 that the function κ can be constructed such that it is tempered relative ∂S of class $\mathcal{C}^m$. Note that in all of these constructions T always remains a regularly situated neighborhood of S.

By $H(x,t) = F_{t,0}(G_0(x))$ and (3.5.4) it therefore remains to show for the proof of the claim that the components of the embeddings $G_0 : T \to M$ and $G_{t,0} : \tilde{T} \to T$ are tempered relative ∂S of class $\mathcal{C}^m$. To achieve this let us represent the vector bundle isomorphisms $\xi : E_0 \to E_1$ and $\zeta : E_1 \to E_1$ in matrix form with the help of local orthonormal frames. From this representation one can immediately read off the temperedness of the matrix components, hence of G_0 and $G_{t,0}$. Without loss of generality we can identify each one of the vector bundles E_i, $i = 0, 1$, with a subvector bundle of $T_{|S}M$ normal to TS; hereby the identification is given like in 3.1.8 (1) by the restricted tangential map $T_{|E_i}\varphi_i : E_i \to TM$. Next we fix two Riemannian metrics over T:

$$\mu_{i,x}(v,w) = \eta_{i,\pi_i(x)}(T_x\varphi_i^{-1}.Q_{i,x}.v, T_x\varphi_i^{-1}.Q_{i,x}.w) + \langle P_{i,x}.v, P_{i,x}.w \rangle, \quad x \in T, v, w \in \mathbb{R}^n, \tag{3.5.7}$$

where η_i denotes the scalar product on E_i and $Q_{i,x}$ the projection $\mathrm{id}_{\mathbb{R}^n} - P_{i,x}$. Additionally let us set $\mu_E = \langle \cdot, \cdot \rangle$.

Now choose for any point of ∂S a ball B with sufficiently small radius < 1. According to Proposition 3.3.12 the three Riemannian metrics μ_i, $i \in \{0, 1, E\}$ then generate locally finite countable families $(\phi_j)_{j\in J}$ and $(f_{jl}^i)_{j\in J, 1\le l\le n}$ consisting of functions resp. sections over T such that (ϕ_j), (f_{jl}^i) satisfy 3.3.12 (1) to (5). In particular every n-tuple $(f_{j1}^i, \cdots, f_{jn}^i)$ thus comprises for every fixed j a μ_i-orthogonal frame over the set $\phi_j \cap B \cap T$. We will construct with respect to this orthogonal frame matrix valued mappings $a_j^i : T \to \mathrm{End}(\mathbb{R}^n)$, $i = 0, 1$, which represent the vector bundle isomorphisms ξ_i over $\operatorname{supp} \phi_j \cap B \cap T$. More precisely we set for $x \in \operatorname{supp} \phi_j \cap B \cap T$

$$a_{j,kl}^i(x) = \frac{\left\langle f_{j(k+\dim S)}^E(x), f_{j(l+\dim S)}^i(x) \right\rangle}{\|f_{j(k+\dim S)}^E(x)\|_{\mu_E} \, \|f_{j(l+\dim S)}^i(x)\|_{\mu_i}}, \qquad k, l = 1, \cdots, n - \dim S,$$

and extend the thus defined matrix valued functions a_j^i to $\mathcal{C}^m$-functions defined on the tubular neighborhood T. By 3.3.12 (4) and because the Riemannian metrics μ_0, μ_1

and μ_E are curvature moderate of order m by (CM7), there exist constants $d, d_m \in \mathbb{N}$ and $D, D_m > 0$ such that for all $j \in J$, $i = 0, 1$

$$\frac{d(x,\partial S)^d}{D} \le \|a_j^i(x)\| \le D\left(1 + \frac{1}{d(x,\partial S)^d}\right), \qquad x \in \operatorname{supp}\phi_j \cap B \cap T,$$
$$\|\partial^\alpha a_j^i(x)\| \le D_m\left(1 + \frac{1}{d(x,\partial S)^{d_m}}\right), \qquad |\alpha| \le m. \tag{3.5.8}$$

hence one can construct $\mathcal{C}^m$-functions $\overline{a}_i^j : T \to \operatorname{End}(\mathbb{R}^n)$ with the property that $\overline{a}_i^j(x)$ is inverse to a_j^i for $x \in \operatorname{supp}\phi_j \cap B \cap T$ and such that after possibly enlarging the constants d, d_m, D, D_m the following estimates are true:

$$\frac{d(x,\partial S)^d}{D} \le \|\overline{a}_j^i(x)\| \le D\left(1 + \frac{1}{d(x,\partial S)^d}\right), \qquad x \in \operatorname{supp}\phi_j \cap B \cap T,$$
$$\|\partial^\alpha \overline{a}_j^i(x)\| \le D_m\left(1 + \frac{1}{d(x,\partial S)^{d_m}}\right), \qquad |\alpha| \le m. \tag{3.5.9}$$

Now we can write the vector bundle isomorphism ξ in the following form using the a_j^i and $\overline{a}_j^i$ and where $v \in E_{0,x}$ and $x \in S$:

$$\xi_x v = \sum_{j\in J} \phi_j(x) \sum_{l,m,k=0}^{n-\dim S} \overline{a}^1_{j,lm}(x)\, a^0_{j,mk}(x) \frac{\mu_0\big(v, f^0_{j(k+\dim S)}(x)\big)}{\|f^0_{j(k+\dim S)}(x)\|_{\mu_0}} \frac{f^1_{j(l+\dim S)}(x)}{\|f^1_{j(l+\dim S)}(x)\|_{\mu_1}}. \tag{3.5.10}$$

This representation, Proposition 3.3.12 and the estimates (3.5.8) and (3.5.9) entail that for $k = 1, \cdots, n$ the functions $T \ni x \mapsto \xi_{\pi_0(x)} Q_{0,\pi_0(x)} e_k \in \mathbb{R}^n$ are tempered relative ∂S of class $\mathcal{C}^m$. Recall the fact (see 3.1.8 (3)) that the map φ_i is given by the exponential function of the Levi–Civita connection of μ_i. Using Proposition 3.3.10 it then follows that

$$G_1 : T \to \mathbb{R}^n, \quad x \mapsto \varphi_1 \circ \xi_{\pi_0(x)} \circ \varphi_0^{-1}(x)$$

is tempered relative ∂S.

Next we have to prove the temperedness relative ∂S of the functions

$$T \to \mathbb{R}^n, \quad x \mapsto \zeta_{\pi_1(x)} Q_{1,\pi_1(x)} e_k.$$

By the polar decomposition A.2.1 the relation $\zeta_x = \sqrt{\xi_x \xi_x^*}$, $x \in S$, holds hence by the analyticity of the square root

$$\big\|\partial_l\big(\zeta_{\pi_1(\cdot)} Q_{1,\pi_1(\cdot)} e_k\big)(x)\big\| \le \frac{1}{2} \frac{\big\|\partial_l\big(\xi_{\pi_1(\cdot)} Q_{1,\pi_1(\cdot)} e_k\big)(x)\big\|}{\|\zeta_{\pi_1(x)}\|}, \qquad k, l = 1, \cdots, n. \tag{3.5.11}$$

A similar estimate follows for every higher partial derivative. More precisely $\partial^\alpha\big(\zeta_{\pi_1(\cdot)} Q_{1,\pi_1(\cdot)} e_k\big)(x)$ is given by a sum of quotients which contain in the nominator higher partial derivatives of $\xi_{\pi_1(\cdot)} Q_{1,\pi_1(\cdot)}$ (multiplied by combinatorial factors) and in the denominator powers of $\|\zeta_{\pi_1(x)}\|$, but both only up to highest order $|\alpha|$. By (3.5.8), (3.5.9) and the representation (3.5.10) there exist on the other hand constants $d' \in \mathbb{N}$ and $D' > 0$ with

$$\frac{d(x,\partial S)^{d'}}{D'} \le \|\zeta_x\| \le D'\left(1 + \frac{1}{d(x,\partial S)^{d'}}\right), \qquad x \in T.$$

Together with 3.5.11 and the temperedness of $\xi_{\pi_1(\cdot)}Q_{1,\pi_0(\cdot)}$ this inequality entails that functions on T given by $\zeta_{\pi_1(\cdot)}Q_{1,\pi_1(\cdot)}e_k$ and $\zeta^{-1}_{\pi_1(\cdot)}Q_{1,\pi_1(\cdot)}e_k$ have to be tempered relative ∂S. Exactly like for G_1 one thus concludes that

$$G_{t,0}(x) = (\varphi_1 \circ ((1-t)\mathrm{id} + t\zeta^{-1}) \circ \varphi_1^{-1})(x), \qquad x \in T,$$

is tempered relative ∂S of class $\mathcal{C}^m$. This finishes the proof of (TU2) in the local case.
2. STEP The proof of the general case will be carried out by first embedding the manifold M as a closed submanifold in Euclidean space $\mathbb{R}^n$ and then applying the just proven case. More precisely one considers the tubular neighborhood $T_M = (E_M, \varepsilon_M, \varphi_M)$ of M in $\mathbb{R}^n$ induced by the Euclidean scalar product. Then the mapping f can be extended canonically to a map on T_M such that $f(\pi_M(x)) = f(x)$ for all $x \in T_M$. Furthermore we then switch from the vector bundles E_i, $i = 0, 1$ to $E_i' := E_i \oplus E_{M|S}$ and define, after possibly changing the ε_i, embeddings $\varphi_i' : T^{\varepsilon_i}_{S\subset E_i'} \to \mathbb{R}^n$ by $\varphi_i'(v, w) = w + \varphi_i(v)$. Then we obtain two tubular neighborhoods $T_i' = (E_i', \varepsilon_i, \varphi_i')$ of S in $\mathbb{R}^n$. Moreover, $\tilde{\psi}$ induces a vector bundle isomorphism $\tilde{\psi}' : E'_{0|U} \to E'_{1|U}$ by $(v, w) \mapsto (\tilde{\psi}(v), w)$, and $\tilde{\psi}' : T'_{0|U} \to T'_{1|U}$ becomes an isomorphism of tubular neighborhoods. Collecting all these data one recognizes that by T_0', T_1' and f the assumptions for the local case in the 1. Step are satisfied. Moreover, this holds for the curvature moderate case (TU2) as well. Firstly we thus obtain a unitary vector bundle isomorphism $\psi' : E_0' \to E_1'$, where it is clear by construction that the restriction of ψ' to $E_{M|S}$ is equal to the identity map, and that ψ' maps the bundle E_0 to E_1. Secondly we obtain an isotopy $H' : T' \times [0,1] \to \mathbb{R}^n$, the components of which are defined by (3.5.4) and where $T' \subset T_M$ is an appropriate tubular neighborhood of S in $\mathbb{R}^n$. Obviously, the function κ contained in (3.5.4) can be chosen such that $\kappa(\pi_M(x)) = \kappa(x)$ for all $x \in \tilde{T}'$, where $\tilde{T}' \subset T_M$ is a further tubular neighborhood of S in $\mathbb{R}^n$. The exponential function appearing in (3.5.4) as well will be defined with respect to the following connection:

$$(\nabla_v w)(x) = P^f_y(D(P^f_y w).v) + Q^f_y(D(Q^f_y w).v) + Q_{M,y}(D(Q_{M,y} w).v),$$
$$v, w \in \mathfrak{X}^{m+1}(\tilde{T}'),\ x \in \tilde{T},\ y = \pi_M(x).$$

Hereby Q_M means the projection onto the vertical bundle of the tubular neighborhood T_M, P^f_x for $x \in M$ means the Euclidean projection from T_xM onto $\ker T_x f$ and Q^f_x the endomorphism $\mathrm{id}_{T_xM} - P^f_x$. This definition of ∇ guarantees that $\exp v$ lies in M for $v \in T_xM$ and that $\exp_x^{-1}(y) \in T_xM$ holds for $x, y \in M$ sufficiently close. In the curvature moderate case the Christoffel symbols of ∇ are curvature moderate. Now one concludes by the definition of the tubular neighborhoods T_i' and the shown properties of ξ', κ and H' that the relation $\pi_M(H'(x,t)) = H'(\pi_M(x,t))$ is true for all $x \in T'$ and $t \in [0,1]$. At this point we can fix the "unslashed" objects: the vector bundle isomorphism $\psi : E_0 \to E_1$ is given by $\psi(v) = \pi_{E_1}(\psi'(v, 0))$, $v \in E_0$, where $\pi_{E_1} : E_1' = E_1 \oplus E_{M|S} \to E_1$ is the canonical projection. The isotopy $H : T \times [0,1] \to M$ is obtained by $T := T' \cap M$ and $H(x,t) := H'(x,t)$ for $x \in T$, $t \in [0,1]$. By construction ψ and H satisfy (TU1). All constructions in the 2. Step do not influence a possible curvature moderate behavior, hence in the curvature moderate case (TU2) holds as well. This finishes the proof. □

3.5.3 Existence of tubular neighborhoods *Let $m \in \mathbb{N}^{>0}$, S be a submanifold of M of class $\mathcal{C}^{m+2}$ and $f : M \to N$ a $\mathcal{C}^{m+2}$-mapping submersive over S. Further let $U \subset S$ be open relative S, $A \subset U$ be closed relative S, and T_0 a tubular neighborhood of U in M compatible with $f_{|U}$. Under these prerequisites the following statements hold:*

(TU3) *There exists a tubular neighborhood T of S compatible with f such that $T_{|A} \sim T_{0|A}$. If $\overline{\pi_0^{-1}(U')} \cap S = U'$ for a neighborhood $U' \subset U$ of A which is closed in S, then T and the isomorphism $(\psi, \delta) : T_{|A} \to T_{0|A}$ can be chosen such that $\varepsilon_{|A} = \varepsilon_{0|A} = \delta$. (cf. MATHER [122, Prop. 6.2])*

(TU4) *If T_0 is a regularly situated neighborhood of A, if both S and T_0 are curvature moderate of order $m+1$ and if f is a submersion strongly curvature moderate of order $m+1$, then the tubular neighborhood T can be chosen to be curvature moderate of order m.*

PROOF: ad (TU3): For the first part of the claim choose an open neighborhood T of S in M such that S is closed in T and choose an arbitrary Riemannian metric θ of class $\mathcal{C}^{m+1}$ on M. After restricting T appropriately one can achieve that the projection valued mapping $P^f : T \to \mathrm{End}(TM)$ onto the kernel bundle of Tf is of class $\mathcal{C}^{m+1}$; hereby the projection P^f_x is assumed to be selfadjoint with respect to θ_x. By the tubular neighborhood $T^\theta = (E^\theta, \varepsilon^\theta, \varphi^\theta)$ of S induced by θ one obtains a further projection valued $\mathcal{C}^{m+1}$-mapping $P^\theta : T \to \mathrm{End}(TM)$ (after possibly shrinking T again); this mapping assigns to every $x \in T$ the θ-orthogonal projection onto the horizontal space of T^θ. The last projection valued section is given by $P^{f,\theta} : T \to \mathrm{End}(TM)$, the θ-orthogonal projection onto the kernel bundle of $P^f - P^\theta$. After shrinking T further $P^{f,\theta}$ is well-defined and of class $\mathcal{C}^{m+1}$. If one now sets $Q^{f,\theta}_x = P^f_x - P^{f,\theta}_x$ then the image of $Q^{f,\theta}_x$ is complementary in T_xM to $\mathrm{im}\, P^\theta_x$, as $f_{|S} : S \to N$ is submersive. Hence there exists for every $x \in T$ a positive definite mapping $A_x \in \mathrm{End}(TM)$ such that $A_{x|\mathrm{im}\, P^\theta_x} = \mathrm{id}_{|\mathrm{im}\, P^\theta_x}$ and $A_x(\mathrm{im}\, Q^{f,\theta}_x) = \mathrm{im}\, Q^\theta_x$, where $Q^\theta_x = \mathrm{id}_{T_xM} - P^\theta_x$. Obviously, $A : T \to \mathrm{End}(TM)$ can be chosen as $\mathcal{C}^m$-section. By the following one obtains a second Riemannian metric μ on T:

$$\mu_x(v, w) = \theta(A_x v, A_x w) \qquad v, w \in T_xM, \quad x \in T.$$

With respect to μ the spaces $\mathrm{im}\, P^\theta_x$ and $\mathrm{im}\, Q^{f,\theta}_x$ are orthogonal to each other. By $N \to S$ we now understand the μ-orthogonal bundle of TS in $T_{|S}M$, by $\exp$ the exponential function of the Levi–Civita connection of μ, and by $T^{\max}_{SCN}$ the neighborhood of the zero section of N as defined in Section 3.2. Hence by definition of μ for every vector $v \in N_x \cap T^{\max}_{SCN}$, $x \in S$ the tangent vector $\dot\gamma(t)$ to the curve $\gamma(t) = \exp(tv)$, $t \in [0,1]$ lies in $Q^{f,\theta}_{\gamma(t)} T_{\gamma(t)}M$. Therefore

$$f(\exp(tv)) = f(x), \qquad v \in N_x \cap T^{\max}_{SCN},\ x \in S,$$

holds for all $t \in [0,1]$. Consequently the triple $\check{T} = (N, \varepsilon^{\max}, \varphi)$ comprises a tubular neighborhood of S in M which is compatible with f, where φ is defined like in 3.2 as the restriction of $\exp_{|T^{\max}_{SCN}}$. Let $V \subset \check{T} := \exp(T^{\max}_{SCN})$ be an open neighborhood of A

such that V lies in $\pi_0^{-1}(U) \cap \check{T}$. By the uniqueness theorem for tubular neighborhoods there exists after possibly shrinking $\check{T}$ an embedding $h : \check{T} \to M$ with support in V such that $h_{|S} = \mathrm{id}_{|S}$ and $(h_*\check{T})_{|A} \sim T_{0|A}$. Hence $T := h_*\check{T}$ satisfies the first part of the claim in (TU3).

Now we come to the second part. Under the assumption that $U' \subset U$ is a closed neighborhood of A in S satisfying the claim $\overline{\pi_0^{-1}(U')} \cap S = U'$ choose further closed neighborhoods V, W and W' of A in S with

$$A \subset V^\circ \subset V \subset U'^\circ \subset U' \subset W^\circ \subset W \subset W'^\circ \subset W' \subset U.$$

According to the first there exists a tubular neighborhood $\tilde{T} = (\tilde{E}, \tilde{\varepsilon}, \tilde{\varphi})$ of S in M and an isomorphism $\tilde{T}_{|W'} \sim_{(\psi,\tilde{\delta})} T_{0|W'}$. By the assumption on U' one finds an open neighborhood O of $S \setminus U'$ in M with $\pi_0^{-1}(V) \cap O = \emptyset$; for instance set $O := M \setminus \overline{\pi_0^{-1}(U')}$. By assumption on the neighborhoods V, U', W and W' it is possible to construct a $\mathcal{C}^m$-mapping $\varepsilon : S \to \mathbb{R}^{>0}$ such that

$$\varepsilon_{|U} \le \varepsilon_0, \quad \varepsilon_{|A} = \varepsilon_{0|A}, \quad \varepsilon_{|W\setminus V} \le \tilde{\delta}_{|W\setminus V}, \quad \varepsilon_{|S\setminus V} \le \tilde{\varepsilon}_{|S\setminus V} \quad \text{and } \tilde{T}^{\varepsilon}_{S\setminus W} \subset O.$$

Now we define $\varphi : T^{\varepsilon}_{S\subset\tilde{E}} \to M$ as the mapping which over the set $\pi_{\tilde{E}}^{-1}(U') \cap T^{\varepsilon}_{S\subset\tilde{E}}$ is equal to $\varphi_0 \circ \psi$ and which coincides over $\pi_{\tilde{E}}^{-1}(S \setminus V) \cap T^{\varepsilon}_{S\subset\tilde{E}}$ with the restriction of $\tilde{\varphi}$. By construction one checks easily that this φ is well-defined and injective. Hence the triple $T = (\tilde{E}, \varepsilon, \varphi)$ together with the isomorphism $(\psi_{|A}, \varepsilon_{|A})$ fulfills the second part of the claim (TU3).

ad (TU4): As soon as the assumptions in (TU4) are given, the projection valued sections P^θ (as S, θ are curvature moderate) and $P^{f,\theta}$ (as f is strongly curvature moderate) have to be curvature moderate of order $m+1$; in particular the common domain T of P^θ and $P^{f,\theta}$ is a regularly situated neighborhood of S. Moreover, as θ is a Riemannian metric on M, the Riemannian metric μ has to be curvature moderate of order $m+1$. Consequently the tubular neighborhood T induced by μ is curvature moderate of order m. By the fact that T_0 is a regularly situated neighborhood of A and by (TU2) in the uniqueness theorem one can achieve that the above chosen embedding $h : T \to M$ is tempered relative ∂S of class $\mathcal{C}^m$. Hence $T = h_*\check{T}$ is curvature moderate of order m, thus satisfies in the curvature moderate case (TU4). This proves the claim. □

3.6 Tubes and control data

3.6.1 Let $(X, \mathcal{S})$ be a locally compact stratified space, and $S \in \mathcal{S}$ one of its strata. By a *tube* of S compatible with $\mathcal{S}$ we understand a triple $T_S = (T_S, \pi_S, \rho_S)$ satisfying the following conditions:

(TB1) T_S is an open neighborhood of S in X such that for every other stratum $R \in \mathcal{S}$ the relation $T_{S,R} := T_S \cap R \neq \emptyset$ implies $R \ge S$.

(TB2) $\pi_S : T_S \to S$ is a continuous retraction of S such that for every stratum $R > S$ the restriction $\pi_{S,R} := \pi_{S|T_{S,R}} : T_{S,R} \to S$ is smooth.

(TB3) $\rho_S : T_S \to \mathbb{R}^{\geq 0}$ is a continuous mapping such that $\rho_S^{-1}(0) = S$ is satisfied and such that for every stratum $R > S$ the restriction $\rho_{S,R} := \rho_{S|T_{S,R}} : T_{S,R} \to \mathbb{R}^{\geq 0}$ is smooth.

(TB4) The mapping $(\pi_{S,R}, \rho_{S,R}) : T_{S,R} \to S \times \mathbb{R}^{>0}$ is a submersion for every pair of strata $R > S$.

3.6.2 Lemma *If the stratum* S *possesses a tube compatible with* $\mathcal{S}$, *then* $\pi_{S,R} : T_{S,R} \to S$ *is surjective and* $\dim R > \dim S$ *for all* $R > S$.

PROOF: That $\pi_{S,R}$ is surjective follows easily from the fact that $S \subset \overline{R}$ and (TB1). The statement about the dimension is an immediate consequence of (TB4). □

3.6.3 Example Let X be a locally compact stratified subspace of a manifold M such that X inherits from M a $\mathcal{C}^m$-structure and let T_S be a tubular neighborhood of the stratum S of class $\mathcal{C}^m$. In case X is Whitney stratified the triple (T_S, π_S, ρ_S) defines according to Proposition 3.4.1, after shrinking T_S, a tube of S in X, which we denote by the letter T. In this case we say that the tube is of *class* $\mathcal{C}^m$ and that it is *induced* by a tubular neighborhood or that it is *normal* of *class* $\mathcal{C}^m$. Using the symbol T both for the tubular neighborhood of S in M and the tube of S in X does in general not lead to any confusion but rather simplifies notation.

If the stratified space $(X, \mathcal{S})$ carries a smooth structure $\mathcal{C}^\infty$, then we call a tube $T_S = (T_S, \pi_S, \rho_S)$ *normal* of *class* $\mathcal{C}^m$, if the following axiom is satisfied:

(TB5) There exists a covering of S by singular charts $x : U \to \mathbb{R}^n$ of X such that π_S and ρ_S are induced by tubular neighborhoods T^x of $x(S \cap U)$ in $\mathbb{R}^n$ of class $\mathcal{C}^m$ that means

$$\pi^x \circ x = x \circ \pi_{S|U},$$
$$\rho^x \circ x = \rho_{S|U},$$

where π^x is the projection and ρ^x the tubular function of T^x.

In this case the functions π_S and ρ_S are $\mathcal{C}^m$-mappings.

Assuming that T_S is a normal tube of S of class $\mathcal{C}^m$ we call T_S *curvature moderate* of *order* m, if the following holds:

(TB6) There exists a covering of S by singular charts $x : U \to \mathbb{R}^n$ of X such that T_S is induced by tubular neighborhoods T^x of $x(S \cap U)$ in $\mathbb{R}^n$ and such that every T^x is curvature moderate of order m.

Two tubes T_S and $\tilde{T}_S$ of S are called *equivalent* over the set $U \subset S$, in symbols $T_{|U} \sim \tilde{T}_{|U}$, if the following holds:

(TB7) There exists a neighborhood $T'_U \subset T_S \cap \tilde{T}_S$ of U, such that

$$\pi_{S|T'_U} = \tilde{\pi}_{S|T'_U},$$
$$\rho_{S|T'_U} = \tilde{\rho}_{S|T'_U}.$$

One checks easily that the equivalence of tubes is an equivalence relation on the set of all tubes of S in X indeed.

3.6.4 *Control data* for a stratified space $(X, \mathcal{S})$ consist of a family $(T_S)_{S \in \mathcal{S}}$ of tubes $T_S = (T_S, \pi_S, \rho_S)$ such that for every pair of strata $R > S$ and all $x \in T_S \cap T_R$ with $\pi_R(x) \in T_S$ the following *control conditions*

(CT1) $$\pi_S \circ \pi_R(x) = \pi_S(x),$$

(CT2) $$\rho_S \circ \pi_R(x) = \rho_S(x)$$

are satisfied. A stratified space on which some control data exist is called a *controllable* space.

If $(X, \mathcal{S})$ carries additionally a $\mathcal{C}^m$-structure and if all tubes of the control data are normal, then $(T_S)_{S \in \mathcal{S}}$ are called *normal control data* of *class* $\mathcal{C}^m$. Some normal control data $(T_S)_{S \in \mathcal{S}}$ are called *curvature moderate* of *order* m, if any two strata of X are regularly situated and if for every stratum S the corresponding tube T_S is curvature moderate of order m.

Two families of control data $(T_S)_{S \in \mathcal{S}}$ and $(\tilde{T}_S)_{S \in \mathcal{S}}$ of a stratified space $(X, \mathcal{S})$ are called *equivalent*, if for every stratum $S \in \mathcal{S}$ the tubes T_S and $\tilde{T}_S$ are equivalent over S. A stratified space $(X, \mathcal{S})$ together with an equivalence class of control data is called a *controlled* (*stratified*) *space*, the corresponding equivalence class a *control structure* for X.

A *morphism* between controlled spaces X and Y is given by a so-called *controlled map* that means a continuous mapping $f : X \to Y$ which is a morphism of stratified spaces and for which control data $(T_S)_{S \in \mathcal{S}}$ and $(T_R)_{R \in \mathcal{R}}$ of $(X, \mathcal{S})$ resp. $(Y, \mathcal{R})$ exist with the following properties: For every connected component S_0 of a stratum $S \in \mathcal{S}$ the relation $f(T_{S_0}) \subset T_{R_{S_0}}$ holds, where $T_{S_0} := \pi_S^{-1}(S_0)$ and R_{S_0} is the stratum of $(Y, \mathcal{R})$ with $f(S_0) \subset R_{S_0}$, and for all $x \in T_{S_0}$ the following conditions are satisfied:

(CT3) $$f \circ \pi_S(x) = \pi_{R_S} \circ f(x),$$

(CT4) $$\rho_S(x) = \rho_{R_S} \circ f(x).$$

If f and appropriate control data fulfill only condition (CT3), then f is called *weakly controlled*. If each of the restrictions $f_{|S_0} : S_0 \to R_{S_0}$ of a controlled mapping $f : X \to Y$ is submersive (resp. immersive), then one says that f is a *controlled submersion* (resp. *controlled immersion*).

Controlled spaces together with the controlled maps as morphisms form a category $\mathfrak{Esp}_{\mathrm{ctr}}$. By associating to every manifold M the trivial control data consisting of the single tube $T_M = (M, \mathrm{id}_M, 0)$, smooth manifolds and $\mathcal{C}^\infty$-mappings form a full subcategory of $\mathfrak{Esp}_{\mathrm{ctr}}$.

If $f : X \to M$ comprises a stratified mapping from X into the manifold M with the canonical stratification then some control data $(T_S)_{S \in \mathcal{S}}$ are called *compatible* with f, if for every stratum S and all $x \in T_S$

(CT5) $$f \circ \pi_S(x) = f(x).$$

Hence the mapping f is compatible with $(T_S)_{S\in\mathcal{S}}$, if and only if it is a morphism of controlled spaces from X to M.

3.6.5 Example Let M be a smooth bounded manifold and $R = \partial M$ its boundary. Choose a smooth collar $k : R \times [0,1[\to U \subset M$ for M. Then k induces control data for M with its natural stratification into the sets $M^\circ = M \setminus R$ and R. Just set $T_R = U$ and define π_R and ρ_R as the uniquely determined functions satisfying

$$k \circ (\pi_R, \rho_R) = \mathrm{id}_U.$$

Together with $T_{M^\circ} = M^\circ$, $\pi_{M^\circ} = \mathrm{id}_{M^\circ}$ and $\rho_{M^\circ} = 0$ we thus obtain normal control data of class $\mathcal{C}^\infty$ for M.

3.6.6 Example Let $f : X \to M$ be a controlled submersion of a controlled space X into a manifold M, and let y be a point of M. Then the fiber $X_y = f^{-1}(y)$ inherits from X in a natural way the structure of a controlled space. By assumption f is a controlled submersion, hence for every stratum S of X the intersection $S_y = f^{-1}(y) \cap S$ is a submanifold of S, so the family $(S_y)_{S\in\mathcal{S}}$ comprises a decomposition and thus a stratification of X_y. Moreover, after choosing control data $(T_S)_{S\in\mathcal{S}}$ of X compatible with f the family $(T_{S_y})_{S\in\mathcal{S}}$ of restricted tubes $T_{S_y} = (T_S \cap X_y, \pi_{S|T_S\cap X_y}, \rho_{S|T_S\cap X_y})$ has to comprise control data for X_y.

3.6.7 Proposition *Let X be a controllable space. Then there exist control data $(T_S)_{S\in\mathcal{S}}$ of X with the following properties:*

(1) *If $R, S \in \mathcal{S}$ and $T_S \cap T_R \neq \emptyset$, then $R < S$, $R = S$ or $R > S$ holds.*

(2) *For every point x of a stratum S there exists an open neighborhood $U \subset S$ such that $\pi_S^{-1}(U)$ has nonempty intersection with only finitely many tubes T_R, $R > S$.*

(3) *For every pair of strata $R > S$ the relation $x \in T_R \cap T_S$ implies $\pi_R(x) \in T_S$.*

(4) *For every stratum S there exists a smooth function $\varepsilon_S : T_S \to \mathbb{R}^{>0}$ such that $T_S^{\varepsilon_S} = T_S$ and such that $(\pi_S, \rho_S) : T_S \to [0, \varepsilon_S[$ is a proper surjective mapping.*

Moreover, such control data satisfy the following relations:

(5) *For every connected component S_0 of a stratum S the preimage $\pi_S^{-1}(S_0)$ is path connected. If S_1 is a second connected component of S disjoint to S_0, then $\pi_S^{-1}(S_0)$ and $\pi_S^{-1}(S_1)$ are disjoint as well.*

(6) *For every pair of strata $R > S$ the map $(\pi_{S,R}, \rho_{S,R}) : T_{S,R} \to]0, \varepsilon_S[$ is a differentiable fibration.*

Control data satisfying conditions (1) to (4) in the proposition are called *proper control data.*

PROOF: We first show (1) by an induction argument. Let $(S_k)_{k\in\mathbb{N}}$ be a denumeration of the strata of X, and $(T_S^{-1})_{S\in\mathcal{S}}$ some control data for X. Suppose that $(T_S^k)_{S\in\mathcal{S}}$ with $k \geq -1$ has been constructed by appropriately shrinking $(T_S^{-1})_{S\in\mathcal{S}}$ and that for all strata S_l with $l \leq k$ and all strata R not comparable with S_l the relation $T^k_{S_l} \cap T^k_R = \emptyset$ holds. As X is locally compact and paracompact there exists a locally finite covering $(U_n)_{n\in\mathbb{N}}$ of S_{k+1} by sets open in X such that $\overline{U_n}$ is compact and has only nonempty intersection with strata $R \geq S_{k+1}$ and $R \leq S_{k+1}$. Then $T^{k+1}_{S_{k+1}} = T^k_{S_{k+1}} \cap \bigcup_{n\in\mathbb{N}} U_n$ is an open neighborhood of S_{k+1}. Moreover, $A = \bigcup_{n\in\mathbb{N}} \overline{U_n}$ is closed and $X \setminus A$ contains all strata not comparable with S_{k+1}. Therefore let us set $T^{k+1}_{S_l} = T^k_{S_l}$ for $l < k$ and $T^{k+1}_{S_l} = T^k_{S_l} \setminus A$ for $l > k$. Restricting the functions π^k_S and ρ^k_S appropriately we thus obtain new control data $(T^{k+1}_S)_{S\in\mathcal{S}}$ which satisfy the above induction assumptions for $k+1$. The family $(T_S)_{S\in\mathcal{S}}$ where $T_S = T^k_S$ for $S = S_k$ then are control data which satisfy (1).

By further restricting the tubes according to Lemma 3.1.2 we immediately obtain some control data $(T_S)_{S\in\mathcal{S}}$ which fulfill (4). In the following we will further shrink $(T_S)_{S\in\mathcal{S}}$ by shrinking the functions ε_S. By this procedure properties (1) and (4) remain true.

Now let us come to (2). As X is locally compact and has a countable basis of its topology there exists a metric d on X compatible with the topology. Let us assume that (2) does not hold for any control data obtained by shrinking $(T_S)_{S\in\mathcal{S}}$. Then there exists a stratum S, a point $x \in S$ and a sequence of pairwise different strata R_n with the following property: For every basis of neighborhoods $(U_n)_{n\in\mathbb{N}}$ of x in S and all sequences of smooth functions $\varepsilon_n : R_n \to \mathbb{R}^{>0}$ and $\delta_n : S \to \mathbb{R}^{>0}$ there exists a point $y_n \in T^{\varepsilon_n}_{R_n} \cap T^{\delta_n}_S \cap \pi_S^{-1}(U_n)$. We can assume that all the sets U_n are relatively compact in S and that $\bigcap_{n\in\mathbb{N}} U_n = \{x\}$. Then according to Lemma 3.1.2 we choose the functions ε_n and δ_n such that

$$T^{\varepsilon_n}_{R_n} \subset B_{\frac{1}{2^n}}(R_n) := \{y \in T_{R_n} \mid d(y, \pi_{R_n}(y)) < \frac{1}{2^n}\} \quad \text{and} \quad T^{\delta_n}_S \subset B_{\frac{1}{2^n}}(S).$$

By definition of the δ_n and U_n and according to the proof of Lemma 3.1.2 the sets $T^{\delta_n}_S \cap \pi_S^{-1}(U_x)$ form a basis of neighborhoods of x in X that means the sequence $(y_n)_{n\in\mathbb{N}}$ converges to x. Now let $B_r(x)$ the ball of radius $r > 0$ around x with respect to the metric d. Choose $N \in \mathbb{N}$ so large that $\frac{1}{2^n} < \frac{r}{2}$ and $y_n \in B_{\frac{r}{2}}(x)$ for all $n > N$. By definition of y_n and ε_n then $\pi_{R_n}(y_n) \in B_r(x)$ follows for $n > N$ that means $B_r(x)$ meets infinitely many different strata which cannot be true. Hence one can shrink $(T_S)_{S\in\mathcal{S}}$ to some control data which we again denote by $(T_S)_{S\in\mathcal{S}}$ such that (2) holds.

To prove (3) we first suppose that the control data $(T_S)_{S\in\mathcal{S}}$ have the following properties:

(3)' For every pair of strata $R > S$ there exist smooth functions $\delta_R : S \to \mathbb{R}^{>0}$ and $\kappa_R : R \to \mathbb{R}^{>0}$ such that the relation $x \in T^{\kappa_R}_R \cap T^{\delta_R}_S$ entails $\pi_R(x) \in T_S$.

With the help of an inductive argument we construct from $(T_S)_{S\in\mathcal{S}}$ control data satisfying (3). To this end we denote by S_k the union of all k-dimensional strata of X and call by slight abuse of language S_k a stratum of X as well. We now suppose that after shrinking $(T_S)_{S\in\mathcal{S}}$ property (3) holds for all strata S_l with $l \leq k$ and every stratum $R > S_l$. Then we choose a locally finite open covering $(U_n)_{n\in\mathbb{N}}$ of $S = S_{k+1}$ by in S open and relatively compact sets $U_n \subset\subset S$, such that only for finitely many U_m the intersection $\pi_S^{-1}(U_m) \cap \pi_S^{-1}(U_n)$ is nonempty. Let further $(\varphi_n)_{n\in\mathbb{N}}$ be a smooth partition of unity of S subordinate to $(U_n)_{n\in\mathbb{N}}$. Then choose for $R > S$ smooth functions

$\delta_R : S \to \mathbb{R}^{>0}$ and $\kappa_R : R \to \mathbb{R}^{>0}$ according to (3)'. After possibly shrinking the U_n the intersection $T_S \cap T_R \cap \pi_S^{-1}(U_n)$ is nonempty only for finitely many strata R, hence there exists smooth functions $\delta_n : S \to \mathbb{R}^{>0}$ with $0 < \delta_{n|U_n} < \delta_{R|U_n}$ for all R with $T_S \cap T_R \cap \pi_S^{-1}(U_n) \neq \emptyset$. Then let us set

$$d_n = \inf \{\delta_m(x) | \, x \in U_m \,\&\, U_n \cap U_m \neq \emptyset\} \quad \text{and} \quad \delta = \sum_{n \in \mathbb{N}} d_n \, \varphi_n.$$

Note that $d_n > 0$ for all n. Hence $\delta > 0$ follows, and by definition $\delta_{|U_n} < (\delta_m)_{|U_n} < (\delta_R)_{|U_n}$ is true for all m and R with $\pi_S^{-1}(U_m) \cap T_R \neq \emptyset$ and $U_m \cap U_n \neq \emptyset$. According to (3)' this implies for all $x \in T_S^\delta \cap T_R^{\kappa_R}$ that for every $n \in \mathbb{N}$ with $x \in U_n$ the relation $x \in T_S^{\delta_R} \cap T_R^{\kappa_R}$ is true. Consequently $\pi_R(x) \in T_S$, which by (CT2) entails that $\rho_S(\pi_R(x)) = \rho_S(x) < \delta(x)$, hence $\pi_R(x) \in T_S^\delta$. Now shrink T_S to T_S^δ and for all $R > S$ shrink T_R to $T_R^{\kappa_R}$. This gives the induction step for $S = S_{k+1}$. Since for every stratum S of X only finitely many strata $R > S$ exist, every tube T_S will be restricted only finitely many times that means we finally obtain by this procedure control data satisfying (3). But it remains to show (3)'. More precisely we will prove that for some control data $(T_S)_{S \in \mathcal{S}}$ fulfilling (1) and (4) the relation (3)' already holds. Let $R > S$ be two strata X. Then we choose a smooth function $\delta_R : S \to \mathbb{R}^{>0}$ with $\delta_R < \varepsilon_S$ and determine for all points $y \in R$ an open neighborhood $U_y \subset R$ and a smooth $\kappa_y : R \to \mathbb{R}^{>0}$, $\kappa_y < \varepsilon_R$, such that $\pi_R^{-1}(U_y) \cap T_R^{\kappa_y} \cap \overline{T}_S^{\delta} = \emptyset$, if $y \notin \overline{T}_S^{\delta_R}$, and $\pi_R^{-1}(U_y) \cap T_R^{\kappa_y} \subset T_S$, if $y \in \overline{T}_S^{\delta_R}$. Then there exists a smooth $\kappa_R : R \to \mathbb{R}^{>0}$ with $T_R^{\kappa_R} \subset \bigcup_{y \in R} \pi_R^{-1}(U_y) \cap T_R^{\kappa_y} \subset T_R$. Then $\pi_R(x) \in U_{\pi_R(x)}$ holds for all $x \in T_R^{\kappa_R} \cap T_S^{\delta_R}$. Thus, by definition of the U_y the relation has to be true $\pi_R(x) \in \overline{T}_S^{\delta_R}$, hence $\pi_R(x) \in T_S$, which proves the claim.

We postpone the proof of (6) until Section 3.9, p. 143. This will not lead to any circular arguments as until then we will not use property (6) of proper control data.

Finally we will show that for every point x of a stratum S the set $W = \overline{T}_S^\varepsilon \cap \pi_S^{-1}(U)$ is path connected, where U is a compact path connected neighborhood of x in S and $\varepsilon : S \to \mathbb{R}^{>0}$ a smooth function with $\varepsilon < \varepsilon_S$. This immediately gives (5). First note that W is compact by (4). Suppose there exists a point $y \in W$ which cannot be connected with x by a continuous path. Let W_y be the set of all points $z \in W$ which can be connected with y by a continuous path in W. Then W_y is closed in W, hence compact and by assumption on y has empty intersection with U. Therefore $d = \inf \{\rho_S(z) | \, z \in W_y\} > 0$, hence there exists a converging sequence $(z_j)_{j \in \mathbb{N}}$ of elements of W_y with $\lim_{j \to \infty} \rho_S(z_j) = d$. Let $z \in W_y$ be the limit of $(z_j)_{j \in \mathbb{N}}$ and $R > S$ the stratum of z. As $(\pi_{S,R}, \rho_{S,R}) : T_{S,R} \to S \times]0, \infty[$ is submersive and R locally path connected, there exists a point $z' \in R \cap W_y$ with $\rho(z') < d$ and $\pi_S(z') = \pi_S(z)$. But this contradicts the minimality of d, hence W is path connected. □

The last part of the proof immediately entails the following.

3.6.8 Corollary *Every controllable stratified space is locally path connected.*

Up to now we do not know whether there exist nontrivial control data besides the ones induced by collars of manifolds with boundary. The following theorem gives an answer to this question.

3.6.9 Theorem *For every Whitney stratified space* X *and every smooth stratified submersion* $f : X \to M$ *to a manifold* M *there exist normal control data* $(T_S)_{S \in \mathcal{S}}$ *on* X *which are compatible with* f. *(cf.* MATHER *[122, Prop. 7.1])*

Let $m \in \mathbb{N}^{>0}$ *and* X *be Euclidean embeddable. If* X *is strongly curvature moderate and* f *strongly curvature moderate over every stratum* S *both of order* $m + (\dim \mathrm{Et}_X(S))^2 - (\dim S)^2 + 1$, *then the normal control data* $(T_S)_{S \in \mathcal{S}}$ *can be chosen to be curvature moderate of order* m.

PROOF: In the first part we will consider the case that X is a Whitney stratified subspace of $\mathbb{R}^n$ and then extend it in the second part to the case of arbitrary Whitney stratified spaces.

1. PART First we suppose that X is a Whitney stratified subspace of $\mathbb{R}^n$ or of a differentiable manifold N. Let X^k be the k-skeleton of X that means the union of all strata of dimension $\leq k$ of X, and $\mathcal{S}^k$ the family of all strata of dimension $\leq k$. Moreover let us set $X^{-1} = \emptyset$ and $\mathcal{S}^{-1} = \emptyset$. Now it will be shown by induction on k that for all X^k there exist control data compatible with f which are curvature moderate in the curvature moderate case.

Let us suppose that for some $k \in \mathbb{N}$ we have a system $(T_S)_{S \in \mathcal{S}^{k-1}}$ of normal tubes $T_S = (T_S, \pi_S, \rho_S)$ in X such that the control conditions (CT1), (CT2) and (CT5) are satisfied. By Proposition 3.4.1 we can suppose that for all $R < S$ with $R, S \in \mathcal{S}^{k-1}$ the mapping

$$(\pi_R, \rho_R)_{|S \cap T_R} : S \cap T_R \longrightarrow R \times \mathbb{R}$$

is submersive. Under the assumptions of the curvature moderate case assume additionally that every one of the tubes T_S, $S \in \mathcal{S}^{k-1}$ is curvature moderate of order $m + (\dim \mathrm{Et}_X(S))^2 - (\dim S)^2$; then by assumption on X every one of the submersions $(\pi_R, \rho_R)_{|S \cap T_R}$ is strongly curvature moderate of order $m + (\dim \mathrm{Et}_X(S))^2 - (\dim S)^2$ over S. Finally suppose that R and S are comparable if and only if $T_R \cap T_S \neq \emptyset$.

As any two strata of X of equal dimension k are not comparable, the following constructions can be performed separately for every stratum of dimension k. So let S be a stratum with $\dim S = k$.

In his proof of the claim JOHN MATHER constructs tubular neighborhoods T_S in two steps. For every $l \leq k$ let U_l be the union of all T_R with $R < S$ and $\dim R \geq l$. Let $S_l = U_l \cap S$. In the first step MATHER defines a tube $T_l = (T_l, \pi_l, \rho_l)$ of S_l in X by a decreasing induction on l. Hereby, the tubes T_R of strata $R < S$ possibly have been shrunk. But as every stratum lies only in the boundary of finitely many strata, the corresponding tubes are shrinked only finitely many times. In a second step the tube T_0 will be then extended to a tube (T_S, π_S, ρ_S) of S in X. In the following we supplement the original argument of MATHER by the curvature moderate case and show that under the corresponding assumptions all constructions can be performed to be curvature moderate.

1. STEP For $l = k$ we have $S_k = \emptyset$ so in this case we are finished. Now suppose that T_{l+1} has been constructed and that the following commutation relations are satisfied: If $R < S$, $\dim R \geq l + 1$, $x \in T_{l+1} \cap T_R$ and $\pi_{l+1}(x) \in T_R$, then

$$(\mathrm{CT})_{l+1} \qquad \begin{aligned} \rho_R \circ \pi_{l+1}(x) &= \rho_R(x), \\ \pi_R \circ \pi_{l+1}(x) &= \pi_R(x). \end{aligned}$$

After possibly shrinking T_{l+1} one can assume that for every $x \in T_{l+1}$ there exists a stratum $Q < S$ of dimension $> l$ such that $x \in T_Q$ and $\pi_{l+1}(x) \in T_Q$. Moreover, in the curvature moderate case we assume that every one of the tubes T_{l+1} is curvature moderate of order $m + (\dim \mathrm{Et}_x(S))^2 - (\dim(S))^2 + 2(l+1)$.

As for any two different l-dimensional strata $R, R' < S$ the relation $T_R \cap T_{R'} = \emptyset$ is true, it suffices to construct the tube T_l seperately over each one of the sets $T_R \cap S$. In other words we want to construct a tubular neighborhood T_S^R of $T_R \cap S$ the restriction of which to $T_R \cap S_{l+1}$ is isomorphic to the restriction of T_{l+1} and such that the following commutation relations are satisfied: for every $x \in T_S^R \cap T_R$ with $\pi_S^R(x) \in T_R$

$$\rho_R \circ \pi_S^R(x) = \rho_R(x),$$
$$\pi_R \circ \pi_S^R(x) = \pi_R(x),$$

where $\pi_S^R := \pi_{T_S^R}$. After shrinking T_R appropriately these commutation relations are already satisfied for all $x \in T_{l+1} \cap T_R$ with $\pi_{l+1}(x) \in T_R$, where π_{l+1} is used instead of π_S^R. To realize this shrink the tube T_R to

$$T_R \setminus \bigcup_{\{Q \mid R<Q<S\}} \pi_Q^{-1}(T_Q \setminus T_R),$$

and denote the thus obtained tube again by T_R. For the chosen $x \in T_{l+1}$ let afterwards $Q < R$ be a stratum with $\dim Q > l$ such that $x \in T_Q$ and $\pi_{l+1}(x) \in T_Q$. Then $\pi_{l+1}(x) \in T_R \cap T_Q$, hence $R < Q < S$. By the just performed shrinking we have $\pi_Q(x) \in T_R$, hence the claimed commutation relations are true:

$$\rho_R \circ \pi_{l+1}(x) = \rho_R \circ \pi_Q \circ \pi_{l+1}(x) = \rho_R \circ \pi_Q(x) = \rho_R(x),$$
$$\pi_R \circ \pi_{l+1}(x) = \pi_R \circ \pi_Q \circ \pi_{l+1}(x) = \pi_R \circ \pi_Q(x) = \pi_R(x).$$

Next recall that by assumption $(\pi_R, \rho_R)_{|S \cap T_R} : S \cap T_R \longrightarrow R \times \mathbb{R}$ is submersive. On the other hand the control conditions (CT_l) mean nothing else than that the tube T_S^R is compatible with the mapping $(\rho_R, \pi_R) : S_{l+1} \cap T_R \to R \times \mathbb{R}$.

Now we shrink each one of the tubular neighborhoods T_Q for $Q < S$, $\dim Q > l$ to a tubular neighborhood T'_Q such that the closure of $T'_Q \cap S$ lies in S_{l+1} and such that in the curvature moderate case T_{l+1} is a regularly situated neighborhood of $T'_Q \cap S$. If one defines S'_{l+1} analogously to S_{l+1}, then the closure of S'_{l+1} lies in S_{l+1}, hence by the existence theorem for tubular neighborhoods there exists a tubular neighborhood T_S^R which satisfies the control conditions (CT_{l+1}) and which is isomorphic to T_{l+1} over S'_{l+1}. In the curvature moderate case $(\pi_R, \rho_R)_{|S \cap T_R}$ is strongly curvature moderate of order

$$m + (\dim \mathrm{Et}_x(R))^2 - (\dim(R))^2 \geq m + (\dim \mathrm{Et}_x(S))^2 - (\dim(S))^2 + 2(l+1) - 1,$$

hence by the existence and uniqueness theorem T_S^R can be chosen curvature moderate of order $m + (\dim \mathrm{Et}_x(S))^2 - (\dim(S))^2 + 2l$. This finishes the induction step with respect to l, so there exists a tubular neighborhood T_0 of S_0, which satisfies (CT_0) for all $R < S$ and which is curvature moderate of order $m + (\dim \mathrm{Et}_x(S))^2 - (\dim(S))^2$ in the curvature moderate case.

2. STEP (CT_0) implies that after possibly shrinking T_0, the tube T_0 is compatible with f. Namely choose T_0 so small that for all $x \in T_0$ there exists some $Q < S$ with $x \in T_Q$ and $\pi_0(x) \in T_Q$. But then

$$f \circ \pi_0(x) = f \circ \pi_Q \circ \pi_0(x) = f \circ \pi_Q(x) = f(x).$$

By applying the existence theorem for tubular neighborhoods there exists analogously to the 1. Step after appropriately shrinking T_0 a tubular neighborhood T_S of S in X which is compatible with f and which satisfies the control conditions. After possibly shrinking T_S again we can assume that the system $(T_S)_{S \in \mathcal{S}^k}$ satisfies the above induction hypothesis that means the induction step is finished. As for every fixed stratum S and sufficiently large k the tube T_S will not be shrinked anymore by the transition to $\mathcal{S}^{k+1}$, we thus finally obtain control data for X with the desired properties.

2. PART After the claim has been shown for the case that X is Euclidean embeddable we now come to the general case. For the construction of control data on X we need an atlas of singular charts $x_j : K^\circ_{j+1} \to O_j \subset \mathbb{R}^{n_j}$, $j \in \mathbb{N}$ which is inductively embedding with respect to a compact exhaustion $(K_j)_{j \in \mathbb{N}}$ of X. Hereby the O_j are open in $\mathbb{R}^{n_j}$ and the $x_j(K^\circ_{j+1})$ are closed in O_j. In the following we will construct for the Whitney stratified spaces $x_j(K^\circ_{j+1}) \subset \mathbb{R}^{n_j}$ control data which are compatible in a certain sense and which altogether will induce the desired control data for X. To this end first construct with the help of a smooth partition of unity and after possibly shrinking the O_j smooth and submersive functions $f_j : O_j \to M$ such that for all $j \in \mathbb{N}$

(1a) $f_j \circ x_j(x) = f(x)$, if $x \in K^\circ_{j+1}$ and

(1b) $f_{j+1}(x) = f_j \circ \pi^{n_{j+1}}_{n_j}(x)$, if $x \in x_{j+1}(K_j)$.

Now the claim is proved, if one can provide for every chart x_j control data $(T^j_S)_{S \in \mathcal{S}}$ consisting of tubes $T^j_S = (T^j_S, \pi^j_S, \rho^j_S)$ around $x_j(S \cap K^\circ_{j+1})$ in $\mathbb{R}^{n_j}$ such that they are induced by tubular neighborhoods $(E^j_S, \varepsilon^j_S, \varphi^j_S)$ and such that the following conditions are satisfied:

(2a) For all $x \in S \cap K_j$ the relation $\varepsilon^{j+1}_S \circ x_{j+1}(x) = \varepsilon^j_S \circ x_j(x)$ holds.

(2b) $\pi^{n_{j+1}}_{n_j}(T^{j+1}_{S \cap K_j}) = T^j_{S \cap K_j}$ holds true, where $T^l_{S \cap K_j} = (\pi^l_S)^{-1}(x_l(S \cap K_j))$ for $l \geq j$.

(2c) For all $x \in T^{j+1}_{S \cap K_j}$ one has $\pi^{n_{j+1}}_{n_j} \circ \pi^{j+1}_S(x) = \pi^j_S \circ \pi^{n_{j+1}}_{n_j}(x)$ and $\rho^{j+1}_S(x) = \rho^j_S \circ \pi^{n_{j+1}}_{n_j}(x)$.

(2d) For $x \in T^j_S$ the equality $f_j \circ \pi^j_S(x) = f(x)$ holds.

Now one constructs control data $(T^0_S)_{S \in \mathcal{S}}$ according to the 1. Part. If $(T^0_S)_{S \in \mathcal{S}}$ to $(T^j_S)_{S \in \mathcal{S}}$ with the desired properties are already given, then one pulls back the tubular neighborhoods $(E^j_S, \varepsilon^j_S, \varphi^j_S)$ via the canonical projections $\pi^{n_{j+1}}_{n_j}$ to tubular neighborhoods $\tilde{T}^{j+1}_S = (\tilde{E}^{j+1}_S, \tilde{\varepsilon}^j_S, \tilde{\varphi}^{j+1}_S)$ of $x_{j+1}(S \cap K^\circ_{j+1})$, where $\tilde{E}^{j+1}_S = E^j_{S|S \cap K^\circ_{j+1}} \oplus \mathbb{R}^{m_j}$ with $m_j = n_{j+1} - n_j$, where $\tilde{\varepsilon}^{j+1}_S = \varepsilon^j_S \circ \pi^{n_{j+1}}_{n_j}{}_{|S \cap K^\circ_{j+1}}$ holds, and where $\tilde{\varphi}^j_S$ comprises an appropriate restriction of $(\varphi, \mathrm{id}_{\mathbb{R}^{m_j}})$. Hereby we have canonically identified in our notation the sets $S \cap K^\circ_{j+1}$ with their images under x_j respectively x_{j+1}.

Recall the arguments given in the 1. Part. Using the uniqueness and existence theorems for tubular neighborhoods several times one now checks that the restrictions $\tilde{T}_S^{j+1}$ can be extended to tubular neighborhoods $T_S^{j+1}{}_{|S\cap K_j}$ satisfying conditions (2a) to (2d). We will not give the details of this somewhat lengthy consideration but mention again that it can be performed exactly like in the 1. Part. □

3.7 Controlled vector fields and integrability

3.7.1 By a *flow* on a metric space X one usually understands a continuous mapping $\gamma : J \to X$, $(x,t) \mapsto \gamma(x,t) = \gamma_x(t)$ which is defined on an open subset $J \subset X \times \mathbb{R}$ and which has the following properties:

(F1) For every $x \in X$ there exist t_x^-, t_x^+ with $-\infty \leq t_x^- < 0 < t_x^+ \leq \infty$ such that $J_x := J \cap (\{x\} \times \mathbb{R})$ is equal to the interval $]t_x^-, t_x^+[$.

(F2) For every $x \in X$ the relation $\gamma_x(0) = x$ holds true.

(F3) If $t \in J_x$ and $s \in J_{\gamma_x(t)}$ then $t + s \in J_x$ follows and $\gamma_x(t+s) = \gamma(\gamma_x(t), s)$.

The set of flows on X is an ordered set by defining the relation $\gamma \leq \tilde{\gamma}$ for two flows $\gamma : J \to X$ and $\tilde{\gamma} : \tilde{J} \to X$ as being equivalent to $J \subset \tilde{J}$ and $\gamma(x,t) = \tilde{\gamma}(x,t)$ for all $(x,t) \in J$.

Now let $X \subset M$ be a submanifold of M and $V : X \to TM$ a vector field tangent to X and which satisfies a local Lipschitz condition. According to the classical theorem of uniqueness and existence of solutions of an ordinary differential equation there exists a unique maximal flow $\gamma : J \to X$ such that for all $x \in X$ and $t \in J_x$

(IF) $$\dot{\gamma}_x(t) = V(\gamma_x(t)) \quad \text{and} \quad \gamma_x(0) = x.$$

Allowing for X an arbitrary stratified subset of the manifold M, while one requires from $V : X \to TM$ that it is a mapping with $V(x) \in T_xS$ for every stratum S and all $x \in S$, then one can ask the question under which conditions on X and V a flow γ exists such that (IF) is satisfied. In other words we want to know under which conditions V is locally integrable. JOHN MATHER has treated this question in his articles [122, 123] in detail. For the case of a Whitney stratified subset $X \subset M$ he could show that every so-called controlled vector field on X has a maximal integral flow. We will prove the corresponding theorem in this section, but before let us introduce some necessary notation.

3.7.2 Definition (MATHER [122, §9]) Let X be a controlled stratified space. A vector field $V : X \to TX$ is called *weakly controlled*, if there exist control data $(T_S)_{S\in\mathcal{S}}$ of X such that for every pair of strata $S < R$

(CT8) $$T\pi_{S,R} \circ V_{|T_{S,R}} = V \circ \pi_{S,R}.$$

If additionally

(CT9) $$T\rho_{S,R} \circ V_{|T_{S,R}} = 0$$

is satisfied, then V is called *controlled*.

A weakly controlled vector field $V : T_S \to TX$ on a tubular neighborhood of the stratum S is called *radial* (with respect to S), if V is controlled over $T_S \setminus S$ and if the relations

(RB1) $$V_{|S} = 0$$

(RB2) $$T\rho_S \circ V_{|T_S \setminus S} = -\frac{\partial}{\partial t}$$

hold true. If besides (RB1) only

(RB3) $$T\rho_S \circ V_{|T_S \setminus S} = -h\frac{\partial}{\partial t}$$

is true with a stratified mapping $h : T_S \to \mathbb{R}^{>0}$, then V is called *conformally radial*.

If X carries additionally a $\mathcal{C}^{m+1}$-structure and if the control structure on X is curvature moderate of order m, then we call a (weakly) controlled vector field V *curvature moderate* of *order* m, if there exist a covering of X by singular charts $x : U \to \mathbb{R}^n$ and control data $(T_S)_{S\in\mathcal{S}}$ curvature moderate of order m such that for these objects the axioms (TB5), (CT8), (and if applicable (CT9)) are satisfied and if the following conditions hold true:

(CM10) For every stratum S the vector field $V_S^x : T_S^x \to \mathbb{R}^n$, $x \mapsto V(x^{-1}(\pi_S^x(x)))$ which is defined on a tubular neighborhood of $x(S \cap U)$ is tempered relative $x(\partial S \cap U)$ of class $\mathcal{C}^m$.

The fundamental existence result for controlled vector fields is the following theorem. Its first part goes back again to MATHER [122, Prop. 9.1].

3.7.3 Theorem *Let X be a controlled space and $f : X \to M$ a controlled submersion in a manifold M. Then there exists for every smooth vector field $W : M \to TM$ a controlled vector field $V : X \to TX$ such that*

(CT10) $$Tf \circ V = W \circ f.$$

If additionally X carries a $\mathcal{C}^{m+1}$-structure and possesses some strongly curvature moderate control data of order $m+1$ and if f is strongly curvature moderate of order $m+1$ over every stratum, then one can choose V curvature moderate of order m.

PROOF: 1. PART By an induction on k MATHER constructs in [122] a controlled vector field $V^k : X^k \to TX^k$ on the k-skeleton X^k such that over X^k the vector field V^k satisfies relation (CT10) and $V^{k+1}_{|X^k} = V^k$. Afterwards he defines $V : X \to TX$ by $V_{|X^k} = V^k$, which gives a controlled vector field satisfying the claim. For the construction of V^k we proceed like in [122] and first choose some control data $(T_S)_{S\in\mathcal{S}}$ compatible with f. Hereby we will use in the following that the skeleton X^k inherits a control structure from X in a canonical way.

If $k = 0$, the claim is trivial. So let us suppose we are given for $k \geq 0$ a vector field V^k on X^k with the desired properties. By possibly shrinking the tubes of the strata S of dimension $\leq k$ one can achieve that the control conditions (CT8) and (CT9) for V^k and all pairs of strata $S \leq R$ with $\dim R \leq k$ are satisfied. After possibly shrinking the tubes again we can achieve according to Proposition 3.6.7 that for all pairs of strata $R \leq S$ the relation

$$\pi_R(T_S \cap T_R) \subset T_R$$

is true, and that for any two not comparable strata S and $\tilde{S}$ of X the tubes T_S and $T_{\tilde{S}}$ are disjoint. Moreover, we shrink the tubes such that for appropriate smooth functions $\varepsilon_S : S \to \mathbb{R}^{>0}$ the mappings $(\pi_S, \rho_S) : T_S^{\varepsilon_S} \to [0, \varepsilon_S[$ are proper, and $T_S = T_S^{\varepsilon_S}$ holds true.

Let us suppose that for every $(k+1)$-dimensional stratum S we are given a smooth vector field $V_S : S \to TS$, such that for all $R < S$, $x \in S \cap T_R^{\varepsilon/2}$ and $z \in S$ the following control conditions are satisfied:

$$(CT8)_R \qquad T\pi_R \circ V_S(x) = V^k \circ \pi_R(x),$$

$$(CT9)_R \qquad T\rho_R \circ V_S(x) = 0,$$

$$(CT10)_S \qquad Tf \circ V_S(z) = W \circ f(z).$$

Now define $V^{k+1} : X^{k+1} \to TX^{k+1}$ by $V^{k+1}_{|X^k} = V^k$ and $V^{k+1}_{|S} = V_S$. Then we obtain a controlled vector field on X^{k+1} which satisfies the induction hypothesis. This would finish the proof.

Hence it remains to construct vector fields V_S with the desired properties. By the assumptions on the control data we can perform the construction of V_S separately for every $(k+1)$-dimensional stratum. Now let y be a point of S. Then define for every stratum $R < S$ a neighborhood $U_{y,R}$ of y in S by

$$U_{y,R} := \begin{cases} S \cap T_R & \text{if } y \in T_R, \\ S \setminus \overline{T_R^{\varepsilon_R/2}} & \text{if } y \notin T_R, \end{cases}$$

and set $U_y := \bigcap_{R<S} U_{R,y}$. Then the family $(U_y)_{y \in S}$ comprises an open covering of S. On the other hand if one denotes by $\mathcal{S}_y$ the set of all strata $R < S$ such that $y \in T_R$ then $\mathcal{S}_y$ is totally ordered by the incidence relation, as the control data have been chosen such that any two strata having nonempty intersection are comparable.

Assume that for a point $y \in S$ the set $\mathcal{S}_y$ is nonempty. Then there exists a largest element R_y in $\mathcal{S}_y$. By the fact that $(\pi_{R_y}, \rho_{R_y}) : S \cap T_{R_y} \to R_y \times \mathbb{R}^{>0}$ is submersive, one can find, after shrinking the neighborhood U_y appropriately, a smooth vector field $V_y : U_y \to TU_y$, such that $(CT8)_{R_y}$ and $(CT9)_{R_y}$ are satisfied for all $x \in U_y$. Moreover, for $z \in U_y$ the condition $(CT10)_S$ holds, as

$$Tf \circ V_S(z) = Tf \circ T\pi_{R_y} \circ V_S(z) = Tf \circ V^k \circ \pi_{R_y}(z) = W \circ f(z).$$

But if the point y does not lie in any one of the tubes T_R with $R < S$, or in other words if $\mathcal{S}_y = \emptyset$, then, after possibly shrinking U_y, choose the vector field $V_y : U_y \to TU_y$ such that for all $z \in U_z$ the condition $(CT10)_S$ is satisfied. By assumption f is a controlled submersion, hence such a V_y exists indeed. After the choice of a partition of unity $(\varphi_j)_{j \in \mathbb{N}}$ subordinate to the covering $(U_y)_{y \in S}$ and of points $y_j \in S$ with $\operatorname{supp} \varphi_j \subset U_{y_j}$ one can finally define V_S as follows:

$$V_S(x) = \sum_{j \in \mathbb{N}} \varphi_j \, V_{y_j}(x), \qquad x \in S. \tag{3.7.2}$$

We show that this V_S satisfies the desired control conditions. Obviously, V_S satisfies the relation $(\mathrm{CT10})_S$, as it is satisfied by all V_y over $\mathcal{U}_y$. If $x \in T_R^{\varepsilon_R/2}$ for some $R < S$, then by construction of $\mathcal{U}_y$ the relation $y_j \in T_R$ follows for all y_j with $x \in \mathcal{U}_j := \mathcal{U}_{y_j}$. This implies $R \leq R_j$, where $R_j := R_{y_j}$, and $\pi_{R_j}(x) \in T_R$. Hence by the above proven fact that V_{y_j} satisfies conditions $(\mathrm{CT8})_{R_j}$ and $(\mathrm{CT9})_{R_j}$ the following equalities have to be true:

$$\begin{aligned} T\pi_R \circ V_S(x) &= \sum_{\{j \mid x \in \mathcal{U}_j\}} \varphi_j(x) \cdot T\pi_R \circ V_{y_j}(x) = \sum_{\{j \mid x \in \mathcal{U}_j\}} \varphi_j(x) \cdot T\pi_R \circ T\pi_{R_j} \circ V_{y_j}(x) \\ &= \sum_{\{j \mid x \in \mathcal{U}_j\}} \varphi_j(x) \cdot T\pi_R \circ V^k \circ \pi_{R_j}(x) = \sum_{\{j \mid x \in \mathcal{U}_j\}} \varphi_j(x) \cdot V^k \circ \pi_R \circ \pi_{R_j}(x) \\ &= \sum_{\{j \mid x \in \mathcal{U}_j\}} \varphi_j(x) \cdot V^k \circ \pi_R(x) = V^k \circ \pi_R(x), \end{aligned}$$

$$\begin{aligned} T\rho_R \circ V_S(x) &= \sum_{\{j \mid x \in \mathcal{U}_j\}} \varphi_j(x) \cdot T\rho_R \circ V_{y_j}(x) = \sum_{\{j \mid x \in \mathcal{U}_j\}} \varphi_j(x) \cdot T\rho_R \circ T\pi_{R_j} \circ V_{y_j}(x) \\ &= \sum_{\{j \mid x \in \mathcal{U}_j\}} \varphi_j(x) \cdot T\rho_R \circ V^k \circ \pi_{R_j}(x) = 0. \end{aligned}$$

This shows the first part of the claim.

2. PART We only sketch the proof that under the assumptions of the curvature moderate case the vector field V can be chosen curvature moderate, as similar constructions already have been used several times.

It is essential for the curvature moderate case that one can chose the partition of unity $(\varphi_j)_{j \in \mathbb{N}}$ like in Lemma 1.7.9 with $K = \partial S$. That this is possible indeed can be proved by using the fact that every tubular neighborhood T_R is a regularly situated neighborhood of R. With the help of a curvature moderate Riemannian metric over T_S one determines V_{y_j} uniquely by requiring that in the case $\mathcal{S}_{y_j} \neq \emptyset$ the relation $V_{y_j}(x) \in \left(\ker T_x(\pi_R, \rho_R)\right)^\perp$ holds for all $x \in \mathcal{U}_{y_i}$ and that in the case $\mathcal{S}_{y_j} = \emptyset$ the relation $V_{y_j}(z) \in \left(\ker T_z f\right)^\perp$ is true for all $z \in \mathcal{U}_{y_i}$. The vector field V_S defined by Eq. 3.7.2 then is tempered relative ∂S of class $\mathcal{C}^m$ in every chart. □

3.7.4 Corollary *Let X be a controlled space. Then there exists for every stratum S a radial vector field $V : T_S \to TX$.*

If additionally X carries a $\mathcal{C}^{m+1}$-structure and possesses some control data which are curvature moderate of order $m+1$, then one can choose V curvature moderate of order m.

PROOF: First choose some control data for X. Then the restricted map $\rho_S : T_S \setminus S \to \mathbb{R}^{>0}$ is a controlled submersion. Then by the preceding theorem there exists a controlled vector field $\tilde{V} : T_S \setminus S \to TX$ such that $T\rho_S \circ \tilde{V} = -\frac{\partial}{\partial t}$. Now define V by $V_{|S} = 0$ and $V_{|T_S \setminus S} = \tilde{V}$, hence V comprises a radial vector field. □

By the last two results we now know that there exist enough controlled vector fields on a stratified space. Therefore we come back to our original matter of concern, the integration of controlled vector fields.

3.7.5 Definition A flow $\gamma : J \to X$ on a stratified space X is called *stratified*, if for all $(x, t) \in J$ the fact that x is an element of the stratum S already implies that $\gamma(x, t) \in S$. If $V : X \to TX$ is a stratified vector field on X, then one calls a stratified flow $\gamma : J \to X$ an *integral flow* of V, if for every stratum S the curves γ_x, $x \in S$ are of class $\mathcal{C}^1$ and if for all $(x, t) \in J \cap (S \times \mathbb{R})$ the condition (IF) is satisfied.

3.7.6 Theorem (cf. MATHER [122, Prop. 10.1]) *If V is a controlled vector field on a controlled space X, then there exists a uniquely determined maximal stratified integral flow $\gamma : J \to X$ of V. Moreover, γ is determined among all stratified integral flows by the following property:*

(IF)$_{\max}$ *For all $x \in X$ with $t_x^+ < \infty$ (resp. $t_x^- > -\infty$) the restricted map*

$$\gamma_{|\{x\}\times[0,t_x^+[} : \{x\} \times [0, t_x^+[\to X \qquad (bzw.\ \gamma_{|\{x\}\times]t_x^-,0]} : \{x\}\times]t_x^-, 0] \to X)$$

is proper.

PROOF: First fix some proper control data on X and for V such that the control conditions (CT8) and (CT9) are satisfied. By assumption on V the restriction $V_{|S}$ is a smooth vector field for every stratum S. By the classical existence and uniqueness theorem on the integrability of vector fields there exists a uniquely determined maximal integral flow $\gamma_S : J_S \to S$ of $V_{|S}$. We now set

$$J := \bigcup_{S \in \mathcal{S}} J_S \quad \text{and} \quad \gamma : J \to X, \quad \gamma_{|J_S} := \gamma_S,$$

and claim that γ is a maximal integral flow of V, that it is determined uniquely and that (IF)$_{\max}$ holds. Under the assumption that J is open and that the continuity of γ has been proved, it follows by definition that γ comprises a stratified integral flow of V. This integral flow is maximal, as for every integral flow $\tilde{\gamma} \geq \gamma$

$$\tilde{\gamma}_{|J\cap(S\times\mathbb{R})} \geq \gamma_{|J\cap(S\times\mathbb{R})} = \gamma_S,$$

which by the maximality of γ_S implies the relation $\tilde{\gamma}_{|J\cap(S\times\mathbb{R})} = \gamma_S$ to be true. Hence $\tilde{\gamma} = \gamma$ follows. An analogous argument proves uniqueness of γ.

To prove (IF)$_{\max}$, that J is open and γ continuous we need the following result.

3.7.7 Lemma *Let x be a point of the stratum S, K a compact neighborhood of x in S and $t_1, t_2 \in \mathbb{R}$ such that $t_x^- < t_1 < t_2 < t_x^+$. For V and $(T_S)_{S\in\mathcal{S}}$ choose $\varepsilon > 0$ with $\varepsilon < \varepsilon_S(\gamma_y(t))$ for $y \in K$ and $t \in [t_1, t_2]$. Set $U := \{y \in T_S \mid \pi_S(y) \in K,\ \rho_S(y) < \frac{\varepsilon}{2}\}$. Then the mapping $\gamma : J \to X$ associated to V fulfills for all $y \in U$ and all $t \in \mathbb{R}$ with $\max(t_y^-, t_1) < t < \min(t_y^+, t_2)$ the following control conditions:*

(CT11) $$\gamma(y, t) \in T_S,$$

(CT12) $$\pi_S(\gamma(y, t)) = \gamma(\pi_S(y), t),$$

(CT13) $$\rho_S(\gamma(y, t)) = \rho_S(y).$$

Proof of the Lemma: Let the point $y \in U$ be an element of the stratum $R > S$. As γ_R is a flow on R, there exists a sufficiently small $t_0 > 0$, such that (CT11) is satisfied for all $t \in \mathbb{R}$ with $|t| < t_0$. By the control conditions (CT8) and (CT9) and after transition to a possibly smaller t_0 the conditions (CT12) and (CT13) are satisfied for $|t| < t_0$. Consequently $\gamma_y(t_0)$ has to be in T_S, as $(\pi_S, \rho_s) : T_S = T_S^{\varepsilon_S} \to [0, \varepsilon_S[$ is proper. But this implies that (CT12) and (CT13) have to be true for $t = t_0$ as well. By the flow property of γ_R the relation $\gamma_y(t_0 + s) \in T_S$ then holds for sufficiently small $s > 0$. Moreover,

$$\begin{aligned}\pi_S(\gamma(y, t_0 + s)) &= \pi_S(\gamma(\gamma(y, t_0), s)) = \gamma(\pi_S(\gamma(y, t_0)), s) \\ &= \gamma(\gamma(\pi_S(y), t_0), s) = \gamma(\pi_S(y), t_0 + s),\end{aligned}$$

and by a similar calculation

$$\rho_S(\gamma(y, t_0 + s)) = \rho_S(y).$$

Hence (CT11), (CT12) and (CT13) are satisfied for $0 \le t < \min(t_y^+, t_2)$. By an analogous argument one shows the claim for negative t. □

Proof of Theorem 3.7.6, continued: Before showing J to be open and γ to be continuous we will prove that γ satisfies $(\text{IF})_{\max}$. Assume that this is not the case. Then there exists $y \in X$ with $t_y^+ < \infty$ (or $t_y^- > -\infty$), a compact set $K \subset X$ and an increasing (resp. decreasing) sequence $(s_j)_{j \in \mathbb{N}} \subset]t_y^-, t_y^+[$ such that $\gamma_y(s_j) \in K$ and $\lim_{j\to\infty} s_j = t_y^+$ (resp. $\lim_{j\to\infty} s_j = t_y^-$). We consider only the first case $s_j \to t_y^+$; the second one can be handled analogously. After selection of a subsequence $\gamma_y(s_j)$ converges to an element $x \in K$. Let S be the stratum of x and R the stratum with $y \in R$. If $R = S$, we obtain a contradiction to the fact that γ_R is the maximal integral flow of V_R, hence we must have $S < R$. Now, if j is chosen sufficiently large, the points $y_j := \gamma_y(s_j)$ all lie in the neighborhood U from Lemma 3.7.7, and for all $k \in \mathbb{N}$ the following relation holds:

$$0 \le s_{j+k} - s_j < t_y^+ - s_j = t_{y_j}^+ \quad \text{and} \quad 0 \le s_{j+k} - s_j < t_2.$$

Then by (CT13)

$$\begin{aligned}0 < \rho_S(y_j) &= \rho_S(\gamma(y_j, s_{j+k} - s_j)) = \rho_S(\gamma(\gamma(y, s_j), s_{j+k} - s_j)) \\ &= \rho_S(\gamma(y, s_{j+k})) = \rho_S(y_{j+k}).\end{aligned}$$

But this means that the sequence $(\rho_S(y_j))_{j\in\mathbb{N}}$ cannot converge to 0 which contradicts $\rho_S(x) = 0$. Hence $(\text{IF})_{\max}$ is true.

Next we will construct for $(x, s) \in J$ an open neighborhood in J. Let t_1, t_2 be such that $t_x^- < t_1 < s < t_2 < t_x^+$. Let us assume that $t_y^+ < t_2$ for some $y \in U$, where U is a neighborhood of x according to Lemma 3.7.7. Then there exists a sequence of elements $s_j \in [0, t_y^+[$ converging to t_y^+. Let $y_j = \gamma(y, t_j)$ for all j. By (CT12) and (CT13) we then have

$$\begin{aligned}\lim_{j\to\infty} \pi_S(y_j) &= \gamma(\pi_S(y), t_y^+), \\ \lim_{j\to\infty} \rho_S(y_j) &= \rho_S(y).\end{aligned}$$

As $(\pi_S, \rho_S) : T_S \to [0, \varepsilon_S[$ is proper, $(y_j)_{j\in\mathbb{N}}$ must have a converging subsequence which contradicts $(\mathrm{IF})_{\max}$. Thus $t_y^+ \geq t_2$ follows and analogously $t_y^- \leq t_1$, which altogether gives $U\times]t_1, t_2[\subset J$, hence J is open. As (CT12) and (CT13) hold as well for all $(y, t) \in U\times]t_1, t_2[$, the map γ must be continuous at (x, s) hence on the whole domain J.

The theorem will be shown, if we can finally prove that every stratified integral flow $\tilde{\gamma}$ of V which satisfies condition $(\mathrm{IF})_{\max}$ coincides with γ. But this is obvious, as every one of the restrictions $\tilde{\gamma}_{|S} : \tilde{J}_{|S} = \tilde{J} \cap (S \times \mathbb{R}) \to S$ has to be maximal by $(\mathrm{IF})_{\max}$, hence coincides with γ_S. □

3.7.8 Corollary *Let* $f : X \to M$ *be a proper controlled mapping between the controlled space* X *and the manifold* M. *Then for any two controlled vector fields* $V : X \to TX$ *and* $W : M \to TM$ *which satisfy the control condition* (CT10) $Tf \circ V = W \circ f$ *the following relation holds:*

$$J_V = (f \times \mathrm{id}_{\mathbb{R}})^{-1}(J_W),$$

where J_W *denotes the domain of the flow of* W *and* J_V *the one of* V.

3.8 Extension theorems on controlled spaces

In this section we will continue the extension theory for smooth functions on stratified spaces, which was introduced in Section 1.7. The main result in 1.7 allows to extend certain jets on the closure $\overline{S}$ of a stratum to a Whitney function on X, but it does not provide general criteria to decide, when a smooth function on S which falls off fast enough near the boundary can be extended to a smooth function on X.

For our extension theory we need an intrinsic notion for the fall off behavior of a smooth function $f \in \mathcal{C}^\infty(S)$ on a stratum S near the boundary ∂S. The corresponding definition 1.7.1 for Whitney functions cannot be transferred directly to smooth functions, as for such functions $f \in \mathcal{C}^\infty(S)$ the higher derivatives $\partial^\alpha f$ are in general not at our disposal. To obtain an appropriate notion of flatness at the boundary, we will extend the definition of FERRAROTTI–WILSON [59, Sec. I] for special singular subspaces in $\mathbb{R}^n$ to the case of arbitrary (A)-stratified spaces.

3.8.1 Definition Let μ be a smooth Riemannian metric on an (A)-stratified space X, and $m \in \mathbb{N}\cup\{\infty\}$. A $\mathcal{C}^m$-function $f : S \to \mathbb{R}$ on a stratum S of X is called *geometrically flat on the boundary* of *order* $c \in \mathbb{R}^{>0}$, if for all natural $k \leq m$ and all $y \in \partial S$

$$\lim_{\substack{x \to y \\ x \in S}} \frac{\|\nabla^k f(x)\|_\mu}{\delta_\mu(x, \partial S)^c} = 0.$$

Hereby ∇ is Levi–Civita connection on tensor fields induced by μ on S, $\|\cdot\|_\mu$ is the norm of a tensor with respect to μ, and δ_μ is the geodesic distance on S induced by μ. If f is geometrically flat on the boundary of arbitrary order, we will say that f is *geometrically flat at the boundary*.

3.8.2 Remark Recall that $\nabla^0 f = f$, $\nabla^1 f = df$ and $\nabla^k f = \nabla(\nabla^{k-1} f)$ for $k \geq 2$. If f is geometrically flat with respect to a smooth Riemannian metric μ, then f is geometrically flat with respect to every other smooth Riemannian metric η on X. The reason is that μ and η are locally equivalent and are both smooth. Hence the notion of geometrically flatness at the boundary is independent of the special choice of a smooth metric on X.

Now we come to our main theorem.

3.8.3 Extension Theorem *Let $m \in \mathbb{N}^{>0}$, X an (A)-stratified space, and μ a smooth Riemannian metric on X. Let S be an (r, l)-regular stratum which is curvature moderate of order $m+3$, and let T_S be the orthogonal projection of the tube of S induced by μ. If then $c \geq c_S^m(r, l)$ is sufficiently large, then one can extend every function $g \in \mathcal{C}^m(S)$ which is geometrically flat at the boundary of order c to a function $f \in \mathcal{C}^m(X)$ such that*

$$f_{|T_S^\delta} = g \circ \pi_{S|T_S^\delta}.$$

Hereby $\delta : S \to \mathbb{R}^{>0}$ is an appropriate (smooth) function.

PROOF: We can suppose that X is closed in an open bounded set $O \subset \mathbb{R}^n$; the general case can be reduced to the statement given in the following with the help of a smooth partition of unity. According to Proposition 3.3.14 and the assumptions on S the components of the projection $\pi := \pi_S$ are tempered of class $\mathcal{C}^m$. Finally recall that the restriction of μ defines a Riemannian metric on S; the corresponding Levi–Civita connection will be denoted by ∇. Under these assumptions let us consider the jet G on $\overline{S}$ which over S is induced by the smooth function $g \circ \pi$ and which vanishes on the boundary ∂S. To prove that G is even a Whitney function we only have to show according to the generalized lemma of Hesténès 1.7.2 that G is flat of order $c_S^m(r, l)$ over the boundary ∂S. To this end let $V_1, V_2, \cdots, V_k, \cdots$ be arbitrary constant vector fields $V_k : T \to \mathbb{R}^n$, in other words this means that all derivatives DV_k vanish. As the functions $T\pi.V_k$ comprise vector fields along π, the covariant derivative $\nabla_{V_j} T\pi.V_k$ along π in direction V_j is defined (see KLINGENBERG [101, Prop. 1.5.5]). Using local coordinates on S and the corresponding Christoffel symbols one checks that the following equalities hold:

$$D(g \circ \pi).V_1 = \langle dg, T\pi.V_1 \rangle,$$

$$D^2(g \circ \pi).(V_1, V_2) = D\langle dg, T\pi.V_2 \rangle.V_1 = \langle \nabla^2 g, T\pi.V_1 \otimes T\pi.V_2 \rangle + \langle dg, \nabla_{V_1} T\pi.V_2 \rangle$$

$$\begin{aligned} D^3(g \circ \pi).(V_1, V_2, V_3) &= \langle \nabla^3 g, T\pi.V_1 \otimes T\pi.V_2 \otimes T\pi.V_3 \rangle + \langle dg, \nabla_{V_1} \nabla_{V_2} T\pi.V_3 \rangle \\ &\quad + \langle \nabla^2 g, \nabla_{V_1} T\pi.V_2 \otimes T\pi.V_3 + T\pi.V_2 \otimes \nabla_{V_1} T\pi.V_3 + T\pi.V_1 \otimes \nabla_{V_2} T\pi.V_3 \rangle . \end{aligned}$$

Analogously one now concludes by an induction argument that for $k \leq m$

$$D^k(g \circ \pi).(V_1, \cdots, V_k) = \sum_{1 \leq j \leq k} \langle \nabla^j g, T_{jk} \rangle,$$

where T_{jk} is an j-times covariant tensor field which can be written as the sum of tensor products of vector fields of the form $\nabla_{V_{i_1}} \cdots \nabla_{V_{i_l}} T\pi.V_{i_0}$ with $l \leq k - j$.

From now on we assume that every one of the vector fields V_i is one of the basis elements $e_1, \cdots, e_n$. As S is curvature moderate of order m and π tempered of class $\mathcal{C}^m$, the relation $\nabla_{V_1} T\pi.V_2(x) = P_{\pi(x)} D(T\pi.V_2)(x).V_1(x)$ implies that $\nabla_{V_1} T\pi.V_2$ is bounded away from ∂S (recall the notion "bounded away from Z" which had been introduced in the proof of 3.3.6 on p. 104). Inductively we thus obtain that the vector fields $\nabla_{V_{i_1}} \cdots \nabla_{V_{i_l}} T\pi.V_{i_0}$, $l \leq m$ are all bounded away from ∂S hence there exists for every point of ∂S a relatively compact neighborhood $W \subset \mathbb{R}^n$ together with constants $C_j > 0$ and $c_j \in \mathbb{N}$ such that

$$\|T_{jk}(x)\| \leq C_j \left(1 + \frac{1}{d(x, \partial S)^{c_j}}\right), \qquad x \in W \cap S. \tag{3.8.2}$$

As by assumption g is geometrically flat of sufficiently large order and as the axiom (RA1) on p. 50 holds true, there exists for a later to be determined constant $d \in \mathbb{N}$ some $C > 0$ such that

$$\|\nabla^l g(x)\| \leq C\, d(x, \partial S)^{c_j + c_S^m(r,l) + d + 1}, \qquad x \in W \cap S,$$

for all $j, l \leq k$. But this implies the existence of appropriate constants $\tilde{C} > 0$ such that

$$\|D^k(g \circ \pi)(x)\| \leq \tilde{C}\, d(x, \partial S)^{c_S^m(r,l) + d + 1}, \qquad x \in W \cap S.$$

So the prerequisites of the generalized lemma of Hesténès are satisfied, hence G is a Whitney function on $\overline{S}$. In particular the following estimates then hold for every $\alpha \in \mathbb{N}^n$ with $|\alpha| \leq m$, if $x, y \in \overline{S}$ converge to some $z \in \overline{S}$:

$$(R_x^m G)^{(\alpha)} = o\big(|x - y|^{m - |\alpha|}\big). \tag{3.8.3}$$

But we are not yet finished with the proof. It remains to show that one can find a smooth function $f : O \to \mathbb{R}$ such that $f = g \circ \pi$ on a neighborhood of S. To construct such an f recall that according to the assumption on S the tubular neighborhood T_S is a regularly situated neighborhood of S. Hence there exists a smooth function $\varphi : \mathbb{R}^n \setminus \partial S \to [0, 1]$ such that φ is tempered relative ∂S of class $\mathcal{C}^\infty$, that $\varphi = 1$ on a neighborhood of S, and that φ vanishes on a neighborhood of $\complement T_S \setminus \partial S$. Now we define $f : \mathbb{R}^n \to \mathbb{R}$ by

$$f(y) = \begin{cases} \varphi(y)\, g(\pi(y)), & \text{if } y \in T, \\ 0, & \text{if } y \in \mathbb{R}^n \setminus T. \end{cases}$$

Choosing d resp. c sufficiently large, then by the temperedness φ the thus defined f comprises a $\mathcal{C}^m$-function on $\mathbb{R}^n$ indeed, hence one on X. As φ is identical to 1 on a neighborhood of S, the map f therefore fulfills the claim of the extension theorem.

□

3.8.4 Remark Ferrarotti–Wilson have proven in [59, Thm III-2.] an analogous extension result for singular subspaces of $\mathbb{R}^n$ with a dense top stratum and differentiability class $m \geq 2$.

3.9 Thom's first isotopy lemma

3.9.1 Definition A stratified mapping $f : X \to M$ from a stratified space X to a manifold M is called *trivial*, if there exists a stratified space Y and an isomorphism of stratified spaces $H : Y \times M \to X$ such that $f \circ H(y, x) = x$ for all $y \in Y$ and $x \in M$.

One says that f is *locally trivial*, if there exists a covering of M by open sets U such that all restricted maps $f_{|f^{-1}(U)} : f^{-1}(U) \to U$ are trivial.

3.9.2 Thom's first isotopy lemma *Let X be a controlled space, M a smooth manifold and $f : X \to M$ a proper controlled submersion. Then f is locally trivial.*

PROOF: As the statement is a local one it suffices to construct in the case $M = \mathbb{R}^n$ an isomorphism of stratified spaces $H : X_0 \times \mathbb{R}^n \to X$ such that $f \circ H(y, v) = v$ for all $y \in X_0$ and $v \in \mathbb{R}^n$. Hereby $X_0 := f^{-1}(0)$ is the fiber of f over 0 according to Example 3.6.6.

Consider the coordinate vector fields $e_1 = \frac{\partial}{\partial v_1}, \cdots, e_n = \frac{\partial}{\partial v_n}$ on $\mathbb{R}^n$ and choose controlled vector fields $E_1, \cdots, E_n$ on X according to Theorem 3.7.3 such that $Tf \circ E_k = e_k \circ f$ for $k = 1, \cdots, n$. As the vector fields e_k are globally integrable, Corollary 3.7.8 entails that the vector fields E_k are globally integrable as well that means for every k the domain J_k of the maximal integral flow γ_k of E_k is equal to $X \times \mathbb{R}$. Hence we can define for all $y \in X_0$ and $v = (v_1, \cdots, v_n) \in \mathbb{R}^n$:

$$H(y, v) = \gamma_n(\gamma_{n-1}(\cdots(\gamma_1(x, v_1), \cdots, v_n).$$

On the other hand we can define a mapping $G : X \to X_0 \times \mathbb{R}^n$ by

$$G(x) = (\gamma_n(\gamma_{n-1}(\cdots(\gamma_1(x, -v_1), \cdots, -v_n), f(x)), \quad (v_1, \cdots, v_n) := f(x), \ x \in X.$$

As for all $x \in X$

$$f(\gamma_k(x, t)) = f(x) + te_k, \tag{3.9.1}$$

$G(x)$ lies in $X_0 \times \mathbb{R}^n$ indeed that means G is well-defined. Moreover, by (3.9.1) the relation $f \circ H(y, v) = v$ holds for $y \in X_0$ and $v \in \mathbb{R}^n$.

As all the γ_k are flows, $G \circ H = \mathrm{id}_{X \times \mathbb{R}^n}$ holds true as well as $H \circ G = \mathrm{id}_X$. Moreover, G and H have to be stratified mappings. Therefore H is a stratified isomorphism with the desired properties. This proves the isotopy lemma. □

With the help of the first isotopy lemma the proof of property (6) in Prop. 3.6.7 now is obvious.

PROOF OF PROP. 3.6.7 (6): For all proper control data $(T_S)_{S \in \mathcal{S}}$ of X and every stratum S the mapping $(\pi_S, \rho_S) : T_S \setminus S \to]0, \varepsilon_S[$ is a proper submersion, hence locally trivial by the isotopy lemma. In particular, this implies that for every stratum $R > S$ the restricted map $(\pi_{S,R}, \rho_{S,R}) : T_{S,R} \to]0, \varepsilon_S[$ is locally trivial, hence a differentiable fibration. □

3.9.3 Corollary *Every Whitney stratified space is locally trivial with cones as typical fibers.*

PROOF: The space X is controllable by Theorem 3.6.9 and $(\pi_S, \rho_S) : T_S \setminus S \to]0, \varepsilon_S[$ is a proper submersion for every stratum S (where the control data $(T_S)_{S\in\mathcal{S}}$ are chosen appropriately), hence the claim follows immediately from the first isotopy lemma. □

As a consequence of local triviality we obtain the theorem that for every controllable stratified space the closed hull of a stratum can be resolved in a certain sense by a manifold-with-corners. This result originates in the work of VERONA [177] and will be proved in the following.

3.9.4 Theorem (VERONA [177, Prop. 2.6]) *Let X be a controllable stratified space of finite dimension such that $\partial X \neq \emptyset$, $\overline{X^\circ} = X$ and such that X° is a manifold of dimension* d. *Then there exist a* (d−1)*-dimensional manifold* Q *(without boundary), some proper control data $(T_S)_{S\in\mathcal{S}}$ of* X*, and a proper continuous mapping $H : Q \times [0,1] \to X$ with the following properties:*

(1) $H(Q\times]0,1[) \subset X^\circ$ *and* $H_{|Q\times]0,1[}$ *is a smooth embedding.*

(2) $H(Q \times \{0\}) = \partial X$.

(3) *The image of* H *is a neighborhood of* ∂X *in* X.

(4) *If* $y \in Q$ *and* $x = H(y, 0)$ *is a point of the stratum* S*, then there exist a neighborhood* $U \subset Q$ *of* y *and a stratum* $R \leq S$ *such that for every* $\tilde{y} \in U$ *and* $t \in [0,1[$ *the relations* $H(\tilde{y}, t) \in T_R$ *and* $\pi_R(H(\tilde{y}, t)) = \pi_R(H(\tilde{y}, 0))$ *hold true.*

3.9.5 Remark VERONA has claimed in [178, Prop. 1.3] that for every Whitney stratified space X the mapping H can be chosen as a $\mathcal{C}^1$-mapping. But in general this does not hold, because otherwise every Whitney space would have a locally finite volume in contradiction to the example constructed by FERRAROTTI [55, 58] of a Whitney stratified space which does not have a locally finite volume. But note that for every subanalytic set the mapping H can be chosen of class $\mathcal{C}^1$, even of class $\mathcal{C}^\omega$ (HIRONAKA [86], see also BIERSTONE–MILMAN [15]).

PROOF: Let $0 \leq d_0 < d_1 < \cdots < d_k = d$ be the sequence of dimensions of strata of X, and k the thus defined *dimension depth* of X. By $\partial X \neq \emptyset$ we have $k \geq 1$. Further let $(T_S)_{S\in\mathcal{S}}$ be some proper control data of X. Finally let S be the union of all strata of smallest dimension d_0. According to Proposition 3.6.7 $\phi := (\pi_{S,X^\circ}, \rho_{S,X^\circ}) : T_S \cap X^\circ \to]0, \varepsilon_S[$ then is a differentiable fibration and $S^\varepsilon := \phi^{-1}([\varepsilon])$ with $\varepsilon := \frac{1}{3}\varepsilon_S$ is a $(d-1)$-dimensional submanifold of X° (see 3.1.1 for the definition of $[\varepsilon]$). Without loss of generality we can assume after possibly shrinking ε_S that $T_S = T_S^{16\varepsilon_S}$ holds true and that $(\pi_S, \rho_S) : T_S \to [0, 16\varepsilon_S[$ is proper. After these agreements we now proceed in several steps.

1. STEP Some helpful smooth mappings.

Let $\varphi : [0,1] \times [0,1] \to [0,1] \times [0,1]$ be a function smoothing the corner according to Lemma C.4.1, and $\chi : \mathbb{R} \to [0,1]$ a further smooth function such that $\chi(s) = 0$ for $s \leq 0$, $\chi'(s) > 0$ for $0 < s < 1$ and $\chi(s) = 1$ for $s \geq 1$. The smooth curve $[0,1] \to \mathbb{R}^2$,

$s \mapsto \chi((s-\frac{1}{2})^2)\varphi(s,1)$ will be abbreviated by the letter c, its components by c_1 and c_2.

2. STEP The integral flow of a radial vector field.
First choose a radial vector field V on X such that

$$\phi_*(V(x)) = \frac{\partial}{\partial t}, \qquad x \in T_S.$$

Moreover, we require $V_{|S} = 0$. Now let $\gamma : J^\circ \to X^\circ$ be the integral flow of $V^\circ := V_{|X^\circ}$. Then for every $x \in X^\circ$ the integral curve γ_x satisfies the differential equation $\frac{\partial}{\partial t}\rho_S(\gamma_x(t)) = -1$ with initial value $\rho_S(x)$, hence

$$\rho_S(\gamma_x(t)) = t + \rho_S(x).$$

Thus by $\gamma_x(-\rho_S(x)) = \pi_S(x)$ one can extend the integral flow to a continuous mapping $\gamma : J \to X$ with

$$J = \{(x,t) \in (X^\circ \cap T_S) \times \mathbb{R} \mid -\rho_S(x) \le t < 2\varepsilon_S(\pi_S(x)) - \rho_S(x)\}.$$

Obviously γ is smooth on $J \cap (X^\circ \times \mathbb{R})$. The function

$$G : T_S^{\varepsilon s} \times [0,2[\to X, \quad (x,t) \mapsto \gamma_x((t-1)\,\rho_S(x))$$

then is continuous and smooth on $(T_S^{\varepsilon s} \setminus S) \times]0,2[$. Moreover, by the properties of the integral flow there exists a smooth function $\delta : X^\circ \to \mathbb{R}^{>0}$ which satisfies

$$\delta(x) \quad \begin{cases} = t, & \text{if } x = G(y,t) \text{ with } y \in S^\varepsilon \text{ and } t \in]0,3/2[, \\ \ge t, & \text{if } x = G(y,t) \text{ with } y \in S^\varepsilon \text{ and } t \in]3/2,2[, \\ \ge 2, & \text{if } x \notin G(S^\varepsilon \times]0,2[). \end{cases}$$

Thus we have a further ingredient for the proof of the theorem which we will now lead by induction on k.

3. STEP: Proof of the claim for $k = 1$.
By the properties of the integral flow γ the manifolds $Q := S^\varepsilon$ and the mapping $H := G_{|S^\varepsilon \times [0,1[}$ satisfy the above conditions (1) to (4). This gives the claim for $k = 1$.

4. STEP: Begin of the induction step, construction of an integral flow.
Let us suppose next that the theorem is true for all spaces of dimension depth $\le k$, and that X is a space of dimension depth $k+1$. With S from the 1. Step let $X' := X \setminus S$. As the following constructions can be performed separately for every connected component of S we can assume without loss of generality that S is connected. By induction hypothesis there exists a $(d-1)$-dimensional manifold Q' and a continuous map $H' : Q' \times [0,2[\to X'$ such that $H'_{|Q' \times [0,1]}$ satisfies the claim with respect to X' and such that $H'_{|Q' \times]0,2[}$ is a diffeomorphism onto its image. In the following constructions we want to glue together the manifolds Q' and S^ε as well as the mappings H' and G in a way such that the resulting objects have the desired properties. The difficulty now lies in smoothing the corner which results from intersecting $H'(Q' \times r)$, $0 < r < 1$ and S^ε.

To achieve this we first show that by (4) there exists a smooth function $r' : Q' \to \mathbb{R}^{>0}$, $r' \leq \frac{2}{3}$ such that for all $(y,t) \in Q' \times [0,1[$ with $t \leq \frac{3}{2} r'(y)$ and $H'(y,0) \in \overline{T_S}^{3\varepsilon_S/2}$ the relations $H'(y,t) \in T_S$ and $\pi_S(H'(y,t)) = \pi_S(H'(y,0))$ as well as $\rho_S(H'(y,t)) = \rho_S(H'(y,0))$ hold true. But then the submanifolds $H'([r'])$ and S^ε of X° are transversal for all $0 < s \leq 1$, and their intersection $B := H'([r']) \cap S^\varepsilon$ comprises a $(d-2)$-dimensional manifold. Now we choose a further smooth function $r : X^\circ \to \mathbb{R}^{>0}$ such that

$$r(x) \begin{cases} = \frac{t}{r'(y)}, & \text{if } x = H'(y,t) \text{ with } (y,t) \in Q' \times]0,3/2], \\ \geq \frac{t}{r'(y)}, & \text{if } x = H'(y,t) \text{ with } (y,t) \in Q' \times]3/2,2[, \\ \geq 2, & \text{if } x \notin H'(Q' \times]0,2[), \end{cases}$$

From now on the function $Q' \times [0,2[\to X$, $(y,t) \mapsto H'(y, t\, r'(y))$ will be denoted by H''. Out of H'' we obtain a vector field $W : U \to TX$, $x = H''(y,t) \mapsto \frac{\partial}{\partial s} H''(y,t+s)\big|_{s=0}$, over the open set $U := H''(Q' \times]0,3/2[)$. The integral flow of W will be denoted by $\zeta : I^\circ \to U$, where

$$I^\circ = \{(x,s) \in U \times \mathbb{R} \,|\, x = H''(y,t) \text{ and } 0 < s+t < \tfrac{3}{2}\}.$$

Now let us bring in the radial vector field $V : X \to TX$ from the 2. Step and its integral flow $\gamma : J^\circ \to Y$. Essential for the following is now the observation that the flows γ and ζ commute, as by assumption on H' and by the definition of V and W the two vector fields V and W commute.

5. STEP: Induction step continued, construction of Q.
Now we can define $Q \subset X^\circ$ as the union $Q = Q_1 \cup Q_2 \cup Q_3$ of the sets

$$Q_1 = \{H'(y, \tfrac{1}{3} r'(y)) \,|\, y \in Q' \text{ and } \delta(H'(y, \tfrac{1}{3} r'(y))) \geq \tfrac{5}{6}\},$$
$$Q_2 = \{H''(G(b, c_2(s)), c_1(s)) \,|\, b \in B \text{ and } s \in]0,1[\},$$
$$Q_3 = \{x \in S^{\varepsilon/81} \,|\, r(x) \geq \tfrac{5}{6}\} = \{x \in G(S^\varepsilon \times \{\tfrac{1}{3}\}) \,|\, r(x) \geq \tfrac{5}{6}\}.$$

Then one realizes that Q comprises even a $d-1$-dimensional submanifold of X°, as its components are submanifolds and as $Q_i \cap Q_j$, $i,j = 1,2,3$ are all open in Q_j. For the detailed proof of this fact use the definition of the functions r and δ, the one of the functions H'' and G and finally the fact that $H''(Q' \times \frac{1}{3})$ and S^ε intersect transversally in B.

6. STEP: End of the induction step, construction of F.
We will provide three functions $H_i : Q_i \times]0,1] \to X$, $i = 1,2,3$ with the desired properties. For $y \in Q_1$ set $H_1(y,t) = \zeta(y, \frac{t}{3} - 1)$. If $y = H''(G(b, c_2(s)), c_1(s))$ with $b \in B$ and $s \in]0,1[$ let $H_2(y,t) = H''(G(b, \varphi_2(s,t)), \varphi_1(s,t))$. Finally define $H_3(y,t) = \gamma(y, \frac{t}{3} - 1)$ for $y \in Q_3$. One now checks easily that the functions H_i and H_j coincide on the intersection $(Q_i \cap Q_j) \times]0,1]$. Hereby one has to use the fact that the integral flows γ and ζ commute. Altogether we thus obtain a smooth function $H : Q \times]0,1] \to X$ the restriction of which to $Q \times]0,1[$ has to be a diffeomorphism onto its image. Using the commutativity of the flows γ and ζ again, one realizes that H can be extended to a continuous function $H : Q \times [0,1] \to X$ and that then the properties (1) to (4) hold true. □

Applying the theorem and using a simple gluing argument one checks immediately the following result.

3.9.6 Corollary *Let* X *be a controlled stratified space and* S *a stratum of dimension* d *with* $\partial \mathsf{S} \neq \emptyset$. *Then there exist a* d*-dimensional manifold-with-boundary* M *and a proper continuous mapping* $\mathsf{f} : \mathsf{M} \to \mathsf{X}$ *such that the following properties hold*

(1) *The equality* $\mathsf{f}(\mathsf{M}^\circ) = \mathsf{S}$ *holds true, and* $\mathsf{f}_{|\mathsf{M}^\circ}$ *is a smooth embedding.*

(2) $\mathsf{f}(\partial \mathsf{M}) = \partial \mathsf{S}$.

3.9.7 A proper continuous mapping $\mathsf{f} : \mathsf{M} \to \mathsf{X}$ for a stratum S of X like in the corollary will be called a *resolution* of S.

Let more generally X be an (A)-stratified space of class $\mathcal{C}^m$ and $\mathsf{S} \subset \mathsf{X}$ a stratum. Then we call a proper $\mathcal{C}^m$-mapping $\mathsf{f} : \mathsf{M} \to \overline{\mathsf{S}}$, where M is a manifold-with-boundary, a *tempered resolution* of *class* $\mathcal{C}^m$, if the following properties hold:

(RT1) $\mathsf{f}_{|\mathsf{M}^\circ} : \mathsf{M}^\circ \to \mathsf{S}$ is a $\mathcal{C}^m$-diffeomorphism.

(RT2) Let μ be a smooth Riemannian metric on X and $\iota : \mathsf{M} \hookrightarrow \mathbb{R}^m$ a proper embedding. Then the components $\overline{f}_1, \cdots, \overline{f}_m$ of the composition $\overline{f} := \iota \circ \mathsf{f}_{|\mathsf{M}^\circ}{}^{-1} : \mathsf{S} \to \mathbb{R}^m$ are *geometrically tempered* of *class* $\mathcal{C}^m$ that means for every $y \in \partial \mathsf{S}$ there exist a neighborhood V and constants $c \in \mathbb{N}$ and $\mathsf{C} > 0$ such that for all $k \leq m$

$$\|\nabla^k f(x)\|_\mu \leq C \left(1 + \frac{1}{\delta_\mu(x, \partial S)^c}\right), \qquad x \in V. \tag{3.9.6}$$

3.9.8 Example By the already mentioned article [86] of HIRONAKA there exists for every stratum of the canonical stratification of a subanalytic set a tempered resolution of class $\mathcal{C}^\infty$. The reason is that according to HIRONAKA there exists for every subanalytic set (at least locally) a bimeromorphic resolution of singularities.

3.10 Cone spaces

According to Corollary 3.9.3 every Whitney space X is locally trivial with cones as typical fiber that means there exists a covering of X by open sets U such that U can be mapped by a stratified homomorphism k onto a cartesian product of the form $(\mathsf{S} \cap \mathsf{U}) \times \mathsf{CL}$, where S is a stratum of X and L a compact Whitney space. But in general one cannot choose k as a diffeomorphism between spaces with a smooth structure as is already shown by the example of Neil's parabola. Though Neil's parabola $\mathsf{X}_{\mathrm{Neil}} \subset \mathbb{R}^2$ is stratified homeomorphic to the edge $\mathsf{X}_{\mathrm{Edge}} = \mathsf{CS}^0 \subset \mathbb{R}^2$, there does not exist a diffeomorphism around the origin of $\mathbb{R}^2$, which (locally around the origin) maps Neil's parabola onto $\mathsf{X}_{\mathrm{Edge}}$, as the legs of $\mathsf{X}_{\mathrm{Neil}}$ touch in higher order then the ones of the edge $\mathsf{X}_{\mathrm{Edge}}$.

A special class of stratified spaces is given by such spaces, for which the homeomorphism k has additional regularity properties like for example that it can be chosen to be smooth or Lipschitz. Thus we obtain different categories of so-called cone spaces which are well suited for further geometric-analytic considerations. In this section we will introduce these cone spaces by a recursive definition.

But before we come to the details let us mention that by $\mathcal{C}^\omega(O)$ (resp. $\mathcal{C}^{1-}(O)$) with $O \subset \mathbb{R}^n$ open we will understand the space of all real analytic functions (resp. of all Lipschitz functions) on O.

3.10.1 Definition Let $m \in \mathbb{N} \cup \{1-, \infty, \omega\}$. A *cone space* of *class* $\mathcal{C}^m$ and *depth* 0 is the topological sum X of countably many smooth (resp. for $m = \omega$ real analytic) connected manifolds together with the stratification the strata S of which are given by the union of connected components of equal dimension. A *cone space* of *class* $\mathcal{C}^m$ and *depth* $d+1$, $d \in \mathbb{N}$ is a stratified space X with smooth (resp. real analytic) structure such that for all $x \in X$ there exist a connected neighborhood U of x, a compact cone space L of class $\mathcal{C}^m$ and depth d and finally a stratified homeomorphism

$$k : U \to (S \cap U) \times CL.$$

If $m \neq 0$ then it is required additionally that L is embedded into a sphere via a fixed smooth global singular chart $l : L \hookrightarrow S^l$ and that k and k^{-1} can be chosen as mappings of class $\mathcal{C}^m$. Hereby, the smooth structure on CL is the one induced by l. In other words this means that the smooth structure is given by the global chart

$$CL \to \mathbb{R}^{l+1}, [t, y] \mapsto t \cdot l(y), \quad t \in [0,1], y \in L.$$

Sometimes we call $l : L \hookrightarrow S^l$ a *link chart* of L and k a *cone chart.*

A *cone space* of *class* $\mathcal{C}^m$ or briefly a $\mathcal{C}^m$*-cone space* then is a stratified space with smooth structure such that for every $x \in X$ there exists a neighborhood U and an integer $d \in \mathbb{N}$ such that U is a cone space of class $\mathcal{C}^m$ and depth d.

3.10.2 Example According to MOSTOWSKI [130] every complex analytic set $X \subset \mathbb{C}^n$ has a so-called *Lipschitz stratification.* Now, for Lipschitz stratifications an isotopy lemma holds, where the local trivializations are bi-Lipschitz, hence complex analytic sets with a Lipschitz stratification according to MOSTOWSKI comprise nontrivial examples for $\mathcal{C}^{1-}$-cone spaces. Moreover, PARUSIŃSKI could show in his article [141] that every subanalytic set possesses a Lipschitz stratification, hence comprises with this stratification a $\mathcal{C}^{1-}$-cone space.

3.10.3 Example Every manifold-with-boundary as well as every manifold-with-corners is a cone space of class $\mathcal{C}^\infty$ as one can show without any difficulties. By a somewhat technically more involved but nevertheless canonical consideration one can prove that every polyhedron X with the smooth structure defined by a fixed triangulation $h : X \to \mathbb{R}^n$ is a cone space of class $\mathcal{C}^\infty$.

3.10.4 Theorem *Every cone space of class* $\mathcal{C}^m$ *with* $m \geq 2$ *is a Whitney space hence satisfies in every singular chart Whitney's condition* (B).

PROOF: We consider the following situation: Assume to be given a compact cone space $L \subset S^l$ of depth d, a stratum $S' \subset L$, an open ball $B \subset \mathbb{R}^n$ and a point $x \in B$. Then $X = B \times CL$ is a cone space of depth $d+1$. Moreover, $B \cong B \times \{0\}$ and

$S := B \times (]0,1[\cdot S')$ are strata of X and $(x,0) \in \mathbb{R}^{n+l+1}$ is a point of X in the stratum B. According to Lemma 1.4.4 and Remark 1.4.6 it suffices to prove that Whitney's condition (B) holds for the pair (B,S) at the point $x \in B$.

So let $(x_k)_{k\in\mathbb{N}}$ be a sequence of points of B with $\lim_{k\to\infty} x_k = x$, $(y_k)_{k\in\mathbb{N}}$ a sequence of points of S with $\lim_{k\to\infty} y_k = x$, and assume that the sequence of secants $\ell_k = \overline{x_k y_k}$ converges in projective space to a line $\ell \subset \mathbb{R}^{n+l+1}$. Now we can represent every y_k uniquely in the form $y_k = (x'_k, t_k \cdot y'_k)$ with $x'_k \in B$, $t_k \in]0,1[$ and $y'_k \in S'$. After transition to subsequences we can assume that the sequence of the y'_k converges to some $y' \in S'$ and that the sequence of normed difference vectors

$$\frac{(x_k - x'_k, -t_k \cdot y'_k)}{\sqrt{\|x_k - x'_k\|^2 + t_k^2}}$$

converges to a vector $(v,w) \in \mathbb{R}^n \times \mathbb{R}^{l+1}$. Then (v,w) spans the line ℓ, and the sequence of the tangent spaces $T_{y_k}S$ converges in the Graßmannian to a subspace $\tau = \mathbb{R}^n \times (\operatorname{span} y' + T_{y'}S')$. By a further transition to subspaces one can achieve, as t_k converges to 0, that either $\lim_{k\to\infty} \frac{\|x_k - x'_k\|}{t_k} = \infty$ holds true or $r := \lim_{k\to\infty} \frac{\|x_k - x'_k\|}{t_k} \in \mathbb{R}^{\geq 0}$. In the first case $w = 0$ follows, in the second $w = -\frac{1}{r+1} y'$. Hence in both cases $(v,w) \in \tau$ resp. $\ell \subset \tau$ is true. This proves the claim. □

3.10.5 Cone metrics In the geometric analysis of singular spaces particular attention is given to metric cones. These spaces have the following shape. Let us be given a compact manifold M with a Riemannian metric μ. Then

$$\eta_{(t,x)} = dt^2 \oplus t^2 \mu_x, \qquad (t,x) \in CM \setminus \{o\} \tag{3.10.1}$$

provides a canonical Riemannian metric on the stratified cone CM. The pair (CM, η) is called a *metric cone* and η its *cone metric* over μ. Now we continue analogously to the definition of a cone space that means we build according to Eq. 3.10.1 the cone metric over η and so on. A Riemannian metric on the cone space X which locally can be constructed by such an iterative processes will be called a *cone metric* for X.

Considering the interesting results already obtained for metric cones the study of more general cone metrics appears promising, though one can expect it to be rather involving.

We close this section with several historical remarks which essentially are taken form LESCH [111]. The study of metric cones was initiated by CHEEGER [38, 39, 41], differential operator of order 1 and 2 on such spaces have been considered among others by BRÜNING–SEELEY [32, 33, 34]. MELROSE [126] and SCHULZE [148, 149] have introduced independently an important class of differential operators on metric cones: the so-called *operators* of *Fuchs type*. These are differential operators of the form

$$t^{-n} \sum_{k=0}^{m} A_k(t) \left(-t\frac{\partial}{\partial t}\right)^k,$$

where $m, n \in \mathbb{N}^{>0}$ and the A_k are smooth families of differential operators on M. The reader can find a detailed exposition of the theory of Fuchs type operators together with further information on metric cones and many references in LESCH [111].

Chapter 4

Orbit Spaces

As already mentioned in the introduction, orbit spaces of certain, or more precisely proper G-actions of a Lie group G give nice examples of stratified spaces with smooth structure. Moreover, orbit spaces play an important role for many considerations in mathematics and mathematical physics. For this reason, they will be treated here in rather detail, where in accordance with the scope of this monograph attention is given primarily on the canonical stratification by orbit types and the construction of the smooth structure. The results of the first three sections of this chapter are standard, at least for the compact case. Thus, we have formulated them from the beginning in the greatest possible generality not only for the case of compact G's but also for the case of proper G-actions. As references for Sections 4.1 to 4.3 serve in particular JÄNICH [95], BREDON [25] and LESCH [110]. Concerning the canonical stratification of an orbit space original references are BIERSTONE [13, 14], SJAMAAR–LERMAN [162], DOVERMANN–SCHULTZ [52, p. 67] and FERRAROTTI [56].

4.1 Differentiable G-Manifolds

4.1.1 Let M be a manifold and G a Lie group. By a *(left) action* of G we mean a smooth mapping

$$\Phi : G \times M \to M, \quad (g, x) \mapsto \Phi(g, x) = \Phi_g(x) = gx$$

such that for all $g, h \in G$ and $x \in M$ the relations $\Phi_g(\Phi_h(x)) = \Phi_{gh}(x)$ and $\Phi_e(x) = x$ hold, e being the identity element of G. By a *right action* of G we mean a smooth mapping

$$\Psi : M \times G \to M, \quad (x, g) \mapsto \Psi(x, g) = \Psi_g(x) = xg,$$

such that $\Phi : G \times M \to M$, $(g, x) \mapsto \Psi(x, g^{-1})$ describes a left action of G.

We often call a manifold M together with a G-action $\Phi : G \times M \to M$ a *differentiable* G-*manifold* or shorter a G-*space*. A left or right action of G on M is said to be *transitive* provided that for all pairs (x, y) of points of M there exists a $g \in G$ with $gx = y$ and $xg = y$, respectively. The G-action is called *effective* or *faithful*, if the relation $\Phi_g = \mathrm{id}_M$ respectively $\Psi_g = \mathrm{id}_M$ is fulfilled, if and only if $g = e$. In other words, this means that the canonical homomorphism of G into the group of diffeomorphisms Diff(M) is injective.

M.J. Pflaum: LNM 1768, pp. 151 - 168, 2001

A *morphism of* G*-actions* or a G*-equivariant mapping* is a differentiable mapping $f : M \to N$ between G-spaces M and N such that for all $g \in G$ and $x \in M$ the equation $f(gx) = gf(x)$ is satisfied. Now, if $\gamma : G \to H$ denotes a smooth homomorphism of Lie groups, we call a smooth mapping $f : M \to N$ from a G-space M into an H-space N *γ-equivariant*, if the diagram

$$\begin{array}{ccc} G \times M & \xrightarrow{\gamma \times f} & H \times N \\ \downarrow & & \downarrow \\ M & \xrightarrow{f} & N \end{array} \tag{4.1.1}$$

commutes. The G-equivariance is therefore equivalent to the id_G-equivariance.

4.1.2 For a point $x \in M$ the set $Gx = \{gx \in M | g \in G\}$ is said to be the *orbit* of x in M. The partition of M into its various orbits then describes an equivalence relation on M; we call the corresponding quotient space of equivalence classes the *orbit space* of M, denoting it by $G\backslash M$. In an analogous way one defines for a manifold N with a right action of G the orbits qG with $q \in N$ and the orbit space N/G. Next, we equip $G\backslash M$ (resp. N/G) with the quotient topology with respect to the canonical projection $\pi : M \to G\backslash M$ (resp. $\pi : N \to N/G$). This makes π into a continuous and open mapping, as for all $U \subset M$ open $\pi^{-1}(\pi(U)) = \bigcup_{g\in G} gU$ is open in M. Usually, the orbit space $G\backslash M$ is not a differentiable manifold, sometimes not even Hausdorff. For a relatively large and most applications sufficient class of G-manifolds – namely those with a so-called proper G-action – the orbit space $G\backslash M$ possesses the structure of a Whitney space. In the next sections we will explain this in more detail and introduce in this paragraph the new notions necessary for this purpose. In the following, if not otherwise mentioned, definitions and results will be given explicitly only for the case of left actions, tacitly assuming that these hold in the "right" case, too.

4.1.3 For each point $x \in M$ define its *isotropy group* or *stabilizer* or *symmetry group* by $G_x = \{g \in G | gx = x\}$. One easily checks that G_x is a subgroup of G, and that for all $g \in G$ the relation $G_{gx} = gG_xg^{-1}$ holds. In other words, this means that the isotropy groups of two points of an orbit are conjugate to each other. Consequently, to each orbit there is a uniquely assigned conjugacy class, namely the conjugacy class (G_y) of the isotropy group G_y of an arbitrary point $y \in Gx$. In the following (G_y) will be called the *type* of the orbit Gx. A G-action of M is said to be *free*, if all the isotropy groups G_x are trivial in the sense of being equal to $\{e\}$. Every free group action is effective, for using the fact that Φ is free it follows immediately from $\Phi_g = \mathrm{id}_M$ that $g = e$. Conversely, not every effective G-action needs to be free.

To every closed subset $H \subset G$ one assigns the following three subspaces of M:

$$M_H := \{x \in M | \ G_x = H\},$$
$$M^H := \{x \in M | \ G_x \supset H\},$$
$$M_{(H)} := \{x \in M | \ G_x \sim H\}.$$

M^H then describes nothing else but the *fixed point set* of H in M.

4.1.4 Differentiating a G-action $\Phi : G\times M \to M$ with respect to the second variable one obtains a G-action on the tangent bundle of M:

$$G \times TM \to TM, \quad (g, v) \mapsto gv = T\Phi_g(v).$$

Conversely, differentiating Φ with respect to the first variable one obtains for every element $\xi \in \mathfrak{g}$ of the Lie algebra of G a canonical vector field on M, the so-called *fundamental vector field* ξ_M of ξ. Explicitly, ξ_M is given by

$$\xi_M(x) := \frac{\partial}{\partial t}\Phi(\exp t\xi, x)\Big|_{t=0}, \qquad x \in M.$$

The G-action Φ yields in a functorial way – apart from the G-action on the tangent bundle – also one on the cotangent bundle, on tensor and exterior products of these bundles and so on. For the sake of completeness we give here the action on T^*M:

$$G \times T^*M \to T^*M, \quad (g, \alpha_x) \mapsto \left(T_{gx}M \ni v \mapsto \langle \alpha_x, g^{-1}v\rangle\right), \qquad \alpha_x \in T_x^*M, \; x \in M.$$

Now the bundle of exterior forms becomes a G-manifold, hence it makes sense to speak of a G-*invariant differential form* on M; this is then a differential form $\alpha \in \Omega^k(M)$ such that $\alpha_{gx} = g\alpha_x$ for all $x \in M$ and $g \in G$. If one finally requires additionally that the contraction of α by each fundamental vector field ξ_M vanishes, i.e. that $i_{\xi_M}\alpha = 0$ holds for every $\xi \in \mathfrak{g}$, then α is said to be a *basic differential form.* The space of basic k-forms on M is denoted by $\Omega^k_{\text{basic}}(G\backslash M)$. The basic differential forms on M can be used for the computation of the cohomology of $G\backslash M$ (see 5.3).

4.2 Proper Group Actions

4.2.1 Definition A G-action $\Phi : G \times M \to M$ is called *proper* if the mapping

$$\Phi_{\text{ext}} : G \times M \to M \times M, \quad (g, x) \mapsto (gx, x)$$

is proper.

4.2.2 Example For a compact Lie group G all G-actions are obviously proper.

4.2.3 Example One might think that all free G-actions are proper. This is, however, not the case as shown by the following action of $\mathbb{R}$ on the torus $S^1 \times S^1$ with the irrational angle $\alpha \in \mathbb{R}/2\pi\mathbb{Z}$:

$$\mathbb{R} \times S^1 \times S^1 \to S^1 \times S^1, \quad (r, e^{i2\pi s}, e^{i2\pi t}) \mapsto (e^{i2\pi(s+r\cos\alpha)}, e^{i2\pi(t+r\sin\alpha)})$$

The following theorem is - aside from the slice theorem proved later on - the starting point of all further investigations concerning proper group actions.

4.2.4 Theorem *Let $\Phi : G \times M \to M$ be a proper group action. Then the following holds:*

(1) *Each orbit* Gx, $x \in M$ *describes a closed submanifold of* M. *Moreover, the canonical mapping*

$$\Phi_x : G/G_x \to M, \quad g\,G_x \mapsto gx$$

yields a diffeomorphism from G/G_x *onto the orbit* Gx.

(2) *The isotropy group* G_x *of any point* $x \in M$ *is compact.*

(3) *The canonical projection* $\pi : M \to G\backslash M$ *is closed. The orbit space* $G\backslash M$ *is Hausdorff, locally compact and endowed with a countable topology.*

(4) *To any covering of* M *by* G*-invariant open sets there exists a subordinate partition of unity by* G*-invariant smooth functions.*

(5) *The algebra* $\mathcal{C}^\infty(M)^G$ *of* G*-invariant smooth functions separates the points of* M.

(6) M *admits a* G*-invariant Riemannian metric.*

PROOF: To begin with, we first show (2). It is possible to write the isotropy group $G_x = \{g \in G \mid gx = x\}$ in the form

$$G_x = \mathrm{pr}_1(\Phi_{\mathrm{ext}}^{-1}(x, x));$$

it is therefore compact since it is the inverse image of the compact set (x, x) under the proper mapping Φ_{ext}.

Let us now prove (1). Since G_x is compact, hence a Lie subgroup of G, G/G_x needs to be a (real analytic) manifold. We first show Φ_x to be an injective immersion. The injectivity is obvious, since from $gx = hx$ it follows immediately $g^{-1}h \in G_x$, hence $gG_x = hG_x$. To show that Φ_x is immersive it suffices to prove that the differential $T_{eG_x}\Phi_x$ is injective, since Φ_x is equivariant with respect to the G-action $G \times G/G_x \to G/G_x$. So, let $v \in T_{eG_x}G/G_x$ be a tangent vector with $T_{eG_x}\Phi_x.v = 0$. Because of the fact that the canonical projection $\pi : G \to G/G_x$ is submersive, there exists a $\xi \in \mathfrak{g} = T_eG$ with $T_e\pi.\xi = v$. Then it follows

$$T_e\Phi(-, x).\xi = T_{eG_x}\Phi_x.T_e\pi.\xi = T_{eG_x}\Phi_x.v = 0,$$

and for the curve $\gamma(t) = \exp t\xi$,

$$\frac{\partial}{\partial t}\Phi(\gamma(t), x)\Big|_{t=s} = \frac{\partial}{\partial t}\Phi(\gamma(t+s), x)\Big|_{t=0} = \left(T_x\Phi(\gamma(s), \cdot) \circ T_e\Phi(-, x)\right)(\xi) = 0,$$

using the fact that $\Phi(\gamma(t+s), x) = \Phi(\gamma(s), \gamma(t)x)$. The result is $\Phi(\gamma(t), x) = x$ for all $t \in \mathbb{R}$, or in other words $\gamma(t) \in G_x$. This implies $\xi \in T_eG_x$, hence $v = T_e\pi.\xi = 0$. Consequently, Φ_x is immersive. Since the mapping $G \times \{x\} \to M \times \{x\}$ is proper, this also holds for Φ_x. Regarded as an injective immersion, Φ_x is therefore an embedding, hence a diffeomorphism onto its image Gx.

Next, we show (3). Let $A \subset M$ be closed. Since Φ_{ext} is proper, $GA \times A = \Phi_{\mathrm{ext}}(G, A)$ needs to be closed in $M \times M$, hence GA is closed in M. Because of $GA = \pi^{-1}(\pi(A))$, $\pi(A)$ is closed in $G\backslash M$, i.e. π is closed. Consider now two different orbits Gx and Gy. M is normal, Gy closed, and therefore there exist two disjoint open neighborhoods U of x and V of Gy. In particular, this means that $\overline{\mathsf{U}} \cap Gy = \emptyset$ and $\pi(y) \notin \pi(\mathsf{U})$. Due to the fact that π is closed $\pi(\mathsf{U})$ and $(G\backslash M)\backslash\pi(\overline{\mathsf{U}})$ then are disjoint open neighborhoods of $\pi(x)$ and $\pi(y)$, respectively. Thus, $G\backslash M$ is Hausdorff. The local compactness and separability of $G\backslash M$ follows directly from the corresponding properties of M.

Before proving (4), first note that by (3) the orbit space $G\backslash M$ is paracompact. Then we assume that $\mathcal{U} = (\mathsf{U}_\iota)_{\iota\in J}$ is a covering of M by G-invariant open sets U_ι. By the paracompactness of $G\backslash M$ there exists a locally finite covering of $G\backslash M$ by open sets V_ι such that $\pi^{-1}(V_\iota) \subset \mathsf{U}_\iota$. Moreover, there is a locally finite smooth partition of unity $(\psi_j)_{j\in\mathbb{N}}$ on M and a mapping $\iota : \mathbb{N} \to J$ in such a way that $\operatorname{supp}\psi_j$ is compact for all $j \in \mathbb{N}$ and such that $\operatorname{supp}\psi_j \subset \pi^{-1}(V_{\iota(j)})$. Choose now a right invariant Haar measure μ on G. By virtue of the hypothesis that the supports $\operatorname{supp}\psi_j$ are compact, there exists for $x \in M$ and $j \in \mathbb{N}$ the integral

$$\psi_j^G(x) = \int_G \psi(gx)\, d\mu(g).$$

An easy argument shows that all the ψ_j^G describe smooth functions on M and that $\operatorname{supp}\psi_j^G \subset \pi^{-1}(V_{\iota(j)})$ holds. On the other hand, the family of supports $(\operatorname{supp}\psi_j^G)_{j\in\mathbb{N}}$ need not be locally finite any more, a lack we intend to remedy by a suitable summation of the functions ψ_j^G. Since $\mathcal{C}^\infty(M)$ is a Fréchet space, there exists a sequence of seminorms $\|\cdot\|_j$ on $\mathcal{C}^\infty(M)$ defining the Fréchet topology such that $\|\cdot\|_j \leq \|\cdot\|_{j+1}$ for all $j \in \mathbb{N}$. We define

$$\tilde{\varphi}_\iota := \sum_{\substack{j\in\mathbb{N}\\ \iota(j)=\iota}} \frac{1}{2^j\,\|\psi_j^G\|_j}\,\psi_j^G.$$

Then even the functions $\tilde{\varphi}_\iota$ are smooth as well, G-invariant and satisfy $\operatorname{supp}\tilde{\varphi}_\iota \subset \pi^{-1}(V_\iota)$. Because of the fact that the covering $(V_\iota)_{\iota\in J}$ is locally finite the family of supports $\operatorname{supp}\tilde{\varphi}_\iota$ is locally finite on its own, hence for all $x \in M$

$$\varphi_\iota(x) = \frac{1}{\tilde{\varphi}(x)}\tilde{\varphi}_\iota(x) \quad \text{with } \tilde{\varphi}(x) = \sum_{\iota\in J} \varphi_\iota(x)$$

is well defined. Now the family $(\varphi_\iota)_{\iota\in J}$ is a locally finite and G-invariant partition of unity subordinate to $\mathcal{U}$.

On (5): Let Gx and Gy be two disjoint orbits. Since $G\backslash M$ is paracompact, hence in particular normal, we can choose two open neighborhoods V_1 and V_2 of $\pi(x)$ and $\pi(y)$, respectively. Setting now $\mathsf{U}_i = \pi^{-1}(V_i)$ for $i = 1, 2$ and $\mathsf{U}_3 = M\backslash(Gx \cup Gy)$, we obtain $(\mathsf{U}_1, \mathsf{U}_2, \mathsf{U}_3)$ as a G-invariant open covering of M, with a G-invariant partition of unity $(\varphi_1, \varphi_2, \varphi_3)$ subordinate to it, existing by virtue of the statements just proven. Then $\varphi_1(x) = \varphi_2(y) = 1$ and $\varphi_2(x) = \varphi_1(y) = 0$ holds, meaning that $\mathcal{C}^\infty(M)^G$ separates the points of M.

Finally, we would like to prove (6). To this end, we first choose an arbitrary Riemannian metric η on M and a compact exhaustion $(K_j)_{j\in\mathbb{N}}$ of M. Afterwards choose for every $j \in \mathbb{N}$ a smooth cut-off function $\chi_j : M \to [0, 1]$ in such a way that $\operatorname{supp}\chi_j \subset K_{j+1}^\circ$ and $\chi_j(x) = 1$ for all $x \in K_j$. By means of the Haar measure on G

already used above we define G-invariant smooth sections $\eta_j : M \to T^*M \otimes_x T^*M$ by

$$\eta_j(x)(v, w) = \int_G \chi_j(gx)\, \eta_{gx}(gv, gw)\, d\mu(g), \qquad x \in M, \;\; v, w \in T_xM.$$

By the assumptions concerning η and the χ_j all the forms $\eta_j(x)$ are symmetric and positive semidefinite. If $x \in GK_j$, $\eta_j(x)$ is even positive definite. Since the family $(U_j)_{j\in\mathbb{N}}$ with $U_j = GK_j^\circ$ describes a G-invariant open covering of M, there exists by (4) a partition of unity $(\varphi_j)_{j\in\mathbb{N}}$ subordinate to $(U_j)_{j\in\mathbb{N}}$. Define $\mu : M \to T^*M \otimes_x T^*M$ by

$$\mu(x) = \sum_{j\in\mathbb{N}} \varphi_j(x)\, \eta_j(x), \quad x \in M.$$

This gives a G-invariant Riemannian metric on M, thus proves the last claim of the theorem. □

4.2.5 In order to prepare the slice theorem, consider for $x \in M$ the normal space $\mathcal{V}_x = T_xM/T_xGx$ of the orbit Gx at x, the so-called *slice* of x. For each element g of the isotropy group G_x the differential $T\Phi_g$ maps the tangent space T_xGx of the orbit Gx again into T_xGx, hence induces an automorphism of $\mathcal{V}_x$. Consequently we obtain the so-called *slice representation*

$$S_x : G_x \to GL(\mathcal{V}_x).$$

Since the homogenous space $G \to G/G_x$ describes a G_x-principal fiber bundle, we obtain an associated bundle $N_x = G \times_{G_x} \mathcal{V}_x$, the *slice bundle* of x. As G_x is compact, there exists on $\mathcal{V}_x$ a G_x-invariant metric, with respect to which one can define the sphere $S\mathcal{V}_x = \{v \in \mathcal{V}_x \mid \|v\| = 1\}$ and the *sphere bundle* $SN_x = G \times_{G_x} S\mathcal{V}_x$. Then the group G_x acts in a natural way on $S\mathcal{V}_x$ such that SN_x is well-defined and becomes a differentiable G-space.

The slice theorem now states only one thing, namely that every G-manifold with a proper G-action locally looks like a neighborhood of the zero section in the slice bundle.

4.2.6 Slice Theorem (Koszul [104, p. 139], Palais [138]) *Let $\Phi : G \times M \to M$ be a proper group action, x a point of M and $\mathcal{V}_x = T_xM/T_xGx$ the normal space to the orbit of x. Then there exists a G-equivariant diffeomorphism from a G-invariant neighborhood of the zero section of $G \times_{G_x} \mathcal{V}_x$ onto a G-invariant neighborhood of Gx such that the zero section is mapped onto Gx in a canonical way.*

Proof: Since the exponential function of a G-invariant metric is again G-invariant, the slice theorem follows immediately from the classical tubular neighborhood theorem 3.1.6. □

4.2.7 Remark In the literature one often calls the uniquely determined zero neighborhood $V \subset \mathcal{V}_x$ with $\varphi(U) \cap (\{e\} \times \mathcal{V}_x) = \{e\} \times V$ the *slice* of x, e being the identity element of G and $\varphi : U \to G \times_{G_x} \mathcal{V}_x$ the G-equivariant diffeomorphism which emerges as in the slice theorem.

4.2.8 Corollary *For every compact subgroup* $H \subset G$ *the stes* M_H, $M_{(H)}$ *and* M^H *are* Σ*-submanifolds of* M. *In other words, this means that each connected component of* M_H, $M_{(H)}$ *and* M^H *is a submanifold of* M. *Moreover, these three sets fulfill the following relation:*

$$M_H = M_{(H)} \cap M^H. \tag{4.2.1}$$

PROOF: Due to the fact that the statement is a local one it suffices by the slice theorem to consider the case that $M = G \times_H \mathcal{V}$, where $H \subset G$ is compact and $\mathcal{V}$ is an H-module. Then the isotropy group of a point $[(g, v)] \in G \times_H \mathcal{V}$ is $G_{[(g,v)]} = gH_v g^{-1}$, where $H_v \subset H$ denotes the isotropy group of the H-manifold $\mathcal{V}$ of v. Indeed, $\tilde{g}[(g, v)] = [(g, v)]$ holds if and only if there exists an $h \in H$ with $\tilde{g}gh^{-1} = g$ and $hv = v$.

By virtue of the lemma below H_v and consequently $G_{[(g,v)]}$ are conjugate to H, if and only if $H_v = H$, i.e. if v lies in the fixed point space $\mathcal{V}^H \subset \mathcal{V}$ of H. Using this, $M_{(H)}$ equals the closed differentiable subbundle $G \times_H \mathcal{V}^H = G/H \times \mathcal{V}^H$. The isotropy group $G_{[(g,v)]} = gH_v g^{-1}$ equals H obviously, if and only if g lies in the normalizer $N_G(H)$ of H in G. Consequently, M_H needs to be the same as the closed differentiable submanifold $N_G(H) \times_H \mathcal{V}^H = N_G(H)/H \times \mathcal{V}^H$, as the normalizer $N_G(H)$ is closed in G, describing therefore a Lie subgroup of G. The relation (4.2.1) is a direct consequence of the definitions of M_H, $M_{(H)}$ and M^H as well as of the following lemma. □

4.2.9 Lemma *Let* G *be a Lie group and* $H \subset G$ *a compact subgroup. Then every closed subgroup* $H_0 \subset H$ *conjugate to* H *is identical to* H.

PROOF: Let $g \in G$ be an element such that $\mathrm{Ad}_g(H) = gHg^{-1} = H_0$. Since Ad_g is a diffeomorphism of G, H_0 needs to be a subgroup of H of the same dimension, meaning that the connected components of the unity of H and H_0 agree. From this it follows that for every $h \in H_0$ the connected component of h in H_0 needs to agree with that of H. Due to the compactness of H and H_0 both of them possess only finitely many connected components and therefore the claim will be given if it can be shown that H and H_0 have the same number of connected components. But noting $\mathrm{Ad}_g(H) = H_0$ this is the case, indeed. □

4.2.10 Proposition *Suppose the Lie group* G *acts in a proper and free way from the left on the manifold* P. *Then there exists a uniquely determined manifold structure on the quotient space* $G\backslash P$ *such that the canonical projection* $\pi : P \to G\backslash P$ *turns into a differentiable fiber bundle with typical fiber* G.

4.2.11 Note and Definition A fiber bundle $P \to N$ occurring as in the proposition by means of a proper free left action of G will be denoted as *opposite* G*-principal bundle.* Usually the structure group of a principal bundle operates from the right on the total space, which is the reason why we have chosen the additive "opposite" to express the structure group acting from the left. Analogously to the case of ordinary principal bundles one can associate to opposite fiber bundles as well: However, they arise from manifolds F on which G operates from the right, and will be denoted by $F\,{}_G{\times} P \to N$.

PROOF: Since the group action is proper, we already know that $G\backslash P$ is a locally compact Hausdorff space with countable topology. With the help of the slice theorem one constructs now local charts for $G\backslash P$. Let $x \in G\backslash P$ and $z \in P$ be a point with $x = Gz$. Due to the fact that the group action is free, there exists a G-equivariant diffeomorphism $\Psi = (\Psi_1, \Psi_2) : U \to G \times V$ from a neighborhood U of z onto a product $G \times V$, $V \subset \mathcal{V}_z$ being a zero neighborhood of the slice to z. Then the map $s : \pi(U) \to \mathcal{V}_z$, $Gy \mapsto \Psi_2(y)$ is well-defined in a neighborhood $\pi(U)$ of x, continuous and a homomorphism onto its image. Any two of those charts of $G\backslash P$ are compatible by virtue of the slice theorem, hence the set of all $s : \pi(U) \to \mathcal{V}_z$ defines a differentiable atlas on $G\backslash P$. Moreover, the projection $\pi : P \to G\backslash P$ describes a fiber bundle, since by construction this is the case locally in charts: $s \circ \pi_{|U} \circ \Psi^{-1} : G \times V \to V \subset \mathcal{V}_z$ is nothing else but the projection onto the second coordinate.

The differentiable structure of $G\backslash P$ is uniquely determined, since by the fiber bundle property of π the sheaf $\mathcal{C}^\infty_{G\backslash P}$ of infinitely times differentiable functions must be the same as $\pi_* \mathcal{C}^\infty_P$, but on the other hand $\mathcal{C}^\infty_{G\backslash P}$ determines the manifold structure. □

4.3 Stratification of the Orbit Space

4.3.1 The set of conjugacy classes of closed subgroups of a Lie group G is ordered, defining $(K) \leq (H)$ as to be equivalent to H being conjugate to a subgroup of K.

4.3.2 Theorem *Let $\Phi : G \times M \to M$ be a proper group action and $G\backslash M$ connected. Then the orbit types of Φ satisfy the following relations:*

(1) *There is a uniquely determined conjugacy class (H°) in G such that $M_{(H^\circ)} \subset M$ is open and dense. Moreover, $G\backslash M_{(H^\circ)}$ is connected.*

(2) *Every compact subgroup $H \subset M$ emerging as isotropy group of an $x \in M$ fulfills $(H) \leq (H^\circ)$. In other words, (H°) is maximal in the ordered set of orbit types of M.*

(3) *For any two compact subgroups $K, H \subset G$ with $(H) < (K)$ the set $M_{(H)} \cap \overline{M_{(K)}}$ is open and closed in $M_{(H)}$.*

4.3.3 Definition (H°) is said to be the *principal orbit type* of M, and $M_{(H^\circ)}$ the *principal orbit bundle*. The orbits lying in $M^\circ := M_{(H^\circ)}$ are called *principal orbits*.

4.3.4 Remark The assumption that $G\backslash M$ is connected does not mean any restriction of generality since an arbitrary G-manifold can be decomposed into the G-manifolds $\pi^{-1}(Z)$, where Z runs through the connected components of $G\backslash M$.

PROOF OF THE THEOREM: The existence of the principal orbit type in (1) will be shown as it is done in JÄNICH [95, Theorem 2.1] by induction by $\dim M$. For $\dim M = 0$ the orbit space $G\backslash M$ consists by assumption only of a single point, hence M has only one orbit. Let now M be n-dimensional. We first consider a slice bundle

$N_x = G \times_{G_x} \mathcal{V}_x$ of $x \in M$ and the sphere bundle $SN_x = G \times_{G_x} S\mathcal{V}_x$ assigned with respect to a G_x-invariant metric. By induction hypothesis the claim is satisfied for SN_x provided that the orbit space $G\backslash SN_x = G_x \backslash S\mathcal{V}_x$ is connected. If, however, it is not connected, then $\dim \mathcal{V}_x = 1$ and $G_x \to GL(\mathcal{V}_x)$ must be the trivial representation. Then the orbit type of each point of SN_x would be (G_x). However, in any case it follows that with $\varepsilon > 0$ the G-space $N_x^\varepsilon = \{[(g, v)] \in G \times_{G_x} \mathcal{V}_x | \, \|v\| < \varepsilon\}$ possesses a principal orbit type in the sense of the theorem. By virtue of the slice theorem and the paracompactness of the orbit space we can now cover M by locally finitely many of such N_x^ε. The fact that $G\backslash M$ is connected gives that any two slice bundles $N_{p_0}^{\varepsilon_0}$ and N_x^ε can be joined with each other by a chain $N_{p_0}^{\varepsilon_0}, \cdots, N_{p_k}^{\varepsilon_k} = N_x^\varepsilon$, i.e. $N_{p_j}^{\varepsilon_j} \cap N_{p_{j+1}}^{\varepsilon_{j+1}} \neq \emptyset$ for $0 \leq j < k$. Hence the principal orbit types (H) of all N_x^ε have to coincide, their union therefore forms an open and dense set $M_{(H)} \subset M$, the quotient $G\backslash M_{(H)}$ of which is connected. So we have proved (1). (2) now follows from (1). Finally, we get to the last claim of the theorem, to (3). We perform the proof according to SJAMAAR [161, Lem. 1.2.21]. Without loss of generality it can be assumed that K is a subgroup of H. By virtue of the slice theorem it suffices to show that for M of the form $M = G \times_H \mathcal{V}$ and a non-empty $M_{(K)}$ the closure of $M_{(K)}$ contains the submanifold $M_{(H)} = G/H \times \mathcal{V}^H$. Be the H-module $\mathcal{V}$ endowed with an H-invariant metric, and let $\mathcal{W}$ be the orthogonal space to $\mathcal{V}^H$. Then M has the representation $M = (G \times_H \mathcal{W}) \times \mathcal{V}^H$, the use of which results in

$$M_{(K)} = G \times_H \mathcal{V}_{(K)} = (G \times_H \mathcal{W}_{(K)}) \times \mathcal{V}^H.$$

Since $\mathcal{W}_{(K)}$ is invariant with respect to multiplication by non-vanishing scalars and by assumption to $M_{(K)}$ must not be empty, the origin of $\mathcal{W}$ lies in $\overline{\mathcal{W}_{(K)}}$. It therefore follows

$$M_{(H)} = G/H \times \mathcal{V}^H \subset (G \times_H \overline{\mathcal{W}_{(K)}}) \times \mathcal{V}^H = \overline{M_{(K)}},$$

which was to be proved. □

4.3.5 To any G-manifold M with G acting properly on it we can now give a stratification of M assigning to each point $x \in M$ the germ $\mathcal{S}_x$ of the set $M_{(G_x)}$. Usually $\mathcal{S}$ is called the *stratification by orbit types*. Due to theorem 4.3.2 $\mathcal{S}$ is a stratification in the sense of definition 1.2.2, indeed, provided that we can furthermore show that the decomposition of M into the submanifolds $M_{(H)}$ is locally finite. Together with the slice theorem, this is, however, a direct consequence of the following lemma.

4.3.6 Lemma *Let $H \subset G$ be a compact subgroup, and $\mathcal{V}$ an H-module. Then the G-space $G \times_H \mathcal{V}$ possesses only finitely many orbit types. In particular, every compact manifold M with a proper G-action on it possesses only finitely many orbit types.*

PROOF: Let us show the second claim by induction on dim M. For $\dim M = 0$ the claim is trivial, because in this case M consists of only finitely many points, hence M has only finitely many orbit types. Now, let M be an n-dimensional manifold. Since M can be covered by virtue of the slice theorem by finitely many open sets of the form $G \times_H \mathcal{V}$ with $\dim \mathcal{V} \leq n$ and $H \subset G$ compact, it suffices to show the induction step for these G-manifolds. Since H is compact and $\mathcal{V}$ a H-module, there is an H-invariant metric on $\mathcal{V}$. Let $S\mathcal{V}$ be the unit sphere with respect to this metric. Then $S\mathcal{V}$ has dimension $< n$, and is a compact H-manifold, moreover. By the induction hypothesis $S\mathcal{V}$ possesses only finitely many orbit types. Due to the proof of corollary 4.2.8 the isotropy group of $[(g, v)] \in G \times_H S\mathcal{V}$ equals $gH_v g^{-1}$, i.e. the number of orbit types of $S\mathcal{V}$, $G \times_H S\mathcal{V}$ and $G \times_H (\mathcal{V} \setminus \{0\})$ agree. Compared to $G \times_H (\mathcal{V} \setminus \{0\})$, the space $G \times_H \mathcal{V}$ has at most the orbit type (H) in addition, hence the induction step follows. Moreover, the proof of the second claim entails that the first claim needs to be true as well. □

4.3.7 Theorem *The stratification by orbit types of a* G*-manifold* M *with proper group action makes* M *into a Whitney stratified space.*

PROOF: It only remains to show that the Whitney condition (B) is satisfied. To this end, let $K \subsetneq H \subset G$ be two (compact) isotropy groups of M, i.e., in other words, $M_{(H)} < M_{(K)}$ may hold. Furthermore, let two sequences $(x_k)_{k\in\mathbb{N}} \subset M_{(K)}$ and $(y_k)_{k\in\mathbb{N}} \subset M_{(H)}$ be given, converging to a $y \in M_{(H)}$, where we additionally assume that in a smooth chart around y the secants $\ell_k = \overline{x_k y_k}$ converge to a straight line ℓ, and the tangent spaces $T_{x_k} M_{(K)}$ converge to a subspace τ. Due to the slice theorem we can assume without loss of generality that

$$M = G \times_H \mathcal{V} = (G \times_H \mathcal{W}) \times \mathcal{V}^H \quad \text{and} \quad y = [(1,0)],$$

where $\mathcal{V}$ denotes a slice of H, $\mathcal{V}^H$ denotes the subspace of H-invariant vectors, and $\mathcal{W} = (\mathcal{V}^H)^\perp$ the orthogonal space with respect to an H-invariant scalar product on $\mathcal{V}$. Let $\mathfrak{g}$ be the Lie algebra of G, $\mathfrak{h}$ the one of H, and $\mathfrak{m}$ the orthogonal space of $\mathfrak{h} \subset \mathfrak{g}$ with respect to an H-invariant scalar product on $\mathfrak{g}$. Via the exponential function on G we obtain a natural smooth chart

$$y : U \to \mathfrak{m} \times \mathcal{V}, \quad y([(\exp \xi, v)]) = (\xi, v), \quad \xi \in \mathfrak{m},\ v \in \mathcal{V},$$

where $U \subset M$ is a suitable open neighborhood of y. We may assume that all elements of the sequences $(x_k)_{k\in\mathbb{N}}$ and $(y_k)_{k\in\mathbb{N}}$ lie in U. Recall now that $M_{(K)} = (G \times_H \mathcal{W}_{(K)}) \times \mathcal{V}^H$ and $M_{(H)} = G/H \times \mathcal{V}^H$. Since $\mathcal{W}_{(K)}$ is invariant with respect to multiplication by non-vanishing scalars, it follows after a possible selection of subsequences

$$(\xi, w, v) := \lim_{k\to\infty} \frac{y(x_k) - y(y_k)}{\|y(x_k) - y(y_k)\|} \in \mathfrak{m} \times \overline{\mathcal{W}_{(K)}} \times \mathcal{V}^H \quad \text{with} \quad \overline{\mathcal{W}_{(K)}} = \mathcal{W}_{(K)} \cup \{0\}.$$

With the representation $y(x_k) = (\xi_k, w_k, v_k) \in \mathfrak{m} \times \mathcal{W}_{(K)} \times \mathcal{V}^H$ we then have (again, after a possible selection of subsequences)

$$\|w\| \lim_{k\to\infty} \frac{w_k}{\|w_k\|} = w.$$

Using once again the invariance of $\mathcal{W}_{(K)}$ with respect to multiplication by scalars gives on the other hand

$$\mathfrak{m} \times \operatorname{span} w \times \mathcal{V}^H \subset \tau = \lim_{k\to\infty} T_{x_k} M_{(K)}.$$

This implies in particular $\ell = \operatorname{span}(\xi, w, v) \subset \tau$, which shows the claim. □

4.3.8 Theorem *Let $f : M \to N$ be a G-equivariant smooth mapping between the manifolds M, N on which G acts in a proper way. Under the additional assumption that f is a stratified submersion with respect to the stratification $\mathcal{S}$ on M by orbit types, there exist G-equivariant control data $(T_S)_{S \in \mathcal{S}}$ compatible with f that means for every stratum S the relation $G \cdot T_S \subset T_S$ holds and*

(KB14) $$\pi_S(gx) = g\,\pi_S(x),$$

(KB15) $$\rho_S(gx) = \rho_S(x),$$

provided that $x \in T_S$ and $g \in G$.

PROOF: For the proof of the claim one needs in first place G-equivariant versions of the existence and uniqueness theorem for tubular neighborhoods. Afterwards one proceeds in accordance with part 1 of the proof of Proposition 3.6.9 and constructs as described there a G-equivariant control data using as ingredients G-equivariant objects only. We already have got a G-equivariant version of the classical tubular neighborhood theorem; this is, in the end, actually the slice theorem with the help of which one shows (almost) word-for-word as in Section 3.1 that the G-equivariant versions of the existence and uniqueness theorem for tubular neighborhoods hold as well. Since the formulation of the various steps of the proof is canonical though somewhat tedious, the proof is left to the reader. □

4.3.9 Though we have just found a natural stratification of M by orbit types, the one of $G\backslash M$ lacks so far. In the following considerations the result will be that the quotients $G\backslash M_{(H)}$ possess a manifold structure in a natural way, where for the proof of this fact Proposition 4.2.10 plays an important role. The manifolds $G\backslash M_{(H)}$ then define the desired stratification of the orbit space.

4.3.10 Theorem *Let H be one of the isotropy groups of a proper G-action on M. Then the quotient group $\Gamma_H = N_G(H)/H$ of the normalizer $N_G(H)$ of H in G acts properly and freely from the left on M_H, i.e. $M_H \to \Gamma_H\backslash M_H$ becomes an opposite Γ_H-principal bundle. Furthermore, the submanifold $M_{(H)}$ can be identified with the associated fiber bundle $G/H \times_{\Gamma_H} M_H \to \Gamma_H\backslash M_H$ by the G-equivariant diffeomorphism*

$$\Psi : G/H \times_{\Gamma_H} M_H \to M_{(H)}, \quad [gH, x] \mapsto gx.$$

(*cf.* BOREL [19, §1] *and* JÄNICH [95, Theorem 1.5])

PROOF: For each element g of the normalizer $N_G(H)$ and each point $x \in M_H$ gx lies again in M_H, hence we have an action $N_G(H) \times M_H \to M_H$. Since, by definition,the isotropy group of any point of M_H is equal to H, the $N_G(H)$-action induces a left action of the quotient group Γ_H on M_H, which has to be free. The action is proper as well, for $N_G(H)$ and M_H are closed subsets of G resp. M, and G acts properly on M by assumption. This gives the first claim of the theorem.

For the proof of the second one first note that Γ_H acts freely from the right on the homogeneous space G/H: $gH \cdot \gamma H := g\gamma H$ is for $g \in G$ and $\gamma \in N_G(H)$ a well-defined product, actually. Consequently, the associated fiber bundle $G/H \times_{\Gamma_H} M_H$ consists of all equivalence classes $[gH, x]$ with respect to the equivalence relation $(gH, \gamma x) \sim (g\gamma H, x)$, $\gamma \in N_G(H)$. Then, one immediately realizes by $g(\gamma x) = (g\gamma)x$ that Ψ is a well-defined differentiable mapping on $G/H \,{}_{\Gamma_H}\!\times M_H$. Since the G-left action commutes with the Γ_H-right action, $G/H \,{}_{\Gamma_H}\!\times M_H$ becomes a differentiable G-space, and Ψ a G-equivariant differentiable mapping. The fact that Ψ is surjective is obvious. It remains to show injectivity. However, from $gx = \tilde{g}y$ it follows $\gamma := g^{-1}\tilde{g} \in N_G(H)$, hence $[gH, x] = [g\gamma H, \gamma^{-1}x] = [\tilde{g}H, y]$. This means nothing else but that Ψ is injective. □

4.3.11 Corollary *Let* M *and* G *as in the preceding theorem. Assigning to each point* Gx *of the orbit space* $G\backslash M$ *the germ of the set* $G\backslash M_{(G_x)} \cong \Gamma_{G_x}\backslash M_{G_x}$ *one obtains a stratification of* $G\backslash M$.

PROOF: The claim is a direct consequence of Theorem 4.3.10 and the fact that the decomposition of M into orbit types describes a stratification. □

4.4 Functional Structure

4.4.1 On the orbit space $G\backslash M$ of a proper G-action one has a canonical sheaf $\mathcal{C}^\infty_{G\backslash M}$ of "smooth" functions. Its sectional spaces are defined in the following way:

$$\mathcal{C}^\infty_{G\backslash M}(U) = \{f \in \mathcal{C}(U) \mid f \circ \pi \in \mathcal{C}^\infty(\pi^{-1}(U))\}, \quad U \subset G\backslash M \text{ open.}$$

For each $U \subset G\backslash M$ open $\mathcal{C}^\infty_{G\backslash M}(U)$ therefore is canonically isomorphic to $\mathcal{C}^\infty(\pi^{-1}(U))^G$, the algebra of G-invariant smooth functions on $\pi^{-1}(U)$. By Theorem 4.2.4, (4) this entails, among other things, that the sheaf $\mathcal{C}^\infty_{G\backslash M}$ is fine.

In this section it will be shown that $\mathcal{C}^\infty_{G\backslash M}$ comes from a canonical smooth structure on the stratified space $G\backslash M$, indeed; in other words it can be defined by a singular atlas in accordance with Section 1.3. To this end, the first and fundamental step is the following classical theorem, attributed to DAVID HILBERT, but probably proven independently by MENAHEM SCHIFFER, too (cf. WEYL [186, Chap. 8, Sec. 14] and BIERSTONE [14]).

4.4.2 Theorem *Let* H *be a compact Lie group, and* $\mathcal{V}$ *a finite dimensional* $\mathbb{R}$*-linear representation space of* H*. Then the algebra* $\mathcal{P}(\mathcal{V})^H$ *of the* H*-invariant polynomials on* $\mathcal{V}$ *is finitely generated.*

A finite generating system of $\mathcal{P}(\mathcal{V})^H$ as in the theorem is usually called a *Hilbert basis* of $\mathcal{P}(\mathcal{V})^H$. If the generating system consists only of homogeneous polynomials, the Hilbert basis is said to be *homogeneous*. A Hilbert basis is called *minimal*, if there is no generating system for $\mathcal{P}(\mathcal{V})^H$ with less elements.

PROOF: We denote by $\mathcal{P}(\mathcal{V})$ the algebra of polynomials on $\mathcal{V}$. In accordance with HILBERT's basis theorem the ideal in $\mathcal{P}(\mathcal{V})$ generated by the H-invariant non-constant polynomials is finitely generated. Therefore there exist H-invariant polynomials $p_1, \cdots, p_k$ generating the ideal. Without loss of generality we can assume that the polynomials p_j are homogeneous, since each H-invariant polynomial can be decomposed into H-invariant components. Let d_0 be the lowest polynomial degree that appears in the generating system $p_1, \cdots, p_k$. Then every non-constant H-invariant polynomial must have at least the degree d_0. We show by induction by the degree $d \geq d_0$ that each element of $\mathcal{P}(\mathcal{V})^G$ is a polynomial in the p_j. Let p be a homogeneous H-invariant polynomial of minimal degree d_0. Then there is a representation of p of the form

$$p = \sum_{j=1}^{k} q_j \cdot p_j, \tag{4.4.1}$$

where $q_j = 0$ whenever $\deg p_j > d$ and $q_j \in \mathbb{C}$ else. This was the initial step of the induction. Let us assume now that for some $d \geq d_0$ every H-invariant q of degree $\leq d$ is a polynomial in the p_j and that $p \in \mathcal{P}(\mathcal{V})^H$ is homogeneous of degree $d+1$. Then, first there exists a representation

$$p = \sum_{j=1}^{k} r_j \cdot p_j, \tag{4.4.2}$$

with $r_j \in \mathcal{P}(\mathcal{V})$ and $\deg r_j < \deg p$. Integrating both sides of this equation over H with respect to the Haar measure μ one obtains a representation of the form (4.4.1), where $q_j \in \mathcal{P}(\mathcal{V})^H$ is given by

$$q_j(v) = \int_G r_j(gv)\, d\mu(v), \quad v \in \mathcal{V},$$

and $\deg q_j < d$ holds. By the induction hypothesis every q_j is a polynomial in $p_1, \cdots, p_k$, hence this holds for p as well. This gives the induction step and therefore completes the proof. □

At this point we recall that the algebra $\mathcal{C}^\infty(M)$ of smooth functions on M possesses a natural Fréchet topology (see appendix C.1), and that via the pullback every smooth function $f : M \to N$ induces a continuous homomorphism $f^* : \mathcal{C}^\infty(N) \to \mathcal{C}^\infty(M)$ of Fréchet algebras. Since $\mathcal{C}^\infty(M)^G$ is a closed subalgebra of $\mathcal{C}^\infty(M)$ with respect to this topology, $\mathcal{C}^\infty_{G\backslash M}$ becomes a sheaf of Frèchet algebras. The theorem following now states that $\mathcal{C}^\infty_{G\backslash M}$ can also be regarded as a topological quotient of the algebra of smooth functions on some $\mathbb{R}^k$ and represents the second important step towards the construction of a smooth structure on $G\backslash M$.

4.4.3 Theorem (SCHWARZ [156], MATHER [124]) *Let H and $\mathcal{V}$ be as in the preceding theorem and $p = (p_1, \cdots, p_k)$ be a Hilbert basis of $\mathcal{P}(\mathcal{V})^H$. Then*

$$p^* : \mathcal{C}^\infty(\mathbb{R}^k) \to \mathcal{C}^\infty(\mathcal{V})^H, \quad f \mapsto f \circ (p_1, \cdots, p_k)$$

is a surjective topologically linear mapping between Fréchet spaces and splits topologically that means there is a topologically linear right inverse $e : \mathcal{C}^\infty(\mathcal{V})^H \to \mathcal{C}^\infty(\mathbb{R}^k)$. Moreover, the mapping

$$\overline{p} : H\backslash\mathcal{V} \to \mathbb{R}^k, \quad Hv \mapsto (p_1(v), \cdots, p_k(v)),$$

induced by p is continuous, injective and proper.

4.4.4 Remark The proof that p^* is surjective comes from SCHWARZ. Then MATHER was able to show that p^* even splits topologically. Moreover, he gave a simplified proof for the result of SCHWARZ.

PROOF: Since the proof of the theorem is very tedious, we refer the reader to the already cited literature [156, 124] or to the monograph of BIERSTONE [14]. □

Now, let $Gx \in G\backslash M$ be a point in the orbit space and U a "slice neighborhood" of x that means it is G-equivariantly diffeomorphic to a neighborhood of the zero section of $G \times_{G_x} \mathcal{V}_x$. Let $\varphi : U \to G \times_{G_x} \mathcal{V}_x$ be the corresponding G-equivariant embedding and $\overline{\varphi} : G\backslash U \to G_x\backslash \mathcal{V}_x$ the canonical quotient map. After choosing a generating system $p = (p_1, \cdots, p_k)$ for $\mathcal{P}(\mathcal{V}_x)^{G_x}$ we can define a singular chart around Gx in the following way:

$$x : G\backslash U \to \mathbb{R}^k, \quad Gz \mapsto \overline{p} \circ \overline{\varphi}(Gz).$$

Since the G_x-invariant functions separate the points of $\mathcal{V}_x$, the map x is injective. Its continuity is obvious; that x is also a homeomorphism onto its image results from the fact that $\overline{p}$ is proper. Since the respective components p and φ are smooth, and the fiber bundle $G \times_{G_x} \mathcal{V}_x \to G/G_x$ as well as $M_{(H)} \to G\backslash M_{(H)}$ for $H \subset G$ possess local sections, the restriction of x onto a stratum of the form $G\backslash(U \cap M_{(H)})$ is smooth. Consequently, we have with x a singular chart at hand, indeed, if it can yet be shown that each of the restrictions $x_{|G\backslash(U\cap M_{(H)})}$ is immersive. In the corresponding proof it will turn out that the family of all such singular charts represents a singular atlas for $G\backslash M$ and that the smooth functions belonging to it are given by $\mathcal{C}^\infty_{G\backslash M}$. For the explicit proof of our claims we now need the following result which makes a statement about the Zariski derivative of $\overline{p}$ (see appendix B.3).

4.4.5 Lemma *Let* H *and* $\mathcal{V}$ *be as above and let* $q = (q_1, \cdots, q_l)$ *be a minimal homogeneous Hilbert basis of* $\mathcal{P}(\mathcal{V})^H$. *Then the Zariski derivative of* $\overline{q}$ *is an isomorphism in the origin that means*

$$T_0^Z \overline{q} : T_0^Z (H\backslash\mathcal{V}) \to T_0^Z \mathbb{R}^l = \mathbb{R}^l$$

is an isomorphism. Consequently, for each Hilbert basis $p = (p_1, \cdots, p_k)$ *of* $\mathcal{P}(\mathcal{V})^H$ *the Zariski derivative*

$$T_v^Z \overline{p} : T_v^Z (H\backslash\mathcal{V}) \to T_v^Z \mathbb{R}^k = \mathbb{R}^k$$

is injective at any point $v \in \mathcal{V}$.

PROOF: The lemma is an immediate consequence of the theorem of SCHWARZ. For a direct proof see BIERSTONE [14, Lem. 2.17] as well. □

We first prove that $x_{|G\backslash(U\cap M_{(G_x)})}$ is an immersion. Via φ the stratum $U \cap M_{(G_x)}$ is mapped G-equivariantly and diffeomorphically onto an open subset of the bundle $G/G_x \times \mathcal{V}_x^{G_x}$, consequently the restriction

$$\overline{\varphi}_{|G\backslash(U\cap M_{(G_x)})} : G\backslash(U \cap M_{(G_x)}) \to G_x\backslash\mathcal{V}_x^{G_x} = \mathcal{V}_x^{G_x}$$

has to be a diffeomorphism onto a zero neighborhood in $\mathcal{V}_x^{G_x}$. On the other hand, by the preceding lemma Lemma 4.4.5 the map $\overline{p}_{|\mathcal{V}_x^{G_x}} : \mathcal{V}_x^{G_x} \to \mathbb{R}^k$ is an immersion and therefore $x_{|G\backslash(U\cap M_{(G_x)})} = \overline{p} \circ \overline{\varphi}_{|G\backslash(U\cap M_{(G_x)})}$ as well.

Assume now, we succeed to show that any two singular charts $x : G\backslash U \to \mathbb{R}^k \subset \mathbb{R}^N$ and $y : G\backslash V \to \mathbb{R}^l \subset \mathbb{R}^N$ as defined above are compatible that means there exists around each point $Gz \in G\backslash(U \cap V)$ a neighborhood W and a diffeomorphism $H : O \to \tilde{O} \subset \mathbb{R}^N$ with $O \subset \mathbb{R}^N$ open such that

$$x_{|W} = H \circ y_{|W} \tag{4.4.4}$$

holds. Then the restrictions $x_{G\backslash(U \cap M_{(H)})}$ are immersive as well. To see this, choose the singular chart y around a point $Gy \in G\backslash U$ with $G_y \sim H$. Since by virtue of the results proven so far $y_{|G\backslash(V \cap M_{(H)})}$ is immersive, Eq. (4.4.4) would then also imply that $x_{|W \cap (G\backslash M_{(H)})}$ hence $x_{|G\backslash(U \cap M_{(H)})}$ are immersive. So, we have to prove the compatibility of the x and y. To this end, it suffices to consider the case that y is defined around a point $Gy \in G\backslash U$ and given by

$$y : G\backslash V \to \mathbb{R}^l, \quad Gz \mapsto \overline{q} \circ \overline{\psi}\,(Gz),$$

where V denotes a slice neighborhood of y, $\psi : V \to G \times_{G_y} \mathcal{V}_y$ the embedding belonging to it and $q = (q_1, \cdots, q_l)$ a minimal homogeneous Hilbert basis for $\mathcal{P}(\mathcal{V}_y)^{G_y}$. The compatibility of x and y is shown when a smooth embedding $H : O \to \mathbb{R}^k$ with $O \subset \mathbb{R}^l$ open can be constructed in such a way that Eq. (4.4.4) is fulfilled for a suitable neighborhood W of Gz. By the theorem of SCHWARZ there are smooth functions $H_1, \cdots, H_k \in \mathcal{C}^\infty(\mathbb{R}^l)$ such that for all v from a zero neighborhood in $\mathcal{V}_y$

$$H_i \circ \overline{q}\,(G_y v) = \pi_i \circ x \circ \overline{\psi}^{-1}\,(G_y v), \quad i = 1, \cdots, k,$$

where $\pi_i : \mathbb{R}^k \to \mathbb{R}$ is the projection onto the i-th coordinate. Then it follows for all Gz from a neighborhood W of Gy

$$H_i \circ y\,(Gz) = H_i \circ \overline{q} \circ \overline{\psi}\,(Gz) = \pi_i \circ x\,(Gz), \quad i = 1, \cdots, k.$$

Now note that the sheaves $\overline{\varphi}^* \mathcal{C}^\infty_{G_x\backslash \mathcal{V}_x}$ and $\mathcal{C}^\infty_{G\backslash V}$ are isomorphic, since φ is a G-invariant diffeomorphism. Consequently, by Lemma 4.4.5, $T^Z_{Gy} x = T^Z \overline{p} \circ T^Z_{Gy} \overline{\varphi}$ is injective. Since $T^Z_{Gy} y$ is by lemma 4.4.5 an isomorphism, too, and $T^Z_{y(Gy)} H \circ T^Z_{Gy} y = T^Z_{Gy} x$ holds, $T^Z_{y(Gy)} H$ has to be injective. Hence, for a suitable open neighborhood $O \subset \mathbb{R}^l$ of $y(Gy)$ the restriction $H := (H_1, \cdots, H_l)_{|O}$ is an embedding. This was the last constituent in the construction of singular atlases $\mathcal{U}$ for $G\backslash M$. Invoking now the chain of equations given by the theorem of SCHWARZ

$$x^* \mathcal{C}^\infty(\mathbb{R}^k) = \overline{\varphi}^* \overline{p}^* \mathcal{C}^\infty(\mathbb{R}^k) = \overline{\varphi}^* \left(\mathcal{C}^\infty(G_x\backslash \mathcal{V}_x)^{G_x}\right) = \mathcal{C}^\infty(G\backslash U)^G,$$

the last claim follows, namely that $\mathcal{C}^\infty_{G\backslash M}$ coincides with the sheaf of smooth functions defined by $\mathcal{U}$. Moreover, even the following holds

4.4.6 Theorem *Let the orbit space* $G\backslash M$ *of a proper* G*-action be supplied with the natural stratification by orbit types. Then the orbit space* $G\backslash M$ *carries a canonical smooth structure the smooth functions of which are given by* $\mathcal{C}^\infty_{G\backslash M}(U) = \mathcal{C}^\infty(\pi^{-1}(U))^G$ *for* $U \subset G\backslash M$ *open. By means of this the orbit space becomes a topologically locally trivial and Whitney stratified space. Moreover, the stratification by orbit types is minimal among all Whitney stratifications of* $G\backslash M$.

4.4.7 Remark The proof that the orbit space of a linear G-action can be Whitney stratified has been given by BIERSTONE [13], see also [14, Thm. 2.5] from the same author.

To carry out the proof of the claims not shown up to now, we need the following

4.4.8 Lemma (BIERSTONE [14, Lem. 2.12]) *Let* $\mathcal{V}^H = \{0\}$ *and* $p = (p_1, \cdots, p_k)$ *be a Hilbert basis. If then* $\gamma :]-\varepsilon, \varepsilon[\to \mathbb{R}^k$ *is a* $\mathcal{C}^1$*-curve in* $X = p(V)$ *with* $\gamma(0) = 0$, *then* $\gamma'(0) = 0$ *holds.*

PROOF: Endow $\mathcal{V}$ with an H-invariant scalar product. Without loss of generality one can assume that p_1 describes the square of the distance from the origin. Let $C > 0$ be a constant such that $|p_i(v)| \leq C$ holds for every i and every unit vector v of $\mathcal{V}$. Let $d_i = \deg p_i$. Then

$$X \subset \{(u_1, \cdots, u_k) \in \mathbb{R}^k \,|\, u_1 \geq 0, \; |u_i| \leq C|u_1|^{d_i/2}, \; i = 2, \cdots, k\}.$$

Obviously, $(\pi_1 \circ \gamma)'(0) = 0$ holds. From $\mathcal{V}^H = \{0\}$ follows $d_i \geq 2$, hence $(\pi_i \circ \gamma)'(0) = 0$ for $i = 2, \cdots, k$. Thus, $\gamma'(0) = 0$. □

PROOF OF THE THEOREM: Only local triviality and that Whitney's condition (B) holds remain to be shown. Since in both cases the properties are local ones, it suffices to carry out the proof for the case that M is a linear G-module $\mathcal{V}$ and G is compact. Equip $\mathcal{V}$ with a G-invariant scalar product μ and a Hilbert basis $p = (p_1, \cdots, p_k)$. Then the orbit space $G\backslash\mathcal{V}$ can via $\overline{p}$ be considered as stratified subspace of $\mathbb{R}^k$. We want to show at first that any pair $R < S$ of strata of $G\backslash\mathcal{V}$ is (A)-regular at every point $Gv \in R \subset G\backslash\mathcal{V}$. Let $\mathcal{W}$ be the orthogonal space of $T_v(Gv)$, and $\mathcal{W}'$ the orthogonal space of $\mathcal{W}^H$ in $\mathcal{W}$, where $H = G_v$ denotes the isotropy group of v. Let $\varphi : U \to (G \times_H \mathcal{W}') \times \mathcal{W}^H$ be a submersive G-invariant embedding in accordance with the slice theorem. Finally, choose a minimal homogeneous Hilbert basis $q = (q_1, \cdots, q_l)$ of $\mathcal{P}(\mathcal{W}')^H$. Then one has the two embeddings

$$G\backslash U \longrightarrow G\backslash\mathcal{V} \xrightarrow{\overline{p}} \mathbb{R}^k \qquad \text{and} \qquad G\backslash U \xrightarrow{\overline{\varphi}} H\backslash\mathcal{W}' \times \mathcal{W}^H \xrightarrow{\overline{q}\times\mathrm{id}} \mathbb{R}^l \times \mathcal{W}^H,$$

resulting in a further embedding obtained by composition

$$i := \overline{p} \circ \overline{\varphi}^{-1} \circ (\overline{q} \times \mathrm{id})^{-1} : Y \to \mathbb{R}^k, \qquad Y := q(\mathcal{W}') \times \mathcal{W}^H$$

By virtue of Lemma 4.4.5 the Zariski derivative $T_0^Z i : T_0^Z Y \to T^Z_{p(v)}\mathbb{R}^k = \mathbb{R}^k$ is injective, consequently there are zero neighborhoods W' and W in $\mathbb{R}^l$ and $\mathcal{W}^H$, respectively, and a smooth embedding $\iota : W' \times W \to \mathbb{R}^k$ such that $\iota_{|Y \cap (W' \times W)} = i_{|Y \cap (W' \times W)}$. The space $Y \subset \mathbb{R}^l \times \mathcal{W}^H$ gets a stratification by the pieces $S_{(K)} = q(\mathcal{W}'_{(K)}) \times \mathcal{W}^H$, where $K \subset H$ runs through the closed subsets of H. Then the origin of $\mathbb{R}^l \times \mathcal{W}^H$ lies in the stratum $S_{(H)} = \{0\} \times \mathcal{W}^H$. Consequently, for every sequence $(w_k)_{k\in\mathbb{N}}$ of points of $S_{(K)} > S_{(H)}$ converging to a $w \in S_{(H)}$ the relation

$$T_w S_{(H)} = \{0\} \times \mathcal{W}^H \subset \lim_{k\to\infty} T_{w_k} S_{(K)}$$

holds. The pair $(S_{(H)}, S_{(K)})$ therefore fulfills the Whitney condition (A), hence, in particular, at the origin. Since the restriction $i = \iota_{|Y} : Y \to G\backslash\mathcal{V}$ describes an isomorphism of stratified spaces and at the same time ι is an embedding, Whitney (A) needs to be satisfied for each stratum $S > R$ at the point $G\nu \in R$. Due to the fact that $G\nu \in G\backslash\mathcal{V}$ has been arbitrary, the stratification of $G\backslash\mathcal{V}$ as a result is (A)-regular.

Next it will be shown that $G\backslash\mathcal{V}$ is even (B)-regular. Since p consists of polynomials, $p(\mathcal{V}) \subset \mathbb{R}^k$ by the theorem of TARSKI–SEIDENBERG is a semi-algebraic and consequently a semi-analytic set. A semi-analytic set $Z \subset \mathbb{R}^k$ possesses by ŁOJASIEWICZ [115] or MATHER [123] a minimal (A)-regular stratification by semi-analytic smooth manifolds. By MATHER [123] this stratification is minimal, too, among all (A)-stratifications of Z by smooth manifolds. Due to Lemma 4.4.8 the stratification of $G\backslash\mathcal{V}$ by orbit types is minimal among all $\mathcal{C}^1$-stratifications. According to the results proven so far the stratification of the orbit space by orbit types has to be minimal among all (A)-regular stratifications of $G\backslash\mathcal{V}$. As a consequence, the corresponding strata of $G\backslash\mathcal{V} \subset \mathbb{R}^k$ and those of $H\backslash\mathcal{W}' \subset \mathbb{R}^l$ are semi-analytic. (They are even semi-algebraic, but we will not need this at this point.) Following ŁOJASIEWICZ [115, p. 103] every smooth semi-analytic manifold is (B)-regular over a point of its closure. As a result, each one of the strata $q(\mathcal{W}'_{(K)})$ with $K \leq H$ is (B)-regular over $\{0\}$. This, on its own, implies that each stratum $S_{(K)}$ is (B)-regular over $S_{(H)}$. By Lemma 1.4.4, $S = \iota(S_{(K)})$ has to be (B)-regular over $R = \iota(S_{(H)})$, too. If K now runs through all closed subgroups of H, S will run through all strata $> R$. Consequently, $G\backslash\mathcal{V}$ is a Whitney stratified space. At the end let us directly show the local triviality of $G\backslash\mathcal{V}$, although this follows immediately from Corollary 3.9.3, too. By virtue of the slice theorem it suffices to prove local triviality around the origin. Let $S \subset \mathcal{V}$ be the unit sphere belonging to μ. Then, because of the G-invariance of μ, S is itself a G-manifold again and moreover compact. In the category of topological spaces the isomorphy $\mathcal{V} \cong CS$ holds. Furthermore, the G-action commutes with the canonical $\mathbb{R}^{>0}$-action on $\mathcal{V}$ and CS, respectively. As a consequence, $G\backslash\mathcal{V}$ and $C(G\backslash S)$ are isomorphic as stratified spaces taht means $G\backslash\mathcal{V}$ is topologically locally trivial around $G0$, where the link is given by $G\backslash S$. This was to show. □

A further consequence of the lemma on page 166 is the following statement about derivations and vector fields on an orbit space.

4.4.9 Proposition (BIERSTONE [14, Prop. 3.9]) *Let the Lie group G act properly on M, and let $\delta \in \mathrm{Der}(\mathcal{C}^\infty(G\backslash M), (G\backslash M))$ be a derivation on the space of smooth functions of $G\backslash M$. Then $\delta \in \mathfrak{X}^\infty(G\backslash M)$ holds, if and only if δ is tangential to every stratum S of codimension 1, hence iff $\delta(f)_{|S} = 0$ for any smooth function f vanishing on such a stratum S. In particular, the relation $\mathrm{Der}(\mathcal{C}^\infty(G\backslash M), (G\backslash M)) = \mathfrak{X}^\infty(G\backslash M)$ is satisfied if and only if $G\backslash M$ does not possess a stratum of codimension 1.*

PROOF: We follow the argument given in [14]. By virtue of the slice theorem it suffices to prove the claim for the case of compact G's and for a linear G-action on a finite dimensional vector space $\mathcal{V}$. Again, due to the slice theorem, it suffices to show that if $\{0\}$ is a stratum of codimension ≥ 2, every derivation $\delta \in \mathrm{Der}(\mathcal{C}^\infty(G\backslash\mathcal{V}), \mathcal{C}^\infty(G\backslash\mathcal{V}))$ being tangential to the strata of codimension 1 vanishes at the origin. By means of an induction argument we can assume that δ is tangential to all strata with codimension less than that of $\{0\}$. Let $p = (p_1, \cdots, p_k)$ be a Hilbert basis for $\mathcal{P}(\mathcal{V})^G$ and

$\overline{p} : G\backslash\mathcal{V} \to X := p(\mathcal{V}) \subset \mathbb{R}^k$ the induced diffeomorphism. Then there is a smooth vector field V on $\mathbb{R}^k$ such that the restriction of V on X equals the derivation $\overline{p}_*(\delta)$. The vector field V generates a local group of diffeomorphisms ϕ_t, where for t sufficiently small ϕ_t is defined on a neighborhood of 0. If the curve $\gamma(t) = \phi_t(0)$ for sufficiently small t lies in X, it follows $\gamma'(0) = 0$ by Lemma 4.4.8. Hence, $V(0) = 0$, implying $\overline{p}_*(\delta) \in \mathfrak{X}^\infty(X)$ and thus $\delta \in \mathfrak{X}^\infty(G\backslash\mathcal{V})$. Assume now on the other hand that there are arbitrary small t with $\phi_t(0) \notin X$, where we can evidently achieve after a possible change to $-\phi_t$ that these t are positive. Since X is closed, there is a neighborhood U of 0 and a $t > 0$ such that $\phi_t(U) = \emptyset$. Let $x \in U \cap X$. Due to the induction hypothesis the curve $\gamma(s) = \phi_s(x)$ lies in X for sufficiently small $s < t$. Let s_0 be the largest $s < t$ such that $\gamma([0, s_0]) \subset X$. Since V is tangential to the strata of $X\backslash\{0\}$, $\phi_{s_0}(x) = 0$ must hold. Consequently, $U \cap X$ is contained in the curve $\phi_{-s}(0)$, $0 \le s \le t$, contradicting the assumption. Finally, we need to show that on an orbit space with a stratum of codimension 1 there exists a derivation δ not being given by a smooth vector field. Again, by virtue of the slice theorem we can restrict ourselves to the situation where the stratum with codimension 1 is given by the origin of a linear G-representation space $\mathcal{V}$. But then $\mathcal{V}$ has the dimension 1, and as a consequence either $G/G_0 = \{e\}$ or $G/G_0 = \mathbb{Z}_2$. Then, in the first case, the orbit space $G\backslash\mathcal{V}$ is diffeomorphic to $\mathbb{R}$, in the second case to $\mathbb{R}^{\geq 0}$. In both cases $\frac{\partial}{\partial x}$ is a derivation that is not induced by a smooth vector field on $G\backslash\mathcal{V}$ for this needs to vanish in G0. □

4.4.10 At the end of this section we would like to present another important class of stratified spaces with smooth structure, the so-called orbifolds. These have been introduced into the mathematical literature by SATAKE [146] under the name *V-manifolds*. Orbifolds represent in a natural way stratified spaces with very mild singularities. Particularly for that reason, orbifolds often allow results and constructions usually known for manifolds only. For example, KAWASAKI [99, 100] succeeded to prove a signature theorem for orbifolds. In the mathematical literature, the exact definition of orbifolds is rather technical (see [146] or also [47, B.2.3.]), in our language of stratified spaces, however, this can be very easily done by means of the orbit spaces provided in this section. A Whitney (A) space $(X, \mathcal{C}^\infty)$ is called an *orbifold*, if there is a covering $\mathcal{U} = (U_j)_{j\in J}$ of X by open sets in such a way that each patch U_j is diffeomorphic to an orbit space of the form $G_j\backslash M_j$, where G_j is a finite group and M_j is a differentiable G_j-manifold. The projections $\pi_j : M_j \to U_j$ belonging to it are named *orbifold charts*. Obviously, it is possible by the slice theorem to choose the manifolds M_j as open zero neighborhoods in a linear G_j-representation space.

Chapter 5

DeRham-Cohomology

5.1 The deRham complex on singular spaces

Considerations on the deRham cohomology of singular spaces have a long tradition. Already in the early years of complex analysis one was interested in the question, what the relation between the (smooth) deRham cohomology of a singular analytic variety and its classical cohomology is (see for example NORGUET [135], 1959). In the year 1967 HERRERA [82] could show by an example that differently to the regular case the deRham cohomology of a singular analytic variety need not coincide with the singular cohomology of the underlying topological space. In another work BLOOM–HERRERA have shown [18] that for every complex analytic space $(\mathsf{X}, \mathcal{O})$ there exists a canonical splitting

$$\mathbf{H}^{\bullet}(\mathsf{X}; \Omega^{\bullet}_{\mathsf{X}}) \cong \mathsf{H}^{\bullet}(\mathsf{X}; \mathbb{C}) \oplus \mathsf{A}^{\bullet},$$

where $\mathbf{H}^{\bullet}(\mathsf{X}; \Omega^{\bullet}_{\mathsf{X}})$ denotes the hypercohomology with values in the sheaf complex $\Omega^{\bullet}_{\mathsf{X}}$ of Kähler differentials of $\mathcal{O}$ and $\mathsf{H}^{\bullet}(\mathsf{X}; \mathbb{C})$ the "classical" complex cohomology of X. The complement $\mathsf{A}^{\bullet}$ to the classical cohomology vanishes for regular X, but if X has singularities then in general $\mathsf{A}^{\bullet} \neq 0$. The reason for that lies mainly in the fact that in the singular case the sequence

$$0 \longrightarrow \mathbb{C}_{\mathsf{X}} \longrightarrow \mathcal{O} \longrightarrow \Omega^1_{\mathsf{X}} \longrightarrow \Omega^2_{\mathsf{X}} \longrightarrow \cdots$$

need not be a resolution of the locally constant sheaf $\mathbb{C}_{\mathsf{X}}$ and this is because in the singular case X need not be holomorphically contractible (see REIFFEN [145]) that means the Poincaré lemma need not hold for holomorphic differential forms on X.

Let us now consider a stratified space X with smooth structure $\mathcal{C}^{\infty}$. Then the question arises what the relation between the (yet to be defined) deRham cohomology of X and the singular cohomology of X is, and whether a kind of deRham theorem can be proved. In this section we define the deRham cohomology of X; in the following sections it will be computed for several different classes of stratified spaces. After the definition of the deRham complex we will give a historical overview about some further approaches to construct cohomology theories on stratified spaces.

5.1.1 Let X be a stratified space with smooth structure $\mathcal{C}^{\infty}$. The sheaf $\mathcal{C}^{\infty}$ of smooth

M.J. Pflaum: LNM 1768, pp. 169 - 181, 2001

functions on X induces according to Section B.3 in Appendix B a complex

$$(\Omega^\bullet, d) : \mathcal{C}^\infty \longrightarrow \Omega^1 \longrightarrow \Omega^2 \longrightarrow \cdots \longrightarrow \Omega^k \longrightarrow \cdots$$

of sheaves $\Omega^k := \Omega^k_{X,\mathbb{R}}$, $k \in \mathbb{N}$. After application of the global section functor we obtain a further complex

$$\mathcal{C}^\infty(X) \longrightarrow \Omega^1(X) \longrightarrow \Omega^2(X) \longrightarrow \cdots \longrightarrow \Omega^k(X) \longrightarrow \cdots,$$

the so-called *deRham complex* of $(X, \mathcal{C}^\infty)$. The cohomology $H^\bullet_{dR}(X)$ of this complex is called the *deRham cohomology* of X.

Besides the already above mentioned studies by HERRERA et al. there exist other approaches for a deRham cohomology or other meaningful cohomology theories on stratified spaces. In the following we explain some of the most important ones, but also refer the reader to the article [20] of BRASSELET for a detailed exposition about the present state of research on deRham theorems for singular varieties.

5.1.2 Controlled differential forms VERONA has introduced in [176, 179] for every controlled space X a complex of so-called *controlled differential forms.* These are families of smooth differential forms $(\alpha_S)_{S\in\mathcal{S}}$, which satisfy a control condition similar to the one for controlled vector fields. Hereby it is not necessary that the family $(\alpha_S)_{S\in\mathcal{S}}$ can be put together to a global continuous differential form, as in particular it is not immediately clear what a globally continuous differential form on a controlled space should be. The important fact now is that the corresponding sheaves of controlled forms comprise a fine resolution of the sheaf of locally constant real functions on X, hence a deRham theorem holds for controlled differential forms. FERRAROTTI [57] has generalized the method of VERONA and introduced a complex of infinitesimally controlled forms which also gives rise to a deRham theorem.

5.1.3 Intersection homology Already POINCARÉ knew that singularities could destroy the particular duality on the homology of manifolds which nowadays is known under his name. Therefore mathematicians have tried to set up a (co)homology theory for for singular spaces which satisfies a kind of Poincaré duality. This has been achieved by the *intersection homology* theory of GORESKY–MCPHERSON [63, 64] which appeared in the mid 80's. In intersection homology one considers the homology of complexes consisting of singular chains which intersect a stratum only in an allowed dimension given by a so-called *perversity.* Hereby a perversity is nothing else than a special integer valued function on the set of strata. A particularly elegant approach is the one via *perverse sheaves*, which comprise complexes of sheaves or more precisely objects in the derived category of sheaves. The Poincaré duality in intersection homology then is an immediate consequence of Verdier's duality in the theory of derived categories (see for example KASHIWARA-SCHAPIRA [98] for derived categories and Verdier duality).

A deRham theorem for intersection homology has been proved by BRASSELET–HECTOR–SARALEGI [22]. More precisely the authors of this article introduce a special class of forms named *intersection forms* and show that integration of intersection forms over chains leads to an isomorphism between the cohomology of intersection forms and intersection homology.

5.1.4 L^2-cohomology At the end of the 80's the importance of L^2-cohomology for the study of singular spaces became apparent by the work of CHEEGER [39] and ZUCKER [194, 193]. Hereby one supplies a stratified space or better the top stratum of it with a Riemannian metric and studies the cohomology of the complex consisting of the L^2-integrable differential forms. By the result [39, Thm. 6.1] of CHEEGER one knows that for a Riemannian pseudomanifold with conic singularities the L^2-cohomology coincides with the intersection homology of middle perversity. CHEEGER–GORESKY–MACPHERSON [42] have extended this result to analytic locally conic varieties and have posed in their work the famous conjecture which says that for any projective algebraic variety with restriction of the Fubini–Study metric as Riemannian metric the L^2-cohomology coincides with the intersection homology of middle perversity. The Cheeger–Goresky–MacPherson conjecture has been shown for the case of isolated singularities by OHSAWA [136]. According to SJAMAAR [161] orbit spaces of Riemannian G-manifolds have as well the property that their L^2-cohomology coincides with the intersection homology of middle perversity. An essential tool for many of these considerations is the sheaf theoretic approach to L^2-cohomology as explained by NAGASE [131, 132]. Some further and intuitive means for a better understanding of intersection homology and its connection to L^2- or more generally to L^q-cohomology is given by the concept of *shadow forms* by BRASSELET–GORESKY–MACPHERSON [21]. We cannot go into this concept at this point but refer the interested reader again to [20], where it is explained in greater detail.

5.2 DeRham cohomology on $\mathcal{C}^\infty$-cone spaces

Before we start with the computation of the deRham cohomology for a $\mathcal{C}^\infty$-cone space let us first recall a classical result which essentially entails the Poincaré lemma on manifolds.

5.2.1 Lemma *Let* M *be a differentiable manifold and* $\iota_t : \mathsf{M} \to \mathsf{M} \times [0,\infty[$ *with* $t \in [0,\infty[$ *the embedding* $x \mapsto (x,t)$. *Define the operator* $\mathsf{K}_{\mathsf{M},t} : \Omega^{l+1}(\mathsf{M} \times [0,t]) \to \Omega^l(\mathsf{M})$ *by*

$$\mathsf{K}_{\mathsf{M},t}(\omega)(y)(v_1,\cdots,v_l) = \int_0^t \omega(y,s)\left(\frac{\partial}{\partial s}, v_1,\cdots,v_l\right) ds,$$

$$\omega \in \Omega^{l+1}(\mathsf{M} \times [0,t]),\ y \in \mathsf{M},\ v_1,\cdots,v_l \in T_y\mathsf{M}.$$

Then $\mathsf{K}_{\mathsf{M},t}$ *satisfies the equality* $d\mathsf{K}_{\mathsf{M},t} + \mathsf{K}_{\mathsf{M},t}d = \iota_t^* - \iota_0^*$, *hence for every smooth homotopy* $\mathsf{H} : \mathsf{M} \times [0,t] \to \mathsf{M}$ *the relation*

$$d\mathsf{K}_{\mathsf{M},t}\mathsf{H}^* + \mathsf{K}_{\mathsf{M},t}\mathsf{H}^*d = \mathsf{H}_t^* - \mathsf{H}_0^* \tag{5.2.1}$$

follows, where $\mathsf{H}_s = \mathsf{H}(\cdot,s)$ *for* $s \in [0,t]$.

PROOF: The claim follows by an easy calculation using Cartan's magic formula $di_V + i_V d = \mathcal{L}_V$, where $\mathcal{L}_V$ is the Lie derivative with respect to the vector field V and i_V the insertion of V. Hereby let V be the vector field on $\mathsf{M} \times [0,t]$ given by

$V(x,s) = \frac{\partial}{\partial s}$. For a further proof of the lemma see HOLMANN–RUMMLER [89, §13].
□

5.2.2 Theorem *Let* $(X, \mathcal{C}^\infty)$ *be a* $\mathcal{C}^\infty$*-cone space. Then the sequence of sheaves*

$$\mathbb{R}_X \longrightarrow \mathcal{C}^\infty \xrightarrow{d_0} \Omega^1 \xrightarrow{d_1} \Omega^2 \xrightarrow{d_2} \cdots \xrightarrow{d_{k-1}} \Omega^k \xrightarrow{d_k} \cdots . \tag{5.2.2}$$

comprises a fine resolution of the sheaf of locally constant real functions on X.

PROOF: We already know by Proposition 2.3.2 that the sheaves Ω^k are fine. Moreover it well-known that $d \circ d = 0$, hence it remains to prove the exactness of the sequence 5.2.2. In other words we have to show that for all $k \in \mathbb{N}$ and all $x \in X$ there exists a basis of neighborhoods W of x such that $\mathcal{H}^k(W) = 0$ for $k > 0$, where $\mathcal{H}^k$ is the cohomology sheaf of $(\Omega^\bullet, d)$ that means the quotient sheaf $\ker d_k / \operatorname{im} d_{k-1}$.

By assumption on X there exists a smooth cone chart $k : U \to (S \cap U) \times CL \subset (S \cap U) \times \mathbb{R}^{l+1}$ around x such that S is the stratum of x and such that S has lowest dimension among the strata of U. Let $l : L \hookrightarrow S^l$ be the link chart for L. As the claim is a local one we can now suppose that k and l are identical embeddings and that

$$X = U = \left\{ y = (y_S, y_{\mathrm{rad}} \cdot y_L) \in S \times \mathbb{R}^{l+1} \,\middle|\, y_{\mathrm{rad}} \in [0,1[\text{ and } y_L \in L \right\}.$$

In this presentation the point x has coordinates $(x, 0)$. Now let $V \subset S$ be a smoothly contractible neighborhood of x, $r \in]0,1[$, and

$$W_{V,r} = \left\{ y \in X \middle| y_S \in V \text{ and } y_{\mathrm{rad}} < r \right\}.$$

By assumption on V there exists a smooth homotopy $H_S : V \times [0,1] \to V$ such that $H_S(y_S, 1) = y_S$ and $H_S(y_S, 0) = x$ for all $y_S \in V$. Then we can extend H_S to a smooth homotopy

$$H : V \times B_r(0) \times [0,1] \to V \times B_r(0), \quad (y_S, y_B, t) \mapsto (H_S(y_S), t \cdot y_B),$$

where $B_r(0)$ is the open ball of radius r around the origin of $\mathbb{R}^{l+1}$. Then $H_1 = \mathrm{id}_{W_{V,r}}$ and $H_0(y) = x$ for all $y \in W_{V,r}$. For the following it is important that H is a homotopy relative $W_{V,r}$, hence $H_t(W_{V,r}) \subset W_{V,r}$ for all $t \in [0,1]$. Moreover, $H_t(R) \subset R$ holds for every stratum R and every $t \in]0,1]$, as the strata $\neq S$ are given by $S \times]0,1[\cdot \tilde{R}$, where $\tilde{R}$ runs through the strata of the link L. Now let $\alpha \in \Omega^k(W_{V,r})$ be a closed k-form. According to Proposition 2.3.7 there exists a smooth form $\omega \in \Omega^k(V \times B_r(0)$ with $k^*\omega = \alpha$. As $d\alpha = 0$, the relation $d\omega(v_0 \otimes \cdots \otimes v_k) = 0$ holds for all tangent vectors $v_0, \cdots, v_k \in T_yR$ of a stratum $R \subset X$. We define the form $\eta \in \Omega^{k-1}(V \times B_r(0))$ by $\eta = KH^*\omega$, where $K := K_{V \times B_r(0),1}$ is the operator from the above lemma, and set $\beta = k^*\eta \in \Omega^{k-1}(W_{V,r})$. Now we claim that

$$d\beta = \alpha, \tag{5.2.3}$$

which entails exactness of the sheaf sequence (5.2.2), as the $W_{V,r}$ form a basis of neighborhoods of x. For the proof of (5.2.3) let us write down how the form $KH^*d\omega$ explicitly looks like:

$$(KH^*d\omega)(y)(v_1 \otimes \cdots \otimes v_k) = \int_0^1 (d\omega)(H(y,t))\big(\dot{H}(y,t) \otimes TH.v_1 \otimes \cdots \otimes TH.v_k\big)\, dt,$$
$$y \in V \times B_r(0),\ v_1, \cdots, v_k \in T_y(V \times B_r(0)).$$

Hereby $\dot{H}$ is the partial derivative of $H(y,t)$ in direction of the variable t and TH the tangent map of H. Let R be the stratum of y and the v_l tangent vectors of R with footpoint y. As $H(y,t) \in R$ holds for $t \in]0,1]$, we must have $(KH^*d\omega)(y)(v_1 \otimes \cdots \otimes v_k) = 0$ by the above considerations, hence $k^*(KH^*d\omega) = 0$ holds true. Now relation (5.2.1) of Lemma 5.2.1 entails by $H_0^*\omega = 0$

$$d\beta = dk^*\eta = k^*dKH^*\omega = k^*(dKH^*\omega + KH^*d\omega) = k^*H_1^*\omega = \alpha.$$

This proves the claim. □

5.2.3 Corollary *The deRham cohomology of a $\mathcal{C}^\infty$-cone space X canonically coincides with its singular cohomology.*

PROOF: For an arbitrary topological space Y and every $k \in \mathbb{N}$ let $S_k(Y)$ be the free Abelian group generated by the k-simplices in Y that means by the continuous maps $\sigma : s_k \to Y$, where s_k denotes the k-th standard simplex (see 1.1.14). Together with the boundary operator $\partial : S_k(Y) \to S_k(Y)$ one thus obtains the well-known singular complex $(S_\bullet(Y), \partial)$ of Y. By $S^k(Y;\mathbb{R})$ we understand the vector space of all homomorphisms from $S_k(Y)$ to $\mathbb{R}$. Together with the coboundary operator δ that is the operator dual to ∂ we thus obtain the singular cochain complex $(S^\bullet(Y;\mathbb{R}), \delta)$. Its cohomology is the singular cohomology of Y. We now want to describe the singular cohomology of the cone space X sheaf theoretically. Hereby we will use constructions given by GODEMENT [60, Ex. 3.9.1]. If $V \subset U \subset X$ are open, then one has a canonical restriction morphism $S^k(U;\mathbb{R}) \to S^k(V;\mathbb{R})$, which commutes with the coboundary operator. Thus $S^k(\,\cdot\,;\mathbb{R})$ becomes a presheaf on X. Let $\mathcal{S}^k(\,\cdot\,;\mathbb{R})$ be its associated sheaf, and $(\mathcal{S}^\bullet(\,\cdot\,;\mathbb{R}), \delta)$ the corresponding sheaf complex. As X is paracompact, the canonical morphism $S^k(X;\mathbb{R}) \to \mathcal{S}^k(X;\mathbb{R})$ is surjective (see [60]), and the cohomology of $(\mathcal{S}^\bullet(X;\mathbb{R}), \delta)$ coincides with the singular cohomology. Now observe that according to [60] $\mathcal{S}^k(\,\cdot\,;\mathbb{R})$ is for every k a soft sheaf, hence in particular acyclic with respect to the right derived functors of the global section functor $\Gamma(X;\,\cdot\,)$. As X is locally path connected and locally contractible, we thus obtain a $\Gamma(X;\,\cdot\,)$-acyclic resolution of the sheaf of locally constant real functions on X:

$$\mathbb{R}_X \longrightarrow \mathcal{S}^0(\,\cdot\,;\mathbb{R}) \longrightarrow \mathcal{S}^1(\,\cdot\,;\mathbb{R}) \longrightarrow \cdots \longrightarrow \mathcal{S}^k(\,\cdot\,;\mathbb{R}) \longrightarrow \cdots.$$

But by the preceding theorem (5.2.2) is a $\Gamma(X;\,\cdot\,)$-acyclic resolution of $\mathbb{R}_X$ as well, hence the singular cohomology that means the cohomology of $(\mathcal{S}^\bullet(X;\mathbb{R}), \delta)$ coincides with the deRham cohomology, i.e. with the cohomology of $(\Omega^\bullet(X), d)$. The canonical morphism from $(\Omega^\bullet(X), d)$ to $(\mathcal{S}^\bullet(X;\mathbb{R}), \delta)$ is obtained like for manifolds by integration of a k-form α over every singular chain $\sigma \in S_k(X)$. □

5.3 DeRham theorems on orbit spaces

In this section we will show that the cohomology of the complex of basic forms on an orbit space of a proper G-action canonically mirrors the singular cohomology of the orbit space. Moreover, we will show that under certain conditions on the dimensions of the strata of the orbit space the sheaf complex of differential forms is quasi isomorphic to the complex of basic forms.

For the case of a compact Lie group G KOSZUL [104] has already claimed and proved that the cohomology of the basic complex coincides with the cohomology of the orbit space. Let us mention that the methods used by KOSZUL are different to the ones given here.

We begin with some simple explanations on the basic complex.

5.3.1 Let M be a G-manifold on which the Lie group G acts properly, let $\pi : \mathsf{M} \to \mathsf{G}\backslash\mathsf{M}$ be the canonical projection and for every open $\mathsf{U} \subset \mathsf{G}\backslash\mathsf{M}$ let $\Omega^k_{\text{basic}}(\mathsf{U})$ be the space of basic k-forms on $\pi^{-1}(\mathsf{U})$. This gives rise to a sheaf Ω^k_{basic} on $\mathsf{G}\backslash\mathsf{M}$. Now let $\alpha \in \Omega^k_{\text{basic}}(\mathsf{U})$ be basic. In particular α then is a differential form on $\pi^{-1}(\mathsf{U})$, hence one can form the differential $d\alpha$. As for every $g \in G$ the relation $\Phi_g{}^* d\alpha = d\Phi_g{}^*\alpha$ is true and for every $\xi \in \mathfrak{g}$ by CARTAN $i_{\xi_M} d\alpha = -d i_{\xi_M}\alpha + \mathcal{L}_{\xi_M}\alpha = 0$, the derivative of a basic form is again basic. Hence we obtain a further sheaf complex on $\mathsf{G}\backslash\mathsf{M}$:

$$(\Omega^\bullet_{\text{basic}}, d) : \mathcal{C}^\infty_{\mathsf{G}\backslash\mathsf{M}} \longrightarrow \Omega^1_{\text{basic}} \longrightarrow \Omega^2_{\text{basic}} \longrightarrow \cdots \longrightarrow \Omega^k_{\text{basic}} \longrightarrow \cdots . \tag{5.3.1}$$

Hereby we canonically identify $\mathcal{C}^\infty_{\mathsf{G}\backslash\mathsf{M}}$ with the sheaf having the sectional spaces $\mathcal{C}^\infty(\pi^{-1}(\mathsf{U}))^{\mathsf{G}}$, where $\mathsf{U} \subset \mathsf{X}$ is open. By application of the global section functor we then obtain the so-called *basic complex* $(\Omega^\bullet_{\text{basic}}(\mathsf{G}\backslash\mathsf{M}), d)$. We will determine its cohomology $\mathsf{H}^\bullet_{\text{basic}}(\mathsf{G}\backslash\mathsf{M})$ and call it the *basic cohomology* of $\mathsf{G}\backslash\mathsf{M}$.

By the universal property (KÄ) of the space of Kähler differentials (see Appendix B) the following diagram commutes:

$$\begin{array}{ccc} \mathcal{C}^\infty(\mathsf{G}\backslash\mathsf{M}) & \xrightarrow{\ d\ } & \Omega^1_{\text{basic}}(\mathsf{G}\backslash\mathsf{M}) \\ {\scriptstyle d}\big\downarrow & \nearrow{\scriptstyle h^1} & \\ \Omega^1(\mathsf{G}\backslash\mathsf{M}) . & & \end{array}$$

The morphism h^1 then induces in a functorial way a morphism of sheaf complexes $h^\bullet : (\Omega^\bullet_{\mathsf{G}\backslash\mathsf{M}}, d) \to (\Omega^\bullet_{\text{basic}}, d)$. It is the goal of the following considerations to prove that under certain assumptions on $\mathsf{G}\backslash\mathsf{M}$ the morphism $h^\bullet$ leads to an isomorphism on the level of cohomology.

5.3.2 Equivariant Poincaré lemma (cf. [185, S. 23], [6, Thm. 6], [75, 2.9]) *Let M be a differentiable manifold on which a Lie group G acts properly and let $\iota : \mathsf{N} \hookrightarrow \mathsf{M}$ be a G-invariant closed submanifold of M. Then there exists an invariant neighborhood U of N such that for every G-invariant closed k-form α with vanishing pullback $\iota^*\alpha$ there exists a $(k-1)$-form β over U such that*

$$d\beta = \alpha_{|\mathsf{U}} \text{ and } \beta_{|\mathsf{N}} = 0.$$

If α is basic, then one can choose β as a basic form as well.

PROOF: As N is a closed invariant submanifold of M, it is the union of (locally finitely many) orbits, hence it suffices to prove the existence of a β with the desired properties only for an invariant neighborhood of a single orbit $\subset N$. Then one can glue together these forms with the help of an invariant smooth partition of unity. So we can assume without loss of generality that N is an orbit Gx. We identify N with the zero section of the bundle $G \times_{G_x} \mathcal{V}_x$, where G_x is the isotropy group of a point $x \in N$ and $\mathcal{V}_x$ the slice at x. For this bundle consider the following homotopy:

$$H : G \times_{G_x} \mathcal{V}_x \times [0,1] \to G \times_{G_x} \mathcal{V}_x, \quad ([(g,v)],t) \mapsto [(g,(1-t)v)].$$

Obviously H then commutes with the G-action, as well as the operator $K := K_{G\times_{G_x}\mathcal{V}_x,1}$ from Lemma 5.2.1. Hence $\beta := KH^*\alpha$ is a G-invariant $(k-1)$-form and satisfies the relation

$$d\beta = -KH^* d\alpha + \alpha - \iota^*\alpha = \alpha.$$

As $\dot{H}(y,t) = 0$ for all $y \in N$, the equality $\beta_{|N} = 0$ is true. Moreover, if α is basic, then we have for every $\xi \in \mathfrak{g}$

$$i_{\xi_M}\beta = -KH^* i_{\xi_M}\alpha = 0,$$

as H and K commute with the G-action. This proves the claim. □

5.3.3 Corollary *The sheaf complex of basic differential forms provides a fine resolution*

$$\mathbb{R}_{G\backslash M} \longrightarrow \mathcal{C}^\infty_{G\backslash M} \longrightarrow \Omega^1_{\text{basic}} \longrightarrow \Omega^2_{\text{basic}} \longrightarrow \cdots \longrightarrow \Omega^k_{\text{basic}} \longrightarrow \cdots .$$

In particular the basic cohomology coincides with the singular cohomology of $G\backslash M$ *in a canonical way.*

PROOF: As $\mathcal{C}^\infty_{G\backslash M}$ is fine and every sheaf Ω^k_{basic} is a $\mathcal{C}^\infty_{G\backslash M}$-module sheaf, all sheaves Ω^k_{basic} must be fine. The equivariant lemma of POINCARÉ implies that the sheaf sequence $\Omega^\bullet_{\text{basic}}$ is exact. The rest of the claim follows by a standard argument analogous to the one given for Corollary 5.2.3. □

5.3.4 Corollary *Let* $Gx \in G\backslash M$ *be a point of the orbit space and* x *the singular chart around* Gx *induced by a Hilbert basis* $p = (p_1, \cdots, p_l)$ *for* $\mathcal{P}(\mathcal{V}_x)^{G_x}$. *Then there exists a contractible neighborhood* $G\backslash U$ *in the domain of* x, *an open set* $O \subset \mathbb{R}^l$, *in which* $x(G\backslash U)$ *is closed, and a smooth homotopy* $\overline{H} : O \times [0,1] \to \mathbb{R}^l$ *relative* $X := x(G\backslash U)$ *such that* $\overline{H}_{0|X} = \mathrm{id}_X$ *and* $\overline{H}_{1|X} = x(Gx) = 0$.

PROOF: The smooth homotopy $H : G \times_{G_x} \mathcal{V}_x \times [0,1] \to G \times_{G_x} \mathcal{V}_x$ which has been constructed in the proof of the equivariant lemma of POINCARÉ is G-equivariant and induces the desired homotopy $\overline{H}$ as one shows by an application of the theorem of SCHWARZ. □

5.3.5 Theorem *Let* M *be a G-manifold on which* G *acts properly and assume that the orbit space* $G\backslash M$ *does not possess strata of codimension* 1. *Then the sheaf complex of differential forms on* $G\backslash M$ *gives rise to a fine resolution*

$$\mathbb{R}_{G\backslash M} \longrightarrow \mathcal{C}^\infty_{G\backslash M} \longrightarrow \Omega^1_{G\backslash M} \longrightarrow \Omega^2_{G\backslash M} \longrightarrow \cdots \longrightarrow \Omega^k_{G\backslash M} \longrightarrow \cdots .$$

The morphism $h^\bullet : (\Omega^\bullet_{G\backslash M}, d) \to (\Omega^\bullet_{\text{basic}}, d)$ *comprises a quasi isomorphism, hence the basic cohomology* $H^\bullet_{\text{basic}}(G\backslash M)$ *and the deRham cohomology of* $G\backslash M$ *coincide. Moreover, both cohomologies are canonically isomorphic to the singular cohomology of* $G\backslash M$.

PROOF: Obviously, all the sheaves in the sequence (besides $\mathbb{R}_{G\backslash M}$) are fine. So, if we can yet show the exactness of the sequence, then it comprises a fine resolution of $\mathbb{R}_{G\backslash M}$. As we already have proved this property for the basic complex and as $h^\bullet$ is a morphism of complexes of sheaves, $h^\bullet$ would then be a quasi isomorphism. Together with Corollary 5.3.3 the rest of the claim would then follow as well. Hence we only need to prove the exactness of $(\Omega^\bullet_{G\backslash M}, d)$. First recall that by assumption on the dimensions of the strata the following relation holds by Proposition 4.4.9:

$$\operatorname{Der}(\mathcal{C}^\infty(G\backslash M), \mathcal{C}^\infty(G\backslash M)) = \mathfrak{X}^\infty(G\backslash M). \tag{5.3.7}$$

Now choose a point $Gx \in G\backslash M$. Let $G\backslash U$ be a neighborhood like in Corollary 5.3.4, $x : G\backslash U \to \mathbb{R}^l$ a singular chart induced by a minimal Hilbert basis p for $\mathcal{P}(\mathcal{V}_x)^{G_x}$ like in 5.3.4 and $O \subset \mathbb{R}^l$ an open neighborhood of 0 such that $X := x(G\backslash U)$ is closed in O. If $\mathcal{J} \subset \mathcal{C}^\infty(O)$ denotes the vanishing ideal of X, then by Proposition 2.2.8 two forms $\omega, \eta \in \Omega^k(O)$ induce the same form $\in \Omega^k(X) \cong \Omega^k(G\backslash U)$, if and only if for all $V_1, \cdots, V_k \in \mathfrak{X}^\infty_{\mathcal{J}}(O)$

$$\omega(V_1, \cdots, V_k) = \eta(V_1, \cdots, V_k).$$

But by (5.3.7) $\mathfrak{X}^\infty_{\mathcal{J}}(O)$ is equal to the space $\mathfrak{X}^\infty_X(O)$ of all smooth vector fields on O which are tangent to every stratum S of X. Let us keep this result in mind for later purposes.

Now let $\alpha \in \Omega^k(G\backslash U)$ be closed. We want to construct a form $\beta \in \Omega^{k-1}(G\backslash U)$ such that $d\beta = \alpha$. The claimed exactness of the above sequence then follows immediately. Let $\overline{H} : O \times [0,1] \to \mathbb{R}^l$ be the homotopy relative $X := x(G\backslash U)$ from Corollary 5.3.4. Using the operator $K := K_{O,1}$ of Lemma 5.2.1 we obtain

$$dK\overline{H}^* + K\overline{H}^* d = \overline{H}_1^* - \overline{H}_0^*. \tag{5.3.8}$$

At this point choose a k-form $\omega \in \Omega^k(O)$ with $x^*\omega = \alpha$. As p has been chosen as a minimal Hilbert basis $d\omega(0) = 0$ holds by Lemma 4.4.5 and $d\alpha = 0$. Hence there exists a form $\eta \in \Omega^{k-1}(O)$ with $d\eta(0) = \omega(0)$. Set $\tilde{\omega} = \omega - d\eta$. As $\overline{H}$ is a homotopy relative $X := x(G\backslash U)$ and $d\tilde{\omega}_{|X} = 0$, Eq. (5.3.8) entails that for every $x \in X$ and all $V_1, \cdots, V_k \in \mathfrak{X}^\infty_X(O)$

$$\begin{aligned}(-dK\overline{H}^*\tilde{\omega}).(V_1, \cdots, V_k)(x) &= \big(\overline{H}_0^*\tilde{\omega} - \overline{H}_1^*\tilde{\omega} + (K\overline{H}^* d\tilde{\omega})\big).(V_1, \cdots, V_k)(x) \\ &= \tilde{\omega}.(V_1, \cdots, V_k)(x) - \tilde{\omega}.(V_1, \cdots, V_k)(0) - (-1)^k K\big((\overline{H}^* d\tilde{\omega}).(V_1, \cdots, V_k)\big)(x) \\ &= \tilde{\omega}.(V_1, \cdots, V_k)(x).\end{aligned}$$

Hereby we have used that $\overline{H}_* V_i \in \mathfrak{X}^\infty_X(O)$, hence $\overline{H}^* d\omega.(V_1, \cdots, V_k)(x) = 0$. Defining β by $\beta := x^*(\eta - K\overline{H}^*\tilde{\omega}) \in \Omega^{k-1}(G\backslash U)$, the equality $d\beta = \alpha$ follows. This proves the claim. □

5.4 DeRham cohomology of Whitney functions

It has been shown in the preceding sections that the deRham cohomology on cone spaces and orbit spaces coincides with the singular cohomology. For arbitrary Whitney spaces this need not be the case. As already mentioned, this fact has been shown by HERRERA, who gave in [82] an example of an analytic variety X such that the deRham cohomology with respect to analytic functions is larger than the singular cohomology of X. On the other hand one knows by the work of GROTHENDIECK [71] and HARTSHORNE [80] on algebraic deRham cohomology (see as well HERRERA–LIEBERMAN [83]) that one can calculate the cohomology of an algebraic or complex analytic variety $(\mathsf{Y}, \mathcal{O}_\mathsf{Y})$ by the deRham cohomology of the formal completion of the structure sheaf $\mathcal{O}_\mathsf{Y}$. This means the following. The variety Y is embedded in some $\mathbb{C}^n$ and inherits the structure sheaf $\mathcal{O}_\mathsf{Y} = \mathcal{O}_{\mathbb{C}^n}/\mathcal{J}$, where $\mathcal{J}$ is the vanishing ideal. Instead of $\mathcal{O}_\mathsf{Y}$ one now regards for natural k the sheaves $\mathcal{O}_{\mathbb{C}^n}/\mathcal{J}^k$ and passes to the inductive limit $\widehat{\mathcal{O}}_\mathsf{Y} := \varinjlim \mathcal{O}_{\mathbb{C}^n}/\mathcal{J}^k$. This limit is called the *formal completion* of $\mathcal{O}_\mathsf{Y}$. Starting from $\widehat{\mathcal{O}}_\mathsf{Y}$ one now forms the sheaf complex $\widehat{\Omega}^\bullet_\mathsf{Y} := \Omega^\bullet(\widehat{\mathcal{O}}_\mathsf{Y})$, applies the global section functor and passes to the cohomology of the thus obtained complex. The resulting cohomology is called *algebraic deRham cohomology* and coincides by [71] and [80] with the cohomology of Y.

Now the reader might compare the concept of formal completion in the algebraic case with the construction of the sheaf of Whitney functions in Section 1.5. The analogy of the two constructions then becomes apparent, hence the conjecture seems reasonable that the deRham cohomology of Whitney functions on a Whitney space gives back the singular cohomology of X. Indeed this will be the case for a curvature moderate Whitney space X, as will be shown in the following. Let us remark that in the proof of this theorem we will use mainly analytic as well as geometric methods.

5.4.1 First we have to explain in some more detail what to understand by the deRham cohomology of Whitney functions on a stratified space X with a smooth structure $\mathcal{C}^\infty$. To this end we choose a covering of X by chart domains $\mathfrak{U} = (\mathsf{U}_j, \mathsf{x}_j)$ and consider the corresponding sheaf $\mathcal{E}^\infty := \mathcal{E}^\infty_{\mathsf{X},\mathfrak{U}}$ of Whitney functions of class $\mathcal{C}^\infty$. According to Section B.3 we then form the sheaf $\Omega^1_{\mathcal{E}^\infty/\mathbb{R}}$ of Kähler differentials. It will be denoted by $\widehat{\Omega}^1_{\mathsf{X},\mathfrak{U}}$ or more briefly by $\widehat{\Omega}^1$.

Out of $\widehat{\Omega}^1$ one can construct for every $k \in \mathbb{N}$ a further sheaf $\widehat{\Omega}^k = \widehat{\Omega}^k_{\mathsf{X},\mathfrak{U}} := \Lambda^k \widehat{\Omega}^1$. Its sections will be called *Whitney k-forms* on X. Together with the Kähler derivative $\mathrm{d} : \widehat{\Omega}^k_{\mathsf{X},\mathfrak{U}} \to \widehat{\Omega}^k_{\mathsf{X},\mathfrak{U}}$ we thus obtain a sheaf complex $\big(\widehat{\Omega}^\bullet, \mathrm{d}\big)$ which after application of the global section functor gives the *Whitney–deRham complex* of $(\mathsf{X}, \mathcal{C}^\infty)$:

$$\mathcal{E}^\infty(\mathsf{X}) \longrightarrow \widehat{\Omega}^1(\mathsf{X}) \longrightarrow \cdots \longrightarrow \widehat{\Omega}^k(\mathsf{X}) \longrightarrow \cdots .$$

We will call the cohomology $\mathrm{H}^\bullet_{\mathrm{WdR}}(\mathsf{X})$ of this complex *Whitney–deRham cohomology* of X.

For the calculation of $\mathrm{H}^\bullet_{\mathrm{WdR}}(\mathsf{X})$ let us first show

5.4.2 Lemma *Under the assumption that the stratified space X can be embedded into some Euclidean space $\mathbb{R}^n$ via a global singular chart x and that the covering $\mathfrak{U}$ is*

equal to (X,x) *the sheaf* $\widehat{\Omega}^k$ *can be identified canonically with* $\mathcal{E}^\infty \otimes_{\mathcal{C}^\infty} \Omega^k_{\mathsf{x}}$. *Hereby* Ω^k_{x} *has been obtained by pullback with the chart* x *that means* $\Omega^1_{\mathsf{x}} = \mathsf{x}^*(\Omega^1_{\mathbb{R}^n})$.

PROOF: To simplify notation we can assume without loss of generality that X is a locally closed stratified subspace of $\mathbb{R}^n$ and that x is the identical embedding. It suffices to prove the claim for $k=1$. Let $\mathcal{J}$ be the sheaf on X consisting of the germs of those smooth functions on $\mathbb{R}^n$ the partial derivatives of which vanish in every order. Then, by the extension theorem of WHITNEY $\mathcal{E}^\infty = \mathcal{C}^\infty_{\mathbb{R}^n|\mathsf{X}}/\mathcal{J}$ holds true. By the second exact fundamental sequence B.1.5 one thus obtains an exact sheaf sequence

$$\mathcal{J}/\mathcal{J}^2 \longrightarrow \mathcal{E}^\infty \otimes_{\mathcal{C}^\infty_{\mathbb{R}^n|\mathsf{X}}} \Omega^k_{\mathbb{R}^n|\mathsf{X}} \longrightarrow \widehat{\Omega}^1 \longrightarrow 0.$$

According to Lemma C.3.3 $\mathcal{J}^2 = \mathcal{J}$, hence the claim follows. □

5.4.3 Theorem *Let* X *be an (A)-stratified space, which for every* $\mathfrak{m} \in \mathbb{N}^{>0}$ *has curvature moderate control data of order* $\mathfrak{m}$ *and which possesses locally tempered resolutions of class* $\mathcal{C}^\mathfrak{m}$. *Let* $\mathfrak{U}$ *be a locally finite covering of* X *by chart domains. Then the sheaf sequence*

$$\mathbb{R}_\mathsf{X} \longrightarrow \mathcal{E}^\infty_{\mathsf{X},\mathfrak{U}} \xrightarrow{d_0} \widehat{\Omega}^1_{\mathsf{X},\mathfrak{U}} \xrightarrow{d_1} \cdots \xrightarrow{d_{k-1}} \widehat{\Omega}^k_{\mathsf{X},\mathfrak{U}} \xrightarrow{d_k} \cdots \qquad (5.4.1)$$

comprises a fine resolution of the sheaf of locally constant real functions on X.

PROOF: That all $\widehat{\Omega}^k_{\mathsf{X},\mathfrak{U}}$ are fine is a consequence of the fact that by Proposition 1.5.4 $\mathcal{E}^\infty_{\mathsf{X},\mathfrak{U}}$ is fine and that the sheaves $\widehat{\Omega}^k_{\mathsf{X},\mathfrak{U}}$ are all $\mathcal{E}^\infty_{\mathsf{X},\mathfrak{U}}$-module sheaves. As $\mathrm{d}\circ\mathrm{d}=0$, only the exactness at every point $\xrightarrow{d_{k-1}} \widehat{\Omega}^k_{\mathsf{X},\mathfrak{U}} \xrightarrow{d_k}$ has to be shown. But this is a consequence of the following lemma. □

5.4.4 Poincaré lemma for Whitney forms *Let* $\mathsf{X} \subset \mathbb{R}^n$ *be an (A)-stratified subspace having for every* $\mathfrak{m} \in \mathbb{N}^{>0}$ *curvature moderate control data* $(\mathsf{T}^\mathfrak{m}_\mathsf{S})_{\mathsf{S}\in\mathcal{S}}$ *of order* $\mathfrak{m}$ *and which possesses for every* $\mathfrak{m}$ *locally tempered resolutions of class* $\mathcal{C}^\mathfrak{m}$. *Let* $\mathcal{E}^\infty$ *be the sheaf of Whitney functions with respect to the embedding* $\mathsf{X} \hookrightarrow \mathbb{R}^n$. *If* x *is a point of* X *and* W *a contractible open neighborhood of* x *then there exists for every closed Whitney form* $\alpha \in \widehat{\Omega}^k(W)$ *a* $\beta \in \widehat{\Omega}^{k-1}(W)$ *fulfilling* $\mathrm{d}\beta = \alpha$.

PROOF: Mainly to fix notation let us recall first the operator $\mathsf{K}_{\mathsf{M},\mathsf{t}}$ from Lemma 5.2.1 which will be needed several times in the following. Then let us fix an $\mathfrak{m} \in \mathbb{N}^{\geq 2}$ and $\tilde{\mathfrak{m}} = \mathfrak{m}+2$. To simplify notation we will write T, π, ρ, ε for $\mathsf{T}^{\tilde{\mathfrak{m}}}_\mathsf{S}$, $\pi^{\tilde{\mathfrak{m}}}_\mathsf{S}$, $\rho^{\tilde{\mathfrak{m}}}_\mathsf{S}$, $\varepsilon^{\tilde{\mathfrak{m}}}_\mathsf{S}$ and T_R for $\mathsf{T}^{\tilde{\mathfrak{m}}}_\mathsf{R}$ and so on. Finally let $\mathfrak{U}_\mathsf{x} \subset \mathsf{S}$ be a contractible relatively compact open neighborhood of x and δ a positive real number with $\delta < \varepsilon(y)$ for all $y \in \mathfrak{U}_\mathsf{x}$. After these preparations we set $\mathsf{O} := \pi^{-1}(\mathfrak{U}_\mathsf{x}) \cap \rho^{-1}([0,\delta[)$ and $W := W_{\mathfrak{U}_\mathsf{x},\delta} := \mathsf{O} \cap \mathsf{X}$. Note that the $W_{\mathfrak{U}_\mathsf{x},\delta}$ are contractible and run through a basis of neighborhoods of x, if $(\mathfrak{U}_\mathsf{x},\delta)$ runs through all admissible $(\mathfrak{U}_\mathsf{x},\delta)$. Now we divide the proof in several steps.

1. STEP Let $\Omega^k_\mathfrak{m}$ be the sheaf defined over O of k-forms ω such that ω and $d\omega$ are of class $\mathcal{C}^\mathfrak{m}$ and let $\widehat{\Omega}^k_\mathfrak{m}$ be the image sheaf of $\Omega^k_\mathfrak{m}$ under the canonical epimorphism $\Omega^k_{\mathcal{C}^\mathfrak{m}_\mathsf{O}/\mathbb{R}} \xrightarrow{\pi^k_\mathfrak{m}} \Omega^k_{\mathcal{E}^\mathfrak{m}_W/\mathbb{R}} = \mathcal{E}^\mathfrak{m}_W \otimes_{\mathcal{E}^\infty_\mathsf{O}} \Omega^k_\mathsf{O}$. It is the goal of the 1. Step to show that for every

closed Whitney form $\alpha \in \widehat{\Omega}^k_m(W)$ there exists a Whitney form $\beta \in \widehat{\Omega}^k_m(W)$ such that $d\beta = \omega$. To this end we first choose a k-form ω in $\Omega^k_m(O)$ having α as image under the canonical epimorphism π^k_m. To simplify notation further we suppose without loss of generality in the 1. Step that $X \subset O$ hence $S = U$, that X possesses only finitely many strata and that all of these are compatible with S. Finally we can assume by the assumptions of the lemma and after possibly shrinking X that every stratum R of X possesses a tempered resolution $f_R : M_R \to \overline{R}$ of class $\mathcal{C}^m$. Under these simplifications let $0 \leq d_0 < d_1 < \cdots < d_d$ be the sequence of dimensions of the strata of X. In particular we then have $\dim S = d_0$. Now we first construct a form $\omega_0 \in \Omega^{k-1}_{\mathcal{C}^m}(O)$ such that

$$\omega - d\omega_0 \in \mathcal{J}^m(S; O) \cdot \Omega^k_m(O). \tag{5.4.2}$$

As U is contractible and O a tubular neighborhood of U in $\mathbb{R}^n$ there exists a $\mathcal{C}^{\tilde{m}}$-homotopy $H : O \times [0,1] \to O$ such that $H(U \times [0,1]) \subset U$, $H_1 = \mathrm{id}_O$ and $H_0(y) = x$ for all $y \in O$. We set $\omega_0 := K_{O,1} H^* \omega \in \mathcal{C}^k_{\mathcal{C}^{\tilde{m}}_O/\mathbb{R}}(O)$. For this ω_0 the relation (5.4.2) is satisfied, as by assumption on ω and H the equality $(K_{O,1} H^* d\omega)_{|S} = 0$ is true, hence by (5.2.1) $(\omega - d\omega_0)_{|S} = 0$. By $d\alpha = 0$ we have $d\omega \in \mathcal{J}^m(S; O) \cdot \Omega^{k+1}_m(O)$ hence $K_{O,1} H^* d\omega \in \mathcal{J}^m(S; O) \cdot \Omega^k_m(O)$. This entails (5.4.2).

Next let us suppose we are given forms $\omega_0, \cdots, \omega_i \in \Omega^{k-1}_m(O)$, $i \leq d$ such that

$$\omega' := \omega - (d\omega_0 + \cdots + d\omega_i) \in \mathcal{J}^m(X^{d_i}; O) \cdot \Omega^k_m(O). \tag{5.4.3}$$

Then we look for a form $\omega_{i+1} \in \Omega^{k-1}_m(O)$, fulfilling

$$\omega' - d\omega_{i+1} \in \mathcal{J}^m(X^{d_{i+1}}; O) \cdot \Omega^k_m(O). \tag{5.4.4}$$

As the following constructions can be performed separately for every connected component of a d_{i+1}-dimensional stratum R of X, we can assume without loss of generality that $X^{d_{i+1}} \setminus X^{d_i}$ consists only of a connected stratum R.

For the construction of ω_{i+1} we now have to make some preparations and must provide some necessary tools. By assumption on the control data there exists a tubular neighborhood $T_{R \subset \mathbb{R}^n} = (E, \varepsilon, \varphi)$ of R in $\mathbb{R}^n$ (which is curvature moderate of order $\tilde{m}$ and) which induces T_R. In the following we identify T_R with $T_{R \subset \mathbb{R}^n}$. As T_R is curvature moderate, T_R has to be a regularly situated neighborhood of R, hence by Lemma 1.7.10 there exists a function $\phi \in \mathcal{M}^\infty(\partial R; \mathbb{R}^n)$ such that $\phi = 1$ over $T_R^{\varepsilon_R/2}$ and such that ϕ vanishes on a neighborhood of $\mathbb{R}^n \setminus (T_R \cup \partial R)$. For every form $\tilde{\omega} \in \mathcal{J}^m(X^{d_i}; O) \cdot \Omega^l_m(O)$ then $\phi \cdot \tilde{\omega}$ lies in $\mathcal{J}^m(X^{d_i}; O) \cdot \Omega^l_m(O)$ as well, has support in $T_R \cup \partial R$ and coincides with $\tilde{\omega}$ on a neighborhood of R. By Corollary 3.7.4 there exists our second tool, namely a with respect to S radial vector field $V : X \to \mathbb{R}^n$ that means

$$(\pi, \rho)_* V = -\frac{\partial}{\partial t}. \tag{5.4.5}$$

As a third tool we choose a collar $k = (k_1, k_2) : M'_R \subset M_R \to N_R \times [0, 1[$ and a decreasing smooth function $\psi : \mathbb{R} \to [0,1]$ which is identical to 1 on $]-\infty, 1/2]$ and which vanishes on $[3/4, \infty[$. Denote by $R_{]0,t]}$, $0 < t < 1$ the open set $f_R k^{-1}(N_R \times]0, t]) \subset R$. These data then induce a homotopy $H : \overline{R} \times [0,1] \to \overline{R}$ by

$$(x,t) \mapsto \begin{cases} f_R k^{-1}\Big(k_1 f_R^{-1}(x), \big(t\psi(k_2 f_R^{-1}(x)) + (1 - \psi(k_2 f_R^{-1}(x)))\big)\, k_2 f_R^{-1}(x)\Big), & \text{if } x \in R_{]0,1]}, \\ x, & \text{else.} \end{cases}$$

Moreover, H can be extended to a mapping $H : T_R \cup \overline{R} \times [0,1] \to \overline{R}$ by requiring $H(x,t) = H(\pi_R(x), t)$ for $x \in T_R$. As $f_R \circ \pi_R$ is tempered relative ∂R of class $\mathcal{C}^{\tilde{m}}$, $H : T_R \times]0,1] \to \overline{R}$ has to be tempered relative ∂R of class $\mathcal{C}^{\tilde{m}}$. By $\omega' \in \mathcal{J}^m(X^{d_i}; O) \cdot \Omega^k_m(O)$ this means that the form $\omega'_{i+1} = \phi \cdot (K_{T_R,1} H^* \omega')$ lies in $\mathcal{J}^m(X^{d_i}; O) \cdot \Omega^{k-1}_m(O)$. According to the definition of H the relation $H_0(R_{]0,1/2]}) \subset \partial R$ holds and $H_1 = \pi_R$, hence by Eq. (5.2.1) and $d\omega'_{|R} = 0$

$$d\omega'_{i+1|R_{]0,1/2]}} = \pi_R^* \omega'_{|R_{]0,1/2]}}, \tag{5.4.7}$$

so even $d\omega'_{i+1} - \pi_R^* \omega' \in \mathcal{J}^m(X^{d_i} \cup R_{]0,1/2]}; O) \cdot \Omega^k_m(O)$. Next we consider the homotopy $F : T_R \times [0,1] \to T_R$, $(x,t) \mapsto \varphi_R(t\varphi_R^{-1}(x))$. Then $F_0 = \pi_R$ and $F_1 = \mathrm{id}_{T_R}$ are true. Using Eq. (5.2.1) again we have for $\check{\omega}_{i+1} := \phi \cdot (K_{T_R,1} F^* \omega')$

$$d\check{\omega}_{i+1|R} = \omega'_{|R} - \pi_R^* \omega'_{|R}. \tag{5.4.8}$$

As F is tempered relative ∂R of class $\mathcal{C}^{\tilde{m}}$, $\check{\omega}_{i+1}$ lies in $\mathcal{J}^m(X^{d_i}; O) \cdot \Omega^{k-1}_m(O)$, and $d\check{\omega}_{i+1} - \omega + \pi_R^* \omega' \in \mathcal{J}^m(X^{d_{i+1}}; O) \cdot \Omega^k_m(O)$ holds true. A third homotopy $G : T_R \times [0,1] \to R$ is given by the integral flow $\gamma : J \to X$ of the radial vector field V. By (5.4.5) and as V is controlled over $X \setminus S$ the relation $J_x \supset [0, \rho(x)[$ holds true and γ_x can be extended to a continuous function on $J_x \cup \{\rho(x)\}$ by setting $\gamma_x(\rho(x)) = \pi(x)$. For the precise definition of G now choose a smooth function $\kappa : R \to [0,1[$ such that $\gamma(x, \kappa(x)\rho(x)) \in R_{]0,1/2[}$, where $\kappa(x) = 0$ for $x \in R_{]0,1/4]}$. Then we set $G(x,t) = \gamma(\pi_R(x), t\kappa(x)\rho(x))$ for $(x,t) \in T_R \times [0,1]$. Hence $G_0 = \pi_R$ follows as well, as $G_1(T_R) \subset R_{]0,1/2[}$ and $G_{t|R_{]0,1/4]}} = \mathrm{id}_{R_{]0,1/4]}}$. Thus $\omega''_{i+1} := \phi \cdot (K_{T_R,1} G^* (\pi_R^* \omega' - d\omega'_{i+1}))$ is a form of $\mathcal{J}^m(X^{d_i}; O) \cdot \Omega^{k-1}_m(O)$. Moreover, by Eq. (5.4.7) ω''_{i+1} satisfies the relations

$$d\omega''_{i+1|R} = G_1^*(\pi_R \omega' - d\omega'_{i+1})_{|R} - (\pi_R \omega' - d\omega'_{i+1})_{|R} = d\omega'_{i+1|R} - \pi_R \omega'_{|R} \tag{5.4.9}$$

and $d\omega''_{i+1} - d\omega'_{i+1} + \pi_R \omega' \in \mathcal{J}^m(X^{d_{i+1}}; O) \cdot \Omega^k_m(O)$. If one sets now

$$\omega_{i+1} := \check{\omega}_{i+1} + \omega'_{i+1} - \omega''_{i+1},$$

then the considerations so far, in particular Equations (5.4.7) to (5.4.9), entail that $d\omega_{i+1|R} = \omega'_{|R}$ and additionally (5.4.4) hold true. Hence the inductive step has been finished and we have

$$\omega - d(\omega_0 + \cdots + \omega_d) \in \mathcal{J}^m(W; O) \cdot \Omega^k_m(O).$$

Let $\beta := \pi_m^{k-1}(\omega_0 + \cdots + \omega_d)$ be the induced Whitney form. Then $\beta \in \Omega^{k-1}_m(W)$ follows, hence $d\beta = \alpha$. This finishes the 1. Step.

2. Step As all the sheaves $\widehat{\Omega}^k_m$ are $\mathcal{E}^\infty$-module sheaves, they are in particular fine, hence by the 1. Step we obtain a fine resolution

$$\mathbb{R}_X \longrightarrow \widehat{\Omega}^0_m \longrightarrow \widehat{\Omega}^1_m \longrightarrow \cdots \longrightarrow \widehat{\Omega}^k_m \longrightarrow \cdots.$$

Thus, by the considerations in the proof of Corollary 5.2.3 the cohomology of the complex

$$\widehat{\Omega}^0_m(X) \longrightarrow \widehat{\Omega}^1_m(X) \longrightarrow \cdots \longrightarrow \widehat{\Omega}^k_m(X) \longrightarrow \cdots$$

coincides with the singular cohomology of X. Hence, over any open contractible set $W \subset X$ there exists for every closed Whitney form $\alpha \in \widehat{\Omega}^k_m(W)$ Whitney form $\beta \in \widehat{\Omega}^{k-1}_m(W)$ with $d\beta = \alpha$.

3. Step Now we are going to prove the claim. Let $\alpha \in \widehat{\Omega}^k(W)$ be a closed Whitney form of class $\mathcal{C}^\infty$ over an open and contractible set $W \subset X$. According to the 2. Step there exists for every natural $m \geq 2$ a Whitney form β_m of class $\mathcal{C}^m$ with $d\beta_m = \alpha$. By the same reason there exist Whitney forms ν'_m of class $\mathcal{C}^{m-1}$ with $d\nu'_m = \beta_m - \beta_{m-1}$. Next choose a compact exhaustion $\mathcal{K} = (K_j)_{j\in\mathbb{N}}$ of W and transfer in a canonical way the seminorms $\| \cdot \|_{K_j,m}$ on $\mathcal{E}^m(W)$, which were defined in Section 1.5, to the spaces $\widehat{\Omega}^k_{\mathcal{C}^m}(W)$ (see also Appendix C.1). As $\mathcal{E}^\infty(W)$ is dense in every $\mathcal{E}^m(W)$ (see for example [118]), $\widehat{\Omega}^k(W)$ has to be dense in every $\widehat{\Omega}^k_m(W)$. Hence there exists for every m a $\nu_m \in \widehat{\Omega}^k(W)$ with

$$\|\beta_m - \beta_{m-1} - d\nu_m\|_{K_m,m-2} < \frac{1}{2^m}.$$

Now set

$$\beta := \beta_2 + \sum_{m\geq 3} (\beta_m - \beta_{m-1} - d\nu_m).$$

As the $\widehat{\Omega}^k(W)$ together with the seminorms $\| \cdot \|_{K_m,m}$ comprise Fréchet spaces, this β is well-defined, hence lies in $\widehat{\Omega}^{k-1}(W)$ and satisfies $d\beta = \alpha$. This proves the claim.
□

5.4.5 Corollary *The Whitney–deRham cohomology of an* (A)*-stratified space* X *having curvature moderate control data of every order* $m \in \mathbb{N}^{>0}$ *and which possesses locally tempered resolutions of class* $\mathcal{C}^m$ *coincides in a canonical way with the singular cohomology of* X*.*

Proof: The proof is similar to the one given for Corollary 5.2.3. □

5.4.6 Remark As in particular subanalytic sets fulfill the prerequisites of the results in this section, the Poincaré lemma holds for Whitney forms on subanalytic sets, hence the Whitney–deRham cohomology on subanalytic sets coincides with the singular cohomology.

Chapter 6

Homology of Algebras of Smooth Functions

Hochschild homology theories and the closely connected cyclic cohomology introduced by ALAIN CONNES [45] have proved to be very useful for the structure theory of algebras (see LODAY [112]). In particular in the framework of noncommutative geometry invented by ALAIN CONNES, where one wants to introduce geometric notions like forms, connections, deRham cohomology and so on for (noncommutative) algebras, these homology theories play an important role. But even for "commutative geometry " Hochschild (co)homology becomes more and more important, because one can prove with its help deep results of geometric analysis like for example the index theorem of ATIYAH–SINGER (cf. NEST–TSYGAN [133]). Moreover, there is hope that it will be possible to formulate and prove appropriate index theorems for singular manifolds with the help of methods of Hochschild homology (cf. MELROSE–NISTOR [128]). Therefore, in this chapter we will study the Hochschild (co)homology of the algebra of smooth functions on a stratified space.

In many cases and in particular in the case of function algebras it turned out that the topological or in other words local version of Hochschild homology is better suited, if one wants to obtain geometric information on the algebras under consideration in the spirit of ALAIN CONNES. For example one can compute the topological Hochschild (co)homology of the algebra of smooth functions on a manifold (see [45, 143, 102, 165]), but one knows only very little about the general Hochschild (co)homology of these spaces. In the following we will first introduce topological Hochschild (co)Homology as a relative homology theory and will then derive some useful properties of this topological homology theory. As a reference for further results on the (co)homology theory of topological algebras see TAYLOR [164].

6.1 Topological algebras and their modules

6.1.1 Let $\mathbb{k}$ be the field of real or complex numbers. A $\mathbb{k}$-algebra $\mathcal{A}$ together with the structure of a topological $\mathbb{k}$-vector space is called a *topological* $\mathbb{k}$*-algebra*, if the product $\cdot : \mathcal{A} \times \mathcal{A} \to \mathcal{A}$ is a a continuous mapping. In case that the underlying topological vector space structure is locally convex and and if for every continuous seminorm $\|\cdot\|$

M.J. Pflaum: LNM 1768, pp. 183 - 199, 2001

on $\mathcal{A}$ there exists a continuous seminorm $\|\cdot\|'$ such that

$$\|ab\| \le \|a\|' \, \|b\|' \qquad \text{for all } a, b \in \mathcal{A},$$

then $\mathcal{A}$ is called a *locally convex (topological)* $\Bbbk$*-algebra*. If additionally $\mathcal{A}$ is a Fréchet space, then $\mathcal{A}$ is called a *Fréchet algebra.* To avoid any misunderstandings let us mention explicitly that we always assume an algebra to possess a unit element.

An $\mathcal{A}$-module $\mathcal{M}$ of a topological $\Bbbk$-algebra $\mathcal{A}$ is called a *topological* $\mathcal{A}$*-module*, if $\mathcal{M}$ has the structure of a topological $\Bbbk$-vector space and if the structure map $(a, m) \mapsto am$ is continuous. In case that $\mathcal{A}$ is a locally convex algebra, $\mathcal{M}$ a locally convex topological $\Bbbk$-vector space, and if for every continuous seminorm $\|\cdot\|$ on $\mathcal{M}$ there exist continuous seminorms $|\cdot|$ on $\mathcal{M}$ and $\|\cdot\|'$ on $\mathcal{A}$ such that

$$\|am\| \le \|a\|' \, |m| \qquad \text{for all } a \in \mathcal{A},\ m \in \mathcal{M},$$

then $\mathcal{M}$ is called a *locally convex topological* $\mathcal{A}$*-module.* If additionally $\mathcal{A}$ and $\mathcal{M}$ are Fréchet spaces, then $\mathcal{M}$ is called a *Fréchet* $\mathcal{A}$*-module.*

For a topological algebra $\mathcal{A}$ and two topological $\mathcal{A}$-modules $\mathcal{M}$ and $\mathcal{N}$ we denote by $\mathrm{Hom}_{\mathcal{A}}(\mathcal{M}, \mathcal{N})$ the set of all continuous $\mathcal{A}$-linear mappings from $\mathcal{M}$ to $\mathcal{N}$. Then the topological $\mathcal{A}$-modules together with the continuous $\mathcal{A}$-linear mappings form a category $\mathcal{A}\text{-}\mathfrak{Mod}_{\mathrm{top}}$. Note hereby that as objects of $\mathcal{A}\text{-}\mathfrak{Mod}_{\mathrm{top}}$ even non-Hausdorff topological $\mathcal{A}$-modules are admitted.

By $\mathrm{tHom}_{\mathcal{A}}(\mathcal{M}, \mathcal{N}) \subset \mathrm{Hom}_{\mathcal{A}}(\mathcal{M}, \mathcal{N})$ we denote the subset of all *topologically* $\mathcal{A}$*-linear homomorphisms* that means the set of all continuous $\mathcal{A}$-linear mappings $f : \mathcal{M} \to \mathcal{N}$ such that the induced mapping $\mathcal{M}/\ker f \to \mathrm{im}\, f \subset \mathcal{N}$ is a topological isomorphism. The composition of two topologically $\mathcal{A}$-linear homomorphisms $f : \mathcal{M} \to \mathcal{N}$ and $g : \mathcal{N} \to \mathcal{P}$ gives again a topological homomorphism $g \circ f \in \mathrm{tHom}_{\mathcal{A}}(\mathcal{M}, \mathcal{P})$, as the following consideration shows. Let $\overline{f} : \mathcal{M}/\ker f \to \mathrm{im}\, f$ and $\overline{g} : \mathcal{N}/\ker g \to \mathrm{im}\, g$ be the induced topological isomorphisms induced by f and g and let $h : \mathrm{im}\, f/\ker g \to \mathcal{M}/(\ker f + f^{-1}(\ker g))$ be the continuous quotient map with $\overline{f}^{-1} : \mathrm{im}\, f \to \mathcal{M}/\ker f$. Then $h \circ \overline{g}_{|\mathrm{im}\,(g \circ f)}$ comprises the $\mathcal{A}$-linear inverse of $\overline{g \circ f} : \mathcal{M}/\ker(g \circ f) \to \mathrm{im}\,(g \circ f)$. Hence the topological $\mathcal{A}$-modules together with the topologically $\mathcal{A}$-linear homomorphisms form a category.

6.1.2 Proposition *Let $\mathcal{A}$ be a topological $\Bbbk$-algebra. Then the category $\mathcal{A}\text{-}\mathfrak{Mod}_{\mathrm{top}}$ of topological $\mathcal{A}$-modules and continuous $\mathcal{A}$-linear maps is additive.*

In case that $\mathcal{A}$ is a locally convex algebra (resp. a Fréchet algebra), then the locally convex (resp. complete locally convex resp. Fréchet-) $\mathcal{A}$-modules form a full additive subcategory $\mathcal{A}\text{-}\mathfrak{Mod}_{\mathrm{lc}}$ (resp. $\mathcal{A}\text{-}\mathfrak{Mod}_{\mathrm{clc}}$ resp. $\mathcal{A}\text{-}\mathfrak{Mod}_{\mathfrak{F}}$) of $\mathcal{A}\text{-}\mathfrak{Mod}_{\mathrm{top}}$.

PROOF: The additivity of $\mathcal{A}\text{-}\mathfrak{Mod}_{\mathrm{top}}$ is clear, because the zero dimensional vector space is a zero object in $\mathcal{A}\text{-}\mathfrak{Mod}_{\mathrm{top}}$, and because continuous $\mathcal{A}$-linear morphisms can be added and multiplied by scalars. The further statements then follow immediately. □

6.1.3 Remark One could conjecture that the category consisting of all Fréchet $\mathcal{A}$-modules and the topologically $\mathcal{A}$-linear homomorphisms is an Abelian category. But this is not the case as the following example shows. Let $\mathcal{A} = \mathbb{C}$ and consider an infinite dimensional separable complex Hilbert space $\mathcal{H}$. Afterwards choose a compact operator $\mathsf{k} : \mathcal{H} \to \mathcal{H}$ of norm < 1 which does not have a closed image. Then $\mathrm{id}_{\mathcal{H}}$ and $\mathsf{f} = \mathrm{id}_{\mathcal{H}} + \mathsf{k}$ are bijective and continuous, and both are topological homomorphisms. But the difference operator $\mathsf{k} = \mathsf{f} - \mathrm{id}_{\mathcal{H}}$ is not a topological homomorphism, as by assumption $\mathrm{im}\,\mathsf{f}$ is not closed.

6.1.4 We want to do homological algebra in the category $\mathcal{A}\text{-}\mathfrak{Mod}_{\mathrm{top}}$. To this end it is necessary that for all morphisms $\mathsf{f} : \mathcal{M} \to \mathcal{N}$ in $\mathcal{A}\text{-}\mathfrak{Mod}_{\mathrm{top}}$ the kernel and cokernel of f exist. If one regards f as a morphism in the category $\mathcal{A}\text{-}\mathfrak{Mod}$ then one already has a kernel $\ker \mathsf{f} \xrightarrow{\mathsf{k}} \mathcal{M}$ and a cokernel $\mathcal{N} \xrightarrow{\mathsf{c}} \operatorname{coker} \mathsf{f}$. We supply $\ker \mathsf{f}$ with the initial topology with respect to k and $\operatorname{coker} \mathsf{f}$ with the final topology with respect to c. Then $\ker \mathsf{f}$ and $\operatorname{coker} \mathsf{f}$ are both topological $\mathcal{A}$-modules, and comprise the kernel resp. cokernel of f in $\mathcal{A}\text{-}\mathfrak{Mod}_{\mathrm{top}}$. Moreover, one obtains the image $\mathrm{im}\,\mathsf{f} \xrightarrow{\mathsf{i}} \mathcal{N}$ of f as the kernel of c. Thus the topological spaces $\ker \mathsf{f}$, $\operatorname{coker} \mathsf{f}$ and $\mathrm{im}\,\mathsf{f}$ satisfy the universal properties of kernel, cokernel and image in the category $\mathcal{A}\text{-}\mathfrak{Mod}_{\mathrm{t}}$. In the following we will therefore call $\ker \mathsf{f}$, $\operatorname{coker} \mathsf{f}$ and $\mathrm{im}\,\mathsf{f}$ the *topological kernel*, *topological cokernel* and *topological image* of f. By the universal properties of $\mathrm{im}\,\mathsf{f}$ the morphism f can be factorized in the form $\mathsf{f} = \mathsf{i} \circ \mathfrak{m}$ with a uniquely determined morphism $\mathfrak{m} : \mathcal{M} \to \mathrm{im}\,\mathsf{f}$. The factorization $\mathsf{f} = \mathsf{i} \circ \mathfrak{m}$ will be called the *canonical factorization* of f. Altogether we thus obtain the following *canonical sequence* of f

$$0 \longrightarrow \ker \mathsf{f} \xrightarrow{\mathsf{k}} \mathcal{M} \xrightarrow{\mathfrak{m}} \mathrm{im}\,\mathsf{f} \xrightarrow{\mathsf{i}} \mathcal{N} \xrightarrow{\mathsf{c}} \operatorname{coker} \mathsf{f} \longrightarrow 0 .$$

If $\mathcal{A}$ is a locally convex algebra and f a continuous $\mathcal{A}$-linear mapping between locally convex $\mathcal{A}$-modules, then $\ker \mathsf{f}$, $\operatorname{coker} \mathsf{f}$ and $\mathrm{im}\,\mathsf{f}$ are again locally convex $\mathcal{A}$-modules in a canonical way and comprise kernel, cokernel and image of f in the category $\mathcal{A}$-$\mathfrak{Mod}_{\mathrm{top}}$. If on the other hand $\mathcal{A}$ is a complete locally convex or even a Fréchet algebra and f a continuous $\mathcal{A}$-linear mapping between complete locally convex (resp. Fréchet) modules, then $\ker \mathsf{f}$ is again complete (resp. Fréchet), but not necessarily $\operatorname{coker} \mathsf{f}$ and $\mathrm{im}\,\mathsf{f}$. Now denote by $\overline{\mathcal{M}}$ the completion of a locally convex $\mathcal{A}$ module $\mathcal{M}$ and define the *complete cokernel* by $\overline{\operatorname{coker}}\,\mathsf{f} := \mathcal{N}/\overline{\mathrm{im}\,\mathsf{f}}$ and the *complete image* by $\overline{\mathrm{im}}\,\mathsf{f} := \overline{\mathrm{im}\,\mathsf{f}}$. Then both $\overline{\operatorname{coker}}\,\mathsf{f}$ and $\overline{\mathrm{im}}\,\mathsf{f}$ are complete locally convex (resp. Fréchet) $\mathcal{A}$-modules. Moreover, $\ker \mathsf{f}$, $\overline{\operatorname{coker}}\,\mathsf{f}$ and $\overline{\mathrm{im}}\,\mathsf{f}$ satisfy the universal properties of kernel, cokernel and image of f in the additive category $\mathcal{A}\text{-}\mathfrak{Mod}_{\mathrm{clc}}$ (resp. $\mathcal{A}\text{-}\mathfrak{Mod}_{\mathfrak{F}}$). Finally let us mention that by the open mapping theorem $\mathrm{im}\,\mathsf{f} = \overline{\mathrm{im}}\,\mathsf{f}$ holds in the Fréchet case, if and only if f is a topological homomorphism between Fréchet $\mathcal{A}$-modules.

As $\mathcal{A}\text{-}\mathfrak{Mod}_{\mathrm{top}}$, $\mathcal{A}\text{-}\mathfrak{Mod}_{\mathrm{lc}}$ and $\mathcal{A}\text{-}\mathfrak{Mod}_{\mathfrak{F}}$ are all additive and possess kernels and cokernels, the notion of exactness of a sequence of topological modules is well-defined. More precisely a sequence

$$\cdots \longrightarrow \mathcal{M}_{k-1} \xrightarrow{\mathsf{f}_{k-1}} \mathcal{M}_k \xrightarrow{\mathsf{f}_k} \mathcal{M}_{k+1} \longrightarrow \cdots$$

of topological (resp. locally convex) $\mathcal{A}$-modules and continuous $\mathcal{A}$-linear mappings is called *exact* at $\mathcal{M}_k$ in $\mathcal{A}\text{-}\mathfrak{Mod}_{\mathrm{top}}$ (resp. $\mathcal{A}\text{-}\mathfrak{Mod}_{\mathrm{lc}}$) or in other words is called *topologically exact*, if $\ker \mathsf{f}_k$ and $\mathrm{im}\,\mathsf{f}_{k-1}$ coincide as topological $\mathcal{A}$-modules. If the modules $\mathcal{M}_k$ are all complete locally convex or Fréchet, then we say that the above sequence is *exact* at $\mathcal{M}_k$ in $\mathcal{A}\text{-}\mathfrak{Mod}_{\text{clc}}$ (resp. $\mathcal{A}\text{-}\mathfrak{Mod}_{\mathfrak{F}}$) or briefly the sequence is *weakly exact* at $\mathcal{M}_k$, if $\ker \mathsf{f}_k = \overline{\operatorname{im}} \, \mathsf{f}_{k-1}$. The notion of topological exactness obviously makes sense as well in $\mathcal{A}\text{-}\mathfrak{Mod}_{\text{clc}}$ and $\mathcal{A}\text{-}\mathfrak{Mod}_{\mathfrak{F}}$, where by the completeness of the kernel $\ker \mathsf{f}_n$ every topological sequence exact at $\mathcal{M}_k$ has to be weakly exact at $\mathcal{M}_k$ as well. Finally we call a sequence of topological vector spaces *topologically* resp. *weakly exact*, if it is topologically resp. weakly exact at every one of its points.

6.2 Homological algebra for topological modules

6.2.1 Let us now consider the category $\mathcal{A}\text{-}\mathfrak{Mod}_{\text{clc}}$ more closely, where it is assumed that $\mathcal{A}$ is a complete locally convex $\mathbb{k}$-algebra. In the following we will introduce a relative homology theory in $\mathcal{A}\text{-}\mathfrak{Mod}_{\text{clc}}$. According to HILTON–STAMMBACH [85, Chap. IX] one needs for the definition of a relative homology theory in an additive category $\mathfrak{A}$ a class $\mathcal{E}$ of epimorphisms in that category with respect to which the projective objects, the so-called $\mathcal{E}$-projective objects have to be defined. Afterwards we will define exactly like in ordinary homology theory $\mathcal{E}$-projective resolutions and $\mathcal{E}$-derived functors, which then will give the desired (co)homology objects.

As epimorphism class, which will fix the relative homology theory in $\mathcal{A}\text{-}\mathfrak{Mod}_{\text{clc}}$, we take the class $\mathfrak{T}$ of all surjective continuous $\mathcal{A}$-linear and $\mathbb{k}$-splitting mappings $\mathsf{e} : \mathcal{M} \twoheadrightarrow \mathcal{N}$. Hereby means $\mathbb{k}$*-splitting*, that there exists a continuous $\mathbb{k}$-linear mapping $\mathsf{s} : \mathcal{N} \to \mathcal{M}$ which satisfies $\mathsf{e} \circ \mathsf{s} = \mathrm{id}_{\mathcal{N}}$ and which sometimes is called a *continuous* $\mathbb{k}$*-section* of e. Let $\mathsf{Proj}(\mathfrak{T}) \subset \mathsf{Ob}(\mathcal{A}\text{-}\mathfrak{Mod}_{\text{clc}})$ be the class of all $\mathfrak{T}$*-projective* or *topologically projective* objects that means the class of all objects $\mathcal{P}$ such that for all morphisms $\mathsf{e} : \mathcal{M} \twoheadrightarrow \mathcal{N}$ from $\mathfrak{T}$ the sequence

$$\operatorname{Hom}(\mathcal{P}, \mathcal{M}) \xrightarrow{\mathsf{e}^*} \operatorname{Hom}(\mathcal{P}, \mathcal{N}) \longrightarrow 0 \tag{6.2.1}$$

is exact. By $C\mathfrak{T}$ we then denote the completion of $\mathfrak{T}$ that means the class of all epimorphisms $\mathsf{e} : \mathcal{M} \twoheadrightarrow \mathcal{N}$ in $\mathcal{A}\text{-}\mathfrak{Mod}_{\text{clc}}$ such that for all topologically projective objects P the sequence (6.2.1) is exact.

Next recall (see Appendix A.3) that the category $\mathcal{A}\text{-}\mathfrak{Mod}_{\text{clc}}$ gets the structure of a tensor category by the functor $\hat{\otimes}_\pi$ of the completed π-tensor product.

6.2.2 Proposition *For every complete locally convex $\mathbb{k}$-algebra $\mathcal{A}$ the following complete locally convex $\mathcal{A}$-modules are topologically projective:*

(1) *every topologically direct summand $\mathcal{P}$ of a topologically projective $\mathcal{A}$-module,*

(2) *every complete locally convex $\mathcal{A}$-module $\mathcal{P}$ of the form $\mathcal{P} = \mathcal{A}\hat{\otimes}_\pi \mathcal{V}$, where $\mathcal{V}$ denotes a complete locally convex topological vector space,*

(3) *in case that $\mathcal{A}$ is a Fréchet algebra, every finitely generated projective $\mathcal{A}$-module $\mathcal{P}$ which has the structure of a Fréchet $\mathcal{A}$-module.*

If conversely $\mathcal{P}$ is a topologically projective $\mathcal{A}$-module, then $\mathcal{P}$ is a topologically direct summand in a complete locally convex $\mathcal{A}$-module of the form $\mathcal{A}\hat{\otimes}_\pi \mathcal{V}$, where $\mathcal{V}$ is complete locally convex.

PROOF: (1) is obvious. Let us show (2). Let $e : \mathcal{M} \to \mathcal{N}$ be a morphism of $\mathfrak{T}$, hence in particular $\mathcal{A}$-linear and $\Bbbk$-splitting. Let $s : \mathcal{N} \to \mathcal{M}$ be a continuous section of e, and $\iota : \mathcal{V} \to \mathcal{P} = \mathcal{A}\hat{\otimes}_\pi\mathcal{V}$ be the canonical injection $v \mapsto 1 \otimes v$. Then one associates to every continuous $\mathcal{A}$-linear mapping $f : \mathcal{P} \to \mathcal{N}$ the mapping $f' = f \circ \iota$ and sets $g' := s \circ f'$. By the universal properties of the tensor product $\hat{\otimes}_\pi$ there exists a uniquely determined continuous $\mathcal{A}$-linear mapping $g : \mathcal{P} \to \mathcal{M}$ such that $g' = g \circ \iota$. As $e \circ g$ is continuously $\mathcal{A}$-linear and

$$e \circ g \circ \iota = e \circ g' = e \circ s \circ f' = f \circ \iota,$$

the relation $e \circ g = f$ follows. Hence $\mathcal{P}$ is topologically projective.

Now we come to (3). Let $(v_1, \cdots, v_k)$ be a generating system of $\mathcal{P}$, and $\mathcal{V}$ the free $\Bbbk$-vector space $\langle v_1, \cdots, v_k \rangle$. Then the mapping

$$f : \mathcal{A} \otimes \mathcal{V} = \mathcal{A}\hat{\otimes}_\pi\mathcal{V} \to \mathcal{P}, \quad a \otimes v_j \mapsto a v_j, \quad j = 1, \cdots, k$$

is continuous, surjective and $\mathcal{A}$-linear. Moreover, the kernel $\mathcal{Q} := \ker f$ is a closed subspace of the Fréchet $\mathcal{A}$-module $\mathcal{A}\otimes\mathcal{V}$ and by the projectivity of $\mathcal{P}$ has a complement $\mathcal{P}'$ which is linearly isomorphic to $\mathcal{P}$. As $\mathcal{Q} \subset \mathcal{A}\otimes\mathcal{V}$ is closed, $\mathcal{P}'$ inherits the structure of a Fréchet $\mathcal{A}$-module of $\mathcal{A}\otimes\mathcal{V}$. By the open mapping theorem $\mathcal{P}' \to \mathcal{P}$ is a topological isomorphism, hence $\mathcal{P}$ is a closed subspace of $\mathcal{A} \otimes \mathcal{V}$. By (1) and (2) the topological projectivity of $\mathcal{P}$ thus follows.

It remains to show that every topologically projective $\mathcal{A}$-module is the topologically direct summand of an $\mathcal{A}$-module of the form $\mathcal{A}\hat{\otimes}_\pi\mathcal{V}$, where $\mathcal{V}$ is complete locally convex. Now set $\mathcal{V} := \mathcal{P}$ and define $e : \mathcal{V} \to \mathcal{P}$ by $\mathcal{A} \otimes v \mapsto a v$. Then e is continuous, $\mathcal{A}$-linear, surjective and possesses a $\Bbbk$-linear continuous splitting $s : \mathcal{P} \to \mathcal{V}$, $v \mapsto 1 \otimes v$. By the topological projectivity of $\mathcal{P}$ there exists a morphism $f : \mathcal{M} \to \mathcal{V}$ of $\mathcal{A}\text{-}\mathfrak{Mod}_{\mathrm{clc}}$ such that $e \circ f = \mathrm{id}_\mathcal{P}$. Now set $\mathcal{P}' := \operatorname{im} e$. Then $\mathcal{P}'$ is a complement of $\ker f$, hence a closed subspace of $\mathcal{A}\hat{\otimes}_\pi\mathcal{V}$. Moreover, $f_{\mathcal{P}'}$ provides a topological isomorphism from $\mathcal{P}'$ to $\mathcal{P}$ with continuous inverse $e : \mathcal{P} \to \mathcal{P}'$. This proves the claim. □

6.2.3 Corollary *The class $\mathfrak{T}$ is projective that means for every object $\mathcal{M}$ of $\mathcal{A}\text{-}\mathfrak{Mod}_{\mathrm{clc}}$ there exists an epimorphism $e : \mathcal{P} \to \mathcal{M}$ of $\mathfrak{T}$ with $\mathcal{P}$ topologically projective.*

PROOF: Define $\mathcal{P} := \mathcal{A}\hat{\otimes}_\pi\mathcal{M}$. Then $\mathcal{P}$ is topologically projective and $e : \mathcal{P} \to \mathcal{M}$, $a \otimes v \to a v$ an epimorphism. □

In the next step we will fix a class of sequences in $\mathcal{A}\text{-}\mathfrak{Mod}_{\mathrm{clc}}$, namely the class of so-called $\mathfrak{T}$-exact sequences and then construct appropriate resolutions of objects.

6.2.4 Definition ([85, IX.1.Def]) A morphism f in $\mathcal{A}\text{-}\mathfrak{Mod}_{\mathrm{clc}}$ is called *$\mathfrak{T}$-admissible*, if it is a topological isomorphism with closed image and if in the canonical decomposition $f = i \circ m$ of f the mapping m in $\mathfrak{T}$ lies in $\mathfrak{T}$. An exact sequence in $\mathcal{A}\text{-}\mathfrak{Mod}_{\mathrm{clc}}$ is called *$\mathfrak{T}$-exact*, if all its morphisms are $\mathfrak{T}$-admissible. A complex in $\mathcal{A}\text{-}\mathfrak{Mod}_{\mathrm{clc}}$

$$\mathbf{C} : \cdots \longrightarrow \mathrm{C}_n \longrightarrow \mathrm{C}_{n-1} \longrightarrow \cdots \longrightarrow \mathrm{C}_0$$

will be called *$\mathfrak{T}$-projective* or *topologically projective*, if every $\mathcal{A}$-module C_n is topologically projective. Moreover, the complex $\mathbf{C}$ is called *$\mathfrak{T}$-acyclic*, if the augmented complex

$$\cdots \longrightarrow \mathsf{C}_k \longrightarrow \mathsf{C}_{k-1} \longrightarrow \cdots \xrightarrow{\partial_1} \mathsf{C}_0 \xrightarrow{\partial_0} \mathrm{H}_0(\mathbf{C}) := \overline{\operatorname{coker} \partial_1} \longrightarrow 0$$

is $\mathfrak{T}$-exact. A *$\mathfrak{T}$-projective* or *topologically projective resolution* of an object $\mathcal{M}$ then is a topologically projective and $\mathfrak{T}$-acyclic complex $\mathbf{C}$ with $\mathrm{H}_0(\mathbf{C}) \cong \mathcal{M}$.

6.2.5 Comparison theorem *Let*

$$\mathbf{C} : \cdots \xrightarrow{\partial_{k+1}} \mathsf{C}_k \xrightarrow{\partial_k} \mathsf{C}_{k-1} \xrightarrow{\partial_{k-1}} \cdots \xrightarrow{\partial_1} \mathsf{C}_0$$

and

$$\mathbf{D} : \cdots \xrightarrow{\delta_{k+1}} \mathsf{D}_k \xrightarrow{\delta_k} \mathsf{D}_{k-1} \xrightarrow{\delta_{k-1}} \cdots \xrightarrow{\delta_1} \mathsf{D}_0$$

be two complexes in $\mathcal{A}$-$\mathfrak{Mod}_{\mathrm{clc}}$, where $\mathbf{C}$ is assumed to be topologically projective and $\mathbf{D}$ $\mathfrak{T}$-acyclic. Then there exists for every continuous $\mathcal{A}$-linear mapping $\mathsf{f} : \mathrm{H}_0(\mathbf{C}) \to \mathrm{H}_0(\mathbf{D})$ a morphism of topological complexes $\mathbf{f} : \mathbf{C} \to \mathbf{D}$ which induces f that means all components $\mathsf{f}_k : \mathsf{C}_k \to \mathsf{D}_k$ of $\mathbf{f}$ are continuous and $\mathrm{H}_0(\mathbf{f}) = \mathsf{f}$ holds true. The homotopy class of $\mathbf{f}$ is determined uniquely by f.

PROOF: We recursively construct the chain map $\mathbf{f}$. To this end we first set $\mathsf{C}_{-1} := \mathrm{H}_0(\mathbf{C})$, $\mathsf{D}_{-1} := \mathrm{H}_0(\mathbf{D})$ and $\mathsf{f}_{-1} := \mathsf{f}$ as well as $\mathsf{C}_{-2} := \mathsf{D}_{-2} := 0$ and $\mathsf{f}_{-2} := 0$. Now suppose that $\mathsf{f}_{-1}, \cdots, \mathsf{f}_{k-1}$ are already defined for $k \in \mathbb{N}$ such that for $l = -1, \cdots, k-1$ the diagram

$$\begin{array}{ccc} \mathsf{C}_l & \xrightarrow{\partial_l} & \mathsf{C}_{l-1} \\ \downarrow{\scriptstyle \mathsf{f}_k} & & \downarrow{\scriptstyle \mathsf{f}_{l-1}} \\ \mathsf{D}_l & \xrightarrow{\delta_l} & \mathsf{D}_{l-1} \end{array} \tag{6.2.2}$$

commutes. Then $\operatorname{im}(\mathsf{f}_{k-1}\partial_k) \subset \ker \delta_{k-1} = \operatorname{im} \delta_k$, hence by the $\mathfrak{T}$-projectivity of C_k there exists a $\mathsf{f}_k : \mathsf{C}_k \to \mathsf{D}_k$ such that the diagram (6.2.2) commutes for $l = k$. As the induction hypothesis holds for $k = 0$, we thus obtain inductively a chain map $\mathbf{f}$ inducing f.

Let $\mathbf{f}$ and $\mathbf{g}$ be two f inducing chain maps from $\mathbf{C}$ to $\mathbf{D}$. We then construct inductively a homotopy $\mathbf{s} = \bigoplus_{k \in \mathbb{N}} \mathsf{s}_k$, $\mathsf{s}_k : \mathsf{C}_k \to \mathsf{D}_{k+1}$ from $\mathbf{f}$ to $\mathbf{g}$. To this end set first $\mathsf{s}_{-1} := \mathsf{s}_{-2} := 0$ and suppose that we already have defined for some $k \in \mathbb{N}$ maps $\mathsf{s}_{-1}, \cdots, \mathsf{s}_{k-1}$ such that for $l = -1, \cdots, k-1$ the equation

$$\mathsf{f}_l - \mathsf{g}_l - \mathsf{s}_{l-1}\partial_l = \delta_{l+1}\mathsf{s}_l \tag{6.2.3}$$

is true. By

$$\delta_k(\mathsf{f}_k - \mathsf{g}_k - \mathsf{s}_{k-1}\partial_k) = (\mathsf{f}_{k-1} - \mathsf{g}_{k-1} - \delta_k \mathsf{s}_{k-1})\partial_k = \mathsf{s}_{k-2}\partial_{k-1}\partial_k = 0$$

the relation $\operatorname{im}(\delta_k(\mathsf{f}_k - \mathsf{g}_k - \mathsf{s}_{k-1}\partial_k)) \subset \ker \delta_k = \operatorname{im} \delta_{k+1}$ then holds as well. As C_k is topologically projective and δ_{k+1} a $\mathfrak{T}$-admissible morphism, there exists a continuous $\mathcal{A}$-linear mapping $\mathsf{s}_n : \mathsf{C}_n \to \mathsf{D}_{k+1}$, such that (6.2.3) is satisfied for $l = k$. As for $k = 0$ the induction hypothesis is true, we thus obtain the desired homotopy $\mathbf{s}$. □

6.2.6 Now we have enough tools to introduce topologically derived functors. To this end let $\mathfrak{A}$ be an Abelian category and $\mathsf{T} : \mathcal{A}\text{-}\mathfrak{Mod}_{\mathrm{clc}} \to \mathfrak{A}$ a (covariant) additive functor. Then choose for every object $\mathcal{M}$ a topologically projective resolution $\mathbf{C}_{\mathcal{M}}$. As by the comparison theorem the homology groups $\mathrm{H}_k(\mathsf{T}\mathbf{C}_{\mathcal{M}})$ of the complex $\mathsf{T}\mathbf{C}_{\mathcal{M}}$ in $\mathcal{A}$ do not depend on the special choice of the resolution $\mathbf{C}_{\mathcal{M}}$, we thus obtain for every $k \in \mathbb{N}$ an additive functor $\mathrm{L}_k\mathsf{T} : \mathcal{A}\text{-}\mathfrak{Mod}_{\mathrm{clc}} \to \mathcal{A}$, $\mathcal{M} \to \mathrm{H}_k(\mathsf{T}\mathbf{C}_{\mathcal{M}})$ called the *k-th left topologically derived functor* of T. If $\mathsf{S} : \mathcal{A}\text{-}\mathfrak{Mod}_{\mathrm{clc}} \to \mathcal{A}$ is a contravariant additive functor, then we analogously define the *k-th right topologically derived functor* $\mathrm{R}^n\mathsf{S}$ by $\mathrm{R}^k\mathsf{S}(\mathcal{M}) = \mathrm{H}^k(\mathsf{S}\mathbf{C}_{\mathcal{M}})$.

For the special case that $\mathsf{T} = \mathcal{M}\hat{\otimes}_{\mathcal{A}}-$ is the functor of the completed tensor product and $\mathsf{S} = \mathrm{Hom}_{\mathcal{A}}(-, \mathcal{M})$ the functor of continuous $\mathcal{A}$-linear mappings, the corresponding derived functors have their own names:

$$\begin{aligned} \mathrm{Tor}_k^{\mathcal{A}}(\mathcal{N}, \mathcal{M}) &:= \mathrm{L}_k\mathsf{T}(\mathcal{N}), \\ \mathrm{Ext}_{\mathcal{A}}^k(\mathcal{N}, \mathcal{M}) &:= \mathrm{R}^k\mathsf{S}(\mathcal{N}), \quad \mathcal{N} \in \mathrm{Ob}(\mathcal{A}\text{-}\mathfrak{Mod}_{\mathrm{clc}}). \end{aligned} \tag{6.2.4}$$

Note that the (co)homology groups $\mathrm{L}_k\mathsf{T}$ and $\mathrm{R}^k\mathsf{S}$ need not be Hausdorff and that each of these groups is Hausdorff if and only if it is complete.

6.3 Continuous Hochschild homology

6.3.1 Let $\mathcal{A}$ be a complete locally convex algebra and $\mathcal{A}^e$ the algebra $\mathcal{A}^e = \mathcal{A}\hat{\otimes}_\pi\mathcal{A}^\circ$. Hereby, $\mathcal{A}^\circ$ is the complete locally convex algebra opposite to $\mathcal{A}$. It has the same underlying topological vector space like $\mathcal{A}$ but the opposite multiplication $(a, b) \mapsto b \cdot a$, where $\cdot$ denotes the multiplication in $\mathcal{A}$. Obviously, $\mathcal{A}$ is an object in the category $\mathcal{A}^e\text{-}\mathfrak{Mod}_{\mathrm{clc}}$, where the $\mathcal{A}^e$-module structure of $\mathcal{A}$ is given by $\mathcal{A}^e \times \mathcal{A} \ni (a \otimes \tilde{a}, b) \mapsto a\, b\, \tilde{a} \in \mathcal{A}$. Now let $\mathcal{M}$ be an object in $\mathcal{A}^e\text{-}\mathfrak{Mod}_{\mathrm{clc}}$ or in other words a *complete locally convex $\mathcal{A}$-bimodule.* Then one defines the *continuous Hochschild homology* of $\mathcal{A}$ with values in $\mathcal{M}$ by

$$\mathrm{H}_n(\mathcal{A}, \mathcal{M}) = \mathrm{Tor}_n^{\mathcal{A}^e}(\mathcal{A}, \mathcal{M}),$$

and the *continuous Hochschild cohomology* of $\mathcal{A}$ with values in $\mathcal{M}$ by

$$\mathrm{H}^n(\mathcal{A}, \mathcal{M}) = \mathrm{Ext}_{\mathcal{A}^e}^n(\mathcal{A}, \mathcal{M}).$$

Instead of $\mathrm{H}_n(\mathcal{A}, \mathcal{A})$ resp. $\mathrm{H}^n(\mathcal{A}, \mathcal{A}^*)$ we often write $\mathrm{HH}_n(\mathcal{A})$ resp. $\mathrm{HH}^n(\mathcal{A})$ and call these vector spaces the *continuous Hochschild homology* resp. *continuous Hochschild cohomology* of $\mathcal{A}$.

6.3.2 To compute the continuous Hochschild (co)homology the *topological Bar resolution* turns out to be very useful. For a complete locally convex algebra $\mathcal{A}$ it is given by the sequence

$$\mathbf{C}_{\mathcal{A}}^{\mathrm{Bar}} : \cdots \xrightarrow{\delta_{k+1}} \mathcal{A}^e\hat{\otimes}_\pi\underbrace{\mathcal{A}\hat{\otimes}_\pi \cdots \hat{\otimes}_\pi\mathcal{A}}_{k\text{--times}} \xrightarrow{\delta_k} \cdots \xrightarrow{\delta_2} \mathcal{A}^e\hat{\otimes}_\pi\mathcal{A} \xrightarrow{\delta_1} \mathcal{A}^e \longrightarrow \mathcal{A} \longrightarrow 0$$

where $\delta_k((a \otimes b) \otimes a_1 \otimes \cdots \otimes a_k)$

$$
\begin{aligned}
&= (aa_1 \otimes b) \otimes (a_2 \otimes \cdots \otimes a_k) + \\
&+ \sum_{j=1}^{k-1} (-1)^j (a \otimes b) \otimes (a_1 \otimes \cdots \otimes a_j a_{j+1} \otimes \cdots \otimes a_k) + \\
&+ (-1)^k (a \otimes a_k b) \otimes (a_1 \otimes \cdots \otimes a_{k-1}), \qquad a, b, a_1, \cdots, a_k \in \mathcal{A}.
\end{aligned}
$$

One checks easily (see for example [112, Sec. 1.1]) that $\delta_{k-1} \circ \delta_k = 0$. Moreover, the mappings

$$s_k : \mathrm{C}^{\mathrm{Bar}}_{\mathcal{A},k} \to \mathrm{C}^{\mathrm{Bar}}_{\mathcal{A},k+1}, \quad (a \otimes b) \otimes a_1 \otimes \cdots \otimes a_k \mapsto (1 \otimes b) \otimes a \otimes a_1 \otimes \cdots \otimes a_k$$

and $s_{-1} : \mathcal{A} \to \mathcal{A}^e$, $a \mapsto (1 \otimes a)$ are continuously $\Bbbk$-linear and induce a contracting homotopy for $\mathrm{C}^{\mathrm{Bar}}_{\mathcal{A}}$ that means one has for all $k \in \mathbb{N}$

$$\delta_{k+1} \circ s_k + s_{k-1} \circ \delta_k = \mathrm{id}.$$

This implies that all δ_k are $\mathfrak{T}$-admissible, but also that $\mathrm{C}^{\mathrm{Bar}}_{\mathcal{A}}$ is acyclic. Hence $\mathrm{C}^{\mathrm{Bar}}_{\mathcal{A}}$ is a topologically projective resolution of $\mathcal{A}$ in $\mathcal{A}^e$-$\mathfrak{Mod}_{\mathrm{clc}}$, if we can yet show that all components $\mathrm{C}^{\mathrm{Bar}}_{\mathcal{A},k} = \mathcal{A}^e \hat{\otimes}_\pi \mathcal{A} \hat{\otimes}_\pi \cdots \hat{\otimes}_\pi \mathcal{A}$ are topologically projective $\mathcal{A}^e$-modules. But this follows directly from the statement (2) in Proposition 6.2.2. If now $\mathcal{M}$ is a complete locally convex $\mathcal{A}^e$-module and if one applies the functors $\mathcal{M} \hat{\otimes}_{\mathcal{A}^e} -$ and $\mathrm{Hom}_{\mathcal{A}^e}(-, \mathcal{M})$ to the Bar complex, then one obtains the *topological Hochschild complex*

$$\mathrm{C}_\bullet(\mathcal{A}, \mathcal{M}) := \mathcal{M} \hat{\otimes}_{\mathcal{A}^e} \mathrm{C}^{\mathrm{Bar}}_{\mathcal{A}} : \cdots \xrightarrow{b_{k+1}} \mathcal{M} \hat{\otimes}_\pi \mathcal{A} \hat{\otimes}_\pi \cdots \hat{\otimes}_\pi \mathcal{A} \xrightarrow{b_k} \cdots \xrightarrow{b_2} \mathcal{M} \hat{\otimes}_\pi \mathcal{A} \xrightarrow{b_1} \mathcal{M} \xrightarrow{b_0} 0$$

and the *topological Hochschild cocomplex*

$$
\begin{aligned}
\mathrm{C}^\bullet(\mathcal{A}, \mathcal{M}) := \mathrm{Hom}_{\mathcal{A}^e}(\mathrm{C}^{\mathrm{Bar}}_{\mathcal{A}}, \mathcal{M}) : \ 0 \longrightarrow \mathcal{M} \xrightarrow{\beta^0} \mathrm{Hom}(\mathcal{A}, \mathcal{M}) \xrightarrow{\beta^1} \cdots \\
\xrightarrow{\beta^{k-1}} \mathrm{Hom}(\mathcal{A} \hat{\otimes}_\pi \cdots \hat{\otimes}_\pi \mathcal{A}, \mathcal{M}) \xrightarrow{\beta^k} \cdots .
\end{aligned}
$$

The homology of $(\mathrm{C}_\bullet(\mathcal{A}, \mathcal{M}), b_\bullet)$ resp. $(\mathrm{C}^\bullet(\mathcal{A}, \mathcal{M}), \beta^\bullet)$ gives as desired the continuous Hochschild (co)homology of $\mathcal{A}$ with values in $\mathcal{M}$. In analogy to ordinary Hochschild homology we call the elements of $\mathcal{Z}_k = \ker b_k$ (resp. $\mathcal{Z}^k = \ker \beta^k$) *continuous Hochschild (co)cycles*, and the elements of $\mathcal{B}_k = \operatorname{im} b_{k+1}$ (resp. $\mathcal{B}^k = \ker \beta^{k-1}$) *continuous Hochschild (co)boundaries.*

6.3.3 Proposition *Let $\mathcal{A}$ be a complete locally convex algebra and $\mathcal{M}$ a complete locally convex $\mathcal{A}^e$-module. Then the low dimensional continuous Hochschild homology groups compute as follows:*

$$
\begin{aligned}
&\mathrm{H}_0(\mathcal{A}, \mathcal{M}) = \mathcal{M}_{\mathcal{A}} := \mathcal{M} \Big/ \Big\{ \sum_{j \in \mathbb{N}} a_j m_j - m_j a_j \Big| \sum a_j \otimes m_j \in \mathcal{A} \hat{\otimes}_\pi \mathcal{M} \Big\}, \\
&\mathrm{H}^0(\mathcal{A}, \mathcal{M}) = \mathcal{M}^{\mathcal{A}} := \{ m \in \mathcal{M} \mid am = ma \text{ for all } a \in \mathcal{A} \}, \\
&\mathrm{H}^1(\mathcal{A}, \mathcal{M}) = \mathrm{Der}_c(\mathcal{A}, \mathcal{M}) / \mathrm{Der}_i(\mathcal{A}, \mathcal{M}).
\end{aligned}
$$

Hereby, $\mathrm{Der_c}(\mathcal{A},\mathcal{M})$ *denotes the space of continuous derivations on* $\mathcal{A}$ *with values in* $\mathcal{M}$*, and* $\mathrm{Der_i}(\mathcal{A},\mathcal{M})$ *the set consisting of all inner derivations that means of all derivations of the form* $a \mapsto \mathrm{ad}(m)(a) = [m,a]$ *with* $m \in \mathcal{M}$*. Moreover, if* $\mathcal{A}$ *is commutative and* $\mathcal{M}$ *a symmetric* $\mathcal{A}^a$*-module, i.e. if* $am = ma$ *holds for all* $a \in \mathcal{A}$ *and* $m \in \mathcal{M}$*, then*

$$\mathrm{HH}_1(\mathcal{A}) = \overline{\Omega}_{\mathcal{A}/\Bbbk} \quad \text{and} \quad \mathrm{H}_1(\mathcal{A},\mathcal{M}) = \mathcal{M}\hat{\otimes}_{\mathcal{A}}\overline{\Omega}_{\mathcal{A}/\Bbbk},$$

where $\overline{\Omega}_{\mathcal{A}/\Bbbk}$ *is the space of topological Kähler differentials.*

For explanation: The space of topological Kähler differentials $\overline{\Omega}_{\mathcal{A}/\Bbbk}$ as defined in Section B.2 carries a canonical complete locally convex topology as shown in B.2. If one supplies the space of continuous derivations $\mathrm{Der_c}(\mathcal{A},\mathcal{M})$ with the strong operator topology that means with the topology of uniform convergence on bounded sets, then $\mathrm{Der_c}(\mathcal{A},\mathcal{M})$ becomes a complete locally convex vector space as well.

PROOF: By definition $\beta^0(m)(a) = am - ma$ and $\beta^1(f)(a \otimes b) = f(ab) - af(b) - f(a)b$ hold for every continuous $\Bbbk$-linear mapping $f : \mathcal{A}\hat{\otimes}_\pi\mathcal{A} \to \mathcal{M}$. That gives the cohomology groups $\mathrm{H}^0(\mathcal{A},\mathcal{M})$ and $\mathrm{H}^1(\mathcal{A},\mathcal{M})$. As $b_1(a \otimes m) = am - ma$, one obtains the homology group $\mathrm{H}_0(\mathcal{A},\mathcal{M})$.

Now let us come to the case that $\mathcal{A}$ is commutative and $\mathcal{M}$ a symmetric $\mathcal{A}$-bimodule. Then $b_1 : \mathcal{M}\hat{\otimes}_\pi\mathcal{A} \to \mathcal{M}$ is trivial, hence $\mathrm{H}_1(\mathcal{A},\mathcal{M})$ is the quotient $(\mathcal{M}\hat{\otimes}_\pi\mathcal{A})/\mathrm{im}\, b_2$. Therefore, the mapping

$$\Theta : \mathrm{H}_1(\mathcal{A},\mathcal{M}) \to \mathcal{M}\hat{\otimes}_{\mathcal{A}}\overline{\Omega}_{\mathcal{A}/\Bbbk}, \quad m \otimes a \mapsto m \otimes da$$

is well-defined and continuous. Together with the closed ideal $\overline{\mathcal{J}}$ of Section B.2 one thus obtains a further continuous mapping

$$\Upsilon : \mathcal{M}\hat{\otimes}_{\mathcal{A}}\overline{\mathcal{J}} \to \mathrm{H}_1(\mathcal{A},\mathcal{M}), \quad m \otimes \sum_{j\in\mathbb{N}} a_j \otimes c_j \mapsto \sum_{j\in\mathbb{N}} a_j m \otimes c_j + \mathrm{im}\, b_2.$$

As for two elements $\sum a_i \otimes c_i \in \overline{\mathcal{J}}$ and $\sum \tilde{a}_j \otimes \tilde{c}_j \in \overline{\mathcal{J}}$ the relation

$$\sum_{i,j} a_i\tilde{a}_j m \otimes c_i\tilde{c}_j + \mathrm{im}\, b_2 = \sum_{i,j} (a_i m \otimes c_i\tilde{a}_j\tilde{c}_j - a_i c_i \tilde{c}_j m \otimes \tilde{a}_j) + \mathrm{im}\, b_2 = 0$$

holds true, Υ factorizes to a continuous $\mathcal{A}$-linear morphism $\overline{\Upsilon} : \mathcal{M}\hat{\otimes}_{\mathcal{A}}\overline{\Omega}_{\mathcal{A}/\Bbbk} \to \mathrm{H}_1(\mathcal{A},\mathcal{M})$, where $\overline{\Omega}_{\mathcal{A}/\Bbbk} = \overline{\mathcal{J}}/\overline{\mathcal{J}}^2$. The two thus defined mappings $\overline{\Upsilon}$ and Θ are inverse to each other, as one checks by some short calculation. □

6.3.4 Proposition *Let* $\mathcal{A}$ *be a commutative, complete locally convex algebra and* $\mathcal{M}$ *a complete locally convex* $\mathcal{A}$*-module. Then the antisymmetrization* $\varepsilon_k : \mathrm{C}^{\mathrm{Bar}}_{\mathcal{A},k} \to \mathrm{C}^{\mathrm{Bar}}_{\mathcal{A},k}$*,*

$$(a \otimes b) \otimes a_1 \otimes \cdots \otimes a_k \mapsto \sum_{\sigma \in S_k} \mathrm{sgn}(\sigma)\left((a \otimes b) \otimes a_{\sigma(1)} \otimes \cdots \otimes a_{\sigma(k)}\right)$$

is continuously $\mathcal{A}^e$*-linear and induces morphisms*

$$\varepsilon_k : \mathrm{C}_k(\mathcal{A},\mathcal{M}) \to \mathrm{C}_k(\mathcal{A},\mathcal{M}) \quad \text{and} \quad \varepsilon^k : \mathrm{C}^k(\mathcal{A},\mathcal{M}) \to \mathrm{C}^k(\mathcal{A},\mathcal{M})$$

on the continuous Hochschild (co)chains such that

$$\mathsf{b}_k \circ \varepsilon_k = 0 \quad \textit{and} \quad \varepsilon^k \circ \beta^{k-1} = 0.$$

PROOF: The continuity of ε_k is clear by definition. We now show that $\mathsf{b}_k \circ \varepsilon_k = 0$; the equality $\varepsilon_k \circ \beta^{k-1} = 0$ follows by an analogous argument. We denote by $\tau_l \in S_k$, $1 \le l < k$ the transposition of l and $l+1$ and by $S_{k,l}$ the set of all permutations σ with $\sigma(l) < \sigma(l+1)$. Then for all $l < k$ the set S_k is the disjoint union of the sets $S_{k,l}$ and $S_{k,l}\tau_l$. Hence

$$\begin{aligned}
&\mathsf{b}\varepsilon_k(\mathfrak{m}\otimes a_1 \otimes \cdots \otimes a_k) = \\
&\quad \sum_{\sigma\in S_k} \Big(a_{\sigma(1)}\mathfrak{m} \otimes a_{\sigma(2)} \otimes \cdots \otimes a_{\sigma(k)} - a_{\sigma(k)}\mathfrak{m} \otimes a_{\sigma(2)} \otimes \cdots \otimes a_{\sigma(k-1)} \otimes a_{\sigma(1)} \Big) \\
&\quad + \sum_{l=1}^{k-1} \sum_{\sigma\in S_{k,l}} (-1)^l \operatorname{sgn}(\sigma) \Big(\mathfrak{m} \otimes a_{\sigma(1)} \otimes \cdots \otimes a_{\sigma(l)} a_{\sigma(l+1)} \otimes \cdots \otimes a_{\sigma(k)} \\
&\qquad\qquad - \mathfrak{m} \otimes a_{\sigma(1)} \otimes \cdots \otimes a_{\sigma(\tau_l(l))} a_{\sigma(\tau_l(l+1))} \otimes \cdots \otimes a_{\sigma(k)} \Big) = 0,
\end{aligned}$$

which proves the claim. □

6.3.5 Proposition *Under the assumptions of the preceding proposition the antisymmetrization operators induce the following continuous mappings:*

$$\begin{aligned}
&[\varepsilon]_k : \mathcal{M}\hat{\otimes}_{\mathcal{A}} \overline{\Omega}^k_{\mathcal{A}/\Bbbk} \to \mathrm{H}_k(\mathcal{A}, \mathcal{M}),\ \mathfrak{m} \otimes da_1 \wedge \cdots \wedge da_k \mapsto [\varepsilon_k(\mathfrak{m} \otimes a_1 \otimes \cdots \otimes a_k)], \\
&[\varepsilon]^k : \overline{\Lambda}^k_{\mathcal{A}} \operatorname{Der}_c(\mathcal{A}, \mathcal{M}) \to \mathrm{H}^k(\mathcal{A}, \mathcal{M}),\ f_1 \wedge \cdots \wedge f_k \mapsto [\varepsilon^k(f_1 \otimes \cdots \otimes f_k)], \\
&[\pi]_k : \mathrm{H}_k(\mathcal{A}, \mathcal{M}) \to \mathcal{M}\hat{\otimes}_{\mathcal{A}} \overline{\Omega}^k_{\mathcal{A}/\Bbbk}, \\
&\qquad\qquad \Big[\sum \mathfrak{m}_{(0)} \otimes a_{(1)} \otimes \cdots \otimes a_{(k)}\Big] \mapsto \sum \mathfrak{m}_{(0)} \otimes da_{(1)} \wedge \cdots \wedge da_{(k)}.
\end{aligned}$$

If $\mathcal{A}$ is finitely generated as a topological algebra and $\mathcal{M}$ a finitely generated topologically projective $\mathcal{A}$-module, then there exists a continuous mapping

$$[\pi]^k : \mathrm{H}^k(\mathcal{A}, \mathcal{M}) \to \overline{\Lambda}^k_{\mathcal{A}} \operatorname{Der}_c(\mathcal{A}, \mathcal{M}),\ \Big[\sum f_{(1)} \otimes \cdots \otimes f_{(k)}\Big] \mapsto \sum f_{(1)} \wedge \cdots \wedge f_{(k)}.$$

For the explanation of the symbols $\overline{\Omega}^k_{\mathcal{A}/\Bbbk}$ and $\overline{\Lambda}^k$ we refer the reader to Section B.2 in the Appendix. Moreover, $[\,\cdot\,]$ denotes the (co)homology class of a continuous Hochschild (co)cycle.

PROOF: By the relation $\mathsf{b}_k \circ \varepsilon_k = 0$ from Proposition 6.3.4, the antisymmetrization operator induces a continuous mapping $\varepsilon_k : \mathcal{M}\hat{\otimes}_{\Bbbk}\overline{\Lambda}^k \mathcal{A} \to \mathrm{H}_k(\mathcal{A}, \mathcal{M})$. To show that this ε_k factorizes by $\mathcal{M}\hat{\otimes}_{\Bbbk}\overline{\Omega}^k_{\mathcal{A}/\Bbbk}$ to a map $[\varepsilon]_k$, we first have to prove that the continuous mappings

$$\mathcal{A} \ni a_l \mapsto [\varepsilon_k(\mathfrak{m} \otimes a_1 \otimes \cdots \otimes a_l \otimes \cdots \otimes a_k)] \in \mathrm{H}_k(\mathcal{A}, \mathcal{M}).$$

are all derivations. For that it suffices to prove (cf. [112, Prop. 1.3.12]) that

$$\varepsilon_k(\mathfrak{m}b \otimes c \otimes a_2 \otimes \cdots \otimes a_k) + \varepsilon_k(\mathfrak{m}c \otimes b \otimes a_2 \otimes \cdots \otimes a_k) - \varepsilon_k(\mathfrak{m} \otimes bc \otimes a_2 \otimes \cdots \otimes a_k)$$

is a (continuous) Hochschild boundary. But this is the case indeed, because after setting $a_0 = b$, $a_1 = c$ and $T_{k+1} = \{\sigma \in S_{k+1} |\ \sigma^{-1}(0) < \sigma^{-1}(1)\}$ we obtain the following relations

$$
\begin{aligned}
& b_{k+1}\Big(\sum_{\sigma\in T_{k+1}} \operatorname{sgn}(\sigma) m \otimes a_{\sigma(0)} \otimes a_{\sigma(1)} \otimes \cdots \otimes a_{\sigma(k)}\Big) = \\
&= \sum_{\sigma\in T_{k+1}} \operatorname{sgn}(\sigma)\Big(m a_{\sigma(0)} \otimes a_{\sigma(1)} \otimes a_{\sigma(k)} + \sum_{l=0}^{k-1} (-1)^{l+1} m \otimes a_{\sigma(0)} \otimes \cdots \otimes a_{\sigma(l)} \otimes a_{\sigma(k)} \\
&\qquad + (-1)^{k+1} m a_{\sigma(1)} \otimes a_{\sigma(0)} \otimes a_{\sigma(2)} \otimes \cdots \otimes a_{\sigma(k)}\Big) \\
&= \varepsilon_k(mb \otimes c \otimes a_2 \otimes \cdots \otimes a_k) + \sum_{\substack{\sigma\in S_{k+1}\\ 0<\sigma^{-1}(0)<\sigma^{-1}(1)}} \operatorname{sgn}(\sigma) m \otimes a_{\sigma(0)} \otimes a_{\sigma(1)} \otimes \cdots \otimes a_{\sigma(k)} \\
&\quad + \sum_{l=0}^{k-1} (-1)^{l+1} \sum_{\substack{\sigma\in S_{k+1}\\ \sigma^{-1}(0)=l,\,\sigma^{-1}(1)=l+1}} m \otimes a_{\sigma(0)} \otimes \cdots \otimes a_{\sigma(l)} a_{\sigma(l+1)} \otimes \cdots \otimes a_{\sigma(k)} \\
&\quad + \varepsilon_k(mc \otimes b \otimes a_2 \otimes \cdots \otimes a_k) \\
&\quad + (-1)^{k+1} \sum_{\substack{\sigma\in S_{k+1}\\ \sigma^{-1}(0)<\sigma^{-1}(1)<k}} \operatorname{sgn}(\sigma) m \otimes a_{\sigma(k)} \otimes a_{\sigma(0)} \otimes \cdots \otimes a_{\sigma(k-1)} \\
&= \varepsilon_k(mb \otimes c \otimes a_2 \otimes \cdots \otimes a_k) + \varepsilon_k(mc \otimes b \otimes a_2 \otimes \cdots \otimes a_k) \\
&\quad - \varepsilon_k(m \otimes bc \otimes a_2 \otimes \cdots \otimes a_k)
\end{aligned} \tag{6.3.3}
$$

For the proof that $[\varepsilon]^k$ is well-defined it suffices to show that for all $f_1, \cdots, f_k \in \mathrm{Der}_c(\mathcal{A}, \mathcal{M})$ the continuous mapping $\Lambda f := \varepsilon^k(f_1 \otimes \cdots \otimes f_k) : \mathcal{A}\hat{\otimes}_\pi \cdots \hat{\otimes}_\pi \mathcal{A} \to \mathcal{M}$ is a Hochschild cocycle. We calculate $\beta^k \varepsilon^k(f_1 \otimes \cdots \otimes f_k)$ by using the fact that all f_l are derivations:

$$
\begin{aligned}
& \beta^k \varepsilon^k (f_1 \otimes \cdots \otimes f_k)(a_1 \otimes \cdots \otimes a_{k+1}) = \\
&= a_1 \Lambda f(a_2 \otimes \cdots \otimes a_{k+1}) + \sum_{l=1}^{k} (-1)^l a_l \Lambda f(a_1 \otimes \cdots \otimes a_{l-1} \otimes a_{l+1} \otimes \cdots \otimes a_{k+1}) \\
&\quad + \sum_{l=1}^{k} (-1)^l a_{l+1} \Lambda f(a_1 \otimes \cdots \otimes a_l \otimes a_{l+2} \otimes \cdots \otimes a_{k+1}) + a_{k+1} \Lambda f(a_1 \otimes \cdots \otimes a_k) \\
&= 0.
\end{aligned}
$$

By a similar argument the morphism $[\pi]_k$ is well-defined (cf. [112, Lem. 1.3.14]).

For the proof of the last part of the claim we suppose that $\mathcal{A}$ is finitely generated as a topological algebra and $\mathcal{M}$ a topologically projective finitely generated $\mathcal{A}$-module. Then $\mathcal{M}$ is the topological direct summand in a topological $\mathcal{A}$-module of the form $\mathcal{A}\hat{\otimes}\mathbb{R}^n$, hence it suffices to prove the claim for the case that $\mathcal{M} = \mathcal{A}$. Now we have

6.3.6 Lemma *For every topologically finitely generated complete locally convex algebra $\mathcal{A}$ the space $\mathrm{Der}_c(\mathcal{A}, \mathcal{A})$ of continuous derivations is finitely generated as an $\mathcal{A}$-module.*

PROOF OF THE LEMMA: By assumption on $\mathcal{A}$ there exists a finite system $(e_1, \cdots, e_n)$ of elements of $\mathcal{A}$ such that the subspace generated by the polynomials in the e_l is dense in $\mathcal{A}$. If there exists a derivation $\delta \in \mathrm{Der_c}(\mathcal{A}, \mathcal{A})$ with $\delta(e_1) \neq 0$, then set

$$\delta_1 = \frac{1}{\delta(e_1)}\delta, \quad \text{and otherwise} \quad \delta_1 = 0.$$

Suppose that one can construct $\delta_1, \cdots, \delta_l$ such that $\delta_k(e_i) = 0$ for $i < k \leq l$ and such that for every continuous derivation δ there exist coefficients $a_1, \cdots, a_l$ such that $\delta_r = \delta - a_1\delta_1 - \ldots - a_l\delta_l$ vanishes on all e_i, $i \leq l$. If there exist δ and $a_1, \cdots, a_l$ with $\delta_r(e_i) = 0$ for $i \leq l$ but $\delta_r(e_{l+1}) \neq 0$, then we set

$$\delta_{l+1} = \frac{1}{\delta_r(e_{l+1})}\delta, \quad \text{and otherwise} \quad \delta_{l+1} = 0.$$

In every case one checks easily that the system $\delta_1, \cdots, \delta_{l+1}$ fulfills the induction hypothesis. As the polynomials in e_k are dense in $\mathcal{A}$, the recursively defined family $\delta_1, \cdots, \delta_n$ generates the $\mathcal{A}$-module $\mathrm{Der_c}(\mathcal{A}, \mathcal{A})$.

PROOF OF 6.3.5, CONTINUED: By $\varepsilon^k \circ \beta^{k-1} = 0$ the map

$$\pi^k : \mathrm{H}^k(\mathcal{A}, \mathcal{M}) \to C^{\mathrm{Bar}}_{\mathcal{A},k}, \quad \left[\sum f_{(1)} \otimes \cdots \otimes f_{(k)}\right] \mapsto \sum f_{(1)} \wedge \cdots \wedge f_{(k)}$$

is well-defined and continuous, hence it remains to show that the image of π^k lies in $\overline{\Lambda}^k_{\mathcal{A}} \mathrm{Der_c}(\mathcal{A}, \mathcal{M})$. By 6.3.3 the mapping

$$a \mapsto \sum_{\sigma \in S_k} \sum f_{(\sigma(1))}(a) \otimes f_{(\sigma(2))}(a_2) \otimes \cdots \otimes f_{(\sigma(k))}(a_k)$$

is a continuous derivation for every Hochschild cocycle $\sum f_{(1)} \otimes \cdots \otimes f_{(k)}$ and all $a_2, \cdots, a_k \in \mathcal{A}$. Next we need a denumeration of the set I consisting of all k-tuples $(i_1, \cdots, i_k)$ of the form $i_1 < i_2 < \cdots < i_k$. For that it suffices to give I the structure of a well-ordered set. One obtains such a well-ordering by defining $(i_1, \cdots, i_k)$ as smaller than $(j_1, \cdots, j_k)$, if $i_1 = j_1, \cdots, i_{l-1} = j_{l-1}$ but $i_l < j_l$ holds for some $l \leq k$. Let $\alpha(j)$, $j \leq \binom{n}{k}$ be the j-the element of I, and $e_\alpha = e_{\alpha_1} \otimes \cdots \otimes e_{\alpha_k}$ for $\alpha = (\alpha_1, \cdots, \alpha_k) \in I$. We now define recursively

$$\begin{aligned}\Delta_1 &= \Lambda f(e_1 \otimes \cdots \otimes e_k)\delta_1 \wedge \cdots \wedge \delta_k,\\ \Delta_{\alpha(j+1)} &= \Delta_{\alpha(j)} + \big(\Lambda f(e_{\alpha(j+1)}) - \Delta_{\alpha(j)}(e_{\alpha(j+1)})\big)\delta_{\alpha_1(j+1)} \wedge \cdots \wedge \delta_{\alpha_k(j+1)},\end{aligned}$$

where $\Lambda f = f_1 \wedge \cdots \wedge f_k$. By an induction argument one concludes that $\Delta_{\alpha\binom{n}{k}} = \Lambda f$. As $\Delta_{\alpha\binom{n}{k}}$ lies by construction in $\overline{\Lambda}^k_{\mathcal{A}} \mathrm{Der_c}(\mathcal{A}, \mathcal{A})$, the proposition now is proven. □

6.3.7 Corollary *Under the assumptions of the proposition* $\mathcal{M}\hat{\otimes}_{\mathcal{A}}\overline{\Omega}^k_{\mathcal{A}/\Bbbk}$ *is in a natural way a topologically direct summand of* $\mathrm{H}_k(\mathcal{A}, \mathcal{M})$. *If* $\mathcal{A}$ *is finitely generated as a topological algebra and* $\mathcal{M}$ *a topologically projective finitely generated complete locally convex* $\mathcal{A}$*-module, then* $\overline{\Lambda}^k_{\mathcal{A}} \mathrm{Der_c}(\mathcal{A}, \mathcal{M})$ *is a canonical topological direct summand of* $\mathrm{H}^k(\mathcal{A}, \mathcal{M})$.

6.4 Hochschild homology of algebras of smooth functions

6.4.1 First let us recall the natural Fréchet topology on the algebra $\mathcal{C}^\infty(M)$ of smooth functions on a manifold M and the natural Fréchet topology on the space $\Omega^k(M)$ of k-forms over M (see Appendix C.1).

Now let us suppose that there exists a global nowhere vanishing vector field V on M. For example this is the case, if M is an open set of some Euclidean space or if M is compact with Euler characteristic 0. Under these assumptions on M ALAIN CONNES has constructed in [45] a topologically projective resolution of $\mathcal{C}^\infty(M)$, which allows the computation of the continuous Hochschild (co)homology of $\mathcal{C}^\infty(M)$. In the following we will give the construction of the resolution according to CONNES [45] and which essentially relies on the Koszul complex. To this end let us choose on M a torsionfree affine connection and denote by exp the corresponding exponential function. Let $\chi : M \times M \to [0,1]$ be a smooth function such that $\exp_y^{-1}(x)$ is defined for all $(x,y) \in \operatorname{supp}\chi$ and such that χ is identical to 1 on a neighborhood of the diagonal. Moreover, let $E_k = \mathrm{pr}_2(\Lambda^k T^*_{\mathbb{C}} M)$ be the complex vector bundle over $M \times M$, which can be obtained by pullback of the k-th exterior product of the complexified cotangent bundle $T^*_{\mathbb{C}} M$ via the projection $\mathrm{pr}_2 : M \times M \to M$ onto the second coordinate. Then E_1^* is the pullback bundle of the complexified tangent bundle $T_{\mathbb{C}} M$ via pr_2, hence

$$Z(x,y) = \chi(x,y)\, \exp_y^{-1}(x) + i\,(1-\chi(x,y))\, V(y) \tag{6.4.1}$$

defines a smooth section Z of E_1^*. It does not vanishes outside the diagonal of $M \times M$. Now we can formulate CONNES' result.

6.4.2 Proposition (CONNES [45], p. 127ff) *Let M be a smooth manifold of dimension n with a nowhere vanishing smooth vector field $V : M \to TM$. Let the bundle E_k and the section $Z \in \mathcal{C}^\infty(E_1^*)$ be defined like above. Then the sequence*

$$\begin{aligned} C^{\mathrm{CK}}_M : 0 \longrightarrow \mathcal{C}^\infty(M \times M, E_n) \xrightarrow{i_Z} \cdots \\ \xrightarrow{i_Z} \mathcal{C}^\infty(M \times M, E_1) \xrightarrow{i_Z} \mathcal{C}^\infty(M \times M) \xrightarrow{\Delta^*} \mathcal{C}^\infty(M) \longrightarrow 0 \end{aligned}$$

comprises a topologically projective resolution of the module $\mathcal{C}^\infty(M)$ over $\mathcal{C}^\infty(M \times M) \cong \mathcal{C}^\infty(M) \hat{\otimes}_\pi \mathcal{C}^\infty(M)$. Hereby i_Z means the canonical insertion of Z in a form and Δ the natural embedding of M as the diagonal of $M \times M$. A continuous chain homotopy from C^{CK}_M to $C^{\mathrm{Bar}}_M := C^{\mathrm{Bar}}_{\mathcal{C}^\infty(M)}$ over the identical map is given by $F_k : C^{\mathrm{CK}}_{M,k} \to C^{\mathrm{Bar}}_{M,k}$, where

$$F_k\omega(x,y,x_1,\cdots,x_k) = \langle Z(x_1,y) \wedge \cdots \wedge Z(x_k,y), \omega(x,y) \rangle \tag{6.4.2}$$

for $x, y, x_i \in M$ and $\omega \in C^{\mathrm{CK}}_{M,k}$.

PROOF: We now give the proof like in [45]. Every one of the modules $\mathcal{M}_k = \mathcal{C}^\infty(M \times M, E_k)$ is finitely generated projective, hence in particular topologically projective. Obviously $i_Z^2 = 0$, hence for the proof that C^{CK}_M is a topologically projective resolution we only have to construct a continuously linear "section" $s_\bullet : C^{\mathrm{CK}}_{M,\bullet} \to C^{\mathrm{CK}}_{M,\bullet+1}$ of i_Z.

To this end we choose a cut-off function $\chi, \chi' : M \times M \to [0,1]$ such that $\chi' = 1$ on the support of χ, $\chi' = 1$ on a neighborhood of the diagonal of $M \times M$ and such that

$$Z(x,y) = \exp_y^{-1}(x) \quad \text{for all } (x,y) \in \operatorname{supp} \chi'.$$

Then there exists a smooth section $\omega' \in \mathcal{C}^\infty(M \times M, E_1)$ with $\langle \omega', Z \rangle_{\operatorname{supp}(1-\chi)} = 1$. For $\omega \in \mathcal{C}^\infty(M \times M, E_k)$ and $x \in M$ one denotes by ω_x the form $\in \Omega^k(M, \mathbb{C})$ such that $\omega_x(y) = \omega_{(x,y)}$ for all $y \in M$. We now define a homotopy

$$H : O \times [0,1] \to M \quad \text{by} \quad H_t(x,y) = H_{x,t}(y) = \exp_x(-t \exp_x^{-1}(y)), \qquad (x,y) \in O.$$

Hereby O is an open neighborhood of the diagonal of $M \times M$ with $\operatorname{supp} \chi' \subset O$. Now we can define

$$s_k(\omega)_x = \chi'_x \int_0^1 H_x^*(d(\chi_x \omega_x)) \frac{dt}{t} + (1 - \chi'_x)\, \omega'_x \wedge \omega_x, \qquad \omega \in \mathcal{C}^\infty(E_k) \tag{6.4.3}$$

According to Lemma 5.2.1 and by using $DH_x(y,t).\frac{\partial}{\partial t} = Z(x,y)$ we obtain for all $\eta \in \mathcal{C}^\infty(E_k)$ with $\operatorname{supp} \eta \subset O$

$$\int_0^1 H_x^*\, d(i_{Z_x} \eta_x) \frac{dt}{t} + i_{Z_x} \int_0^1 H_x^*\, d\eta_x \frac{dt}{t} = K_{M,1} H_x^* \eta_x + H_x^* K_{M,1} \eta_x = \eta_x.$$

But then $s_{k-1} i_{Z_x} \omega_x + i_{Z_x} s_k \omega_x = \omega_x$. Hence $C^{\mathrm{CK}}_{M,\bullet}$ is a topologically projective resolution of $\mathcal{C}^\infty(M)$ indeed.

Next we must show that by $F_\bullet$ there is given a chain map. To this end we compute for $\omega \in \mathcal{C}^\infty(M \times M, E_k)$:

$$\begin{aligned}
(\delta_k F_k \omega)(x, y, x_1, \cdots, x_{k-1}) &= F\omega(x, y, x, x_1 \cdots, x_k) \\
&\quad - \sum_{i=1}^{k-1} (-1)^i F\omega(x, y, x_1, \cdots, x_i, x_i, \cdots, x_{k-1}) + (-1)^k F\omega(x, y, x_1, \cdots, x_{k-1}, y) \\
&= \langle Z(x,y) \wedge Z(x_1, y) \wedge \cdots \wedge Z(x_{k-1}, y), \omega(x,y) \rangle \\
&= (F i_Z \omega)(x, y, x_1, \cdots, x_{k-1}),
\end{aligned}$$

hence F is a chain map. □

According to the work [165] of Teleman and [182] of Wasserman it is possible to calculate for algebras of smooth functions the "global" Hochschild homology from the "local" homology. In the following we will explain this in more detail according to [165]. To this end let X be an (A)-stratified space which is embeddable in a geodesically complete Riemannian manifold (M, μ) and assume X inherits the smooth structure from M. Let d_μ be the geodesic distance on M and $\chi : \mathbb{R}^{\geq 0} \to [0,1]$ a smooth function with support in $[0,1]$ and $\chi = 1$ over $[0, 1/2]$. Then the function

$$\delta_k : X^{k+1} \to \mathbb{R}^{\geq 0}, (x_0, \cdots, x_k) \mapsto d_\mu^2(x_0, x_1) + d_\mu^2(x_1, x_2) + \cdots + d_\mu^2(x_{k-1}, x_k) + d_\mu^2(x_k, x_0)$$

is smooth and measures the distance to the diagonal Δ_{k+1} of X^{k+1}. For $r > 0$ denote by U_r the so-called r-*neighborhood* of the diagonal $\{(x_0, x_1, \cdots, x_k) \mid \delta_k(x_0, \cdots, x_k) < r\}$. According to Teleman the function $\delta_{k,r} := \chi_r \circ \delta_k$ with $\chi_r(t) = \chi(r/t)$ has the following properties:

(1) Multiplication by $\delta_{\bullet,r}$ commutes with the Hochschild boundary operator, as $\delta_{\bullet,r}$ comprises a cyclic homomorphisms in the sense of CONNES [45].

(2) The support of $\delta_{k,r}$ lies in $\mathcal{U}_r$.

(3) The map $\delta_{k,r}$ is identical to 1 over the neighborhood $\mathcal{U}_{r/2}$.

Now one can decompose every Hochschild cycle (or analogously every cocycle) $c \in \mathcal{M}\hat{\otimes}_{\mathcal{C}^\infty(X)} C^{\mathrm{Bar}}_{X,k}$ in the form $c = \delta_{k,r}c + (1-\delta_{k,r})c$. Then the second summand in this decomposition turns out to be acyclic, hence does not contribute to homology. To make this precise we define for a complete locally convex $\mathcal{C}^\infty(M)$-module $\mathcal{M}$ the following complexes:

$$C^r_{X,\mathcal{M},\bullet} := (1-\delta_{\bullet,r})\cdot \mathcal{M}\hat{\otimes}_{\mathcal{C}^\infty(X)} C^{\mathrm{Bar}}_X \quad \text{and} \quad C_r^{X,\mathcal{M},\bullet} := (1-\delta_{\bullet,r})\cdot \mathrm{Hom}_{\mathcal{C}^\infty(X)}(C^{\mathrm{Bar}}_X, \mathcal{M}).$$

The inductive limit with $r \to 0$ will be denoted by $C^0_{X,\mathcal{M},\bullet}$ resp. $C_0^{X,\mathcal{M},\bullet}$. Now, by Theorem 3.2 of TELEMAN [165] we obtain immediately the following

6.4.3 Localization theorem (*cf.* [165, Sec. 3]) *The complexes* $C^0_{X,\mathcal{M},\bullet}$ *and* $C_0^{X,\mathcal{M},\bullet}$ *are acyclic.*

The localization theorem entails that for every locally finite covering $(\mathcal{U}_j)_{j\in\mathbb{N}}$ of X and every Hochschild cycle $c \in \mathcal{M} \otimes_{\mathcal{C}^\infty(X)} C^{\mathrm{Bar}}_{X,k}$ there exist Hochschild cycles c_j with support in $\mathcal{U}_j^{k+1}$ such that c and $\sum c_j$ are homologous. An analogous argument obviously holds for Hochschild cocycles. Thus the computation of Hochschild (co)homology of algebras of smooth functions can be reduced to the one of smooth functions on locally closed stratified subspaces of $\mathbb{R}^n$.

Next we consider the antisymmetrization operators of Proposition 6.3.4.

6.4.4 Lemma *Let M be a manifold and $\mathcal{M}$ a complete locally convex $\mathcal{C}^\infty(M)$-module. Then every Hochschild (co)homology class in $\mathcal{M}$ can be represented by an antisymmetric (co)cycle.*

PROOF: By the localization theorem it suffices to prove the claim for the case that M is an open set in $\mathbb{R}^n$. Then according to Proposition 6.4.2 we have for $\mathcal{C}^\infty(M)$ the topologically projective resolution $C^{\mathrm{CK}}_{M,\bullet}$ and the chain map $F : C^{\mathrm{CK}}_{M,\bullet} \to C^{\mathrm{Bar}}_{M,\bullet}$. Let $[c]$ be a Hochschild homology class in $\mathcal{M}$, represented by the cycle $c \in C^{\mathrm{Bar}}_{M,k}$. Then there exists an element $\omega \in \mathcal{C}^\infty(M \times M, E_k)$, such that c and $F\omega$ are homologous. But by definition of F $F\omega$ is antisymmetric and represents $[c]$ by 6.4.2. Hence the claim holds in the homology case; analogously one shows the claim in the cohomology case. □

6.4.5 Theorem *Let M be a manifold and $\mathcal{M}$ a complete locally convex $\mathcal{C}^\infty$-module. Then one has the canonical isomorphy* $H_\bullet(\mathcal{C}^\infty(M), \mathcal{M}) = \Omega^\bullet(M)\hat{\otimes}_{\mathcal{C}^\infty(M)}\mathcal{M}$. *Moreover, if $\mathcal{M}$ is finitely generated projective, then* $H^\bullet(\mathcal{C}^\infty(M), \mathcal{M}) = \mathcal{C}^\infty(M, \Lambda^\bullet TM)\hat{\otimes}_{\mathcal{C}^\infty(M)}\mathcal{M}$. *In particular* $H_\bullet(\mathcal{C}^\infty(M), \mathcal{C}^\infty(M)) = \Omega^\bullet(M)$ *and* $H^\bullet(\mathcal{C}^\infty(M), \mathcal{C}^\infty(M)) = \mathcal{C}^\infty(M, \Lambda^\bullet TM)$.

PROOF: The claim follows immediately from the preceding lemma and Proposition 6.3.5, if one recalls that the derivations on $\mathcal{C}^\infty(M)$ are given by the smooth vector fields on M. □

It is the goal of the following considerations to transfer the results about the Hochschild homology of smooth functions on manifolds to certain cone spaces of class $\mathcal{C}^\infty$. To this end we consider a compact (A)-stratified subspace $L \subset S^{n-1}$ with the smooth structure inherited from S^{n-1} and a smooth manifold S, on which there exists a nowhere vanishing vector field V. Additionally we assume that the cone $CL \subset \mathbb{R}^n$ is convex. Let X be the stratified space $S \times CL \subset S \times \mathbb{R}^n$ with the induced smooth structure. Now we want to transfer the topologically projective resolution of $\mathcal{C}^\infty(M)$ defined above to the stratified case. We define:

$$C^{CK}_{X,k} := \mathcal{C}^\infty(X) \hat{\otimes} \mathcal{C}^\infty(X, \Lambda^k T^*_{\mathbb{C}} \mathbb{R}^n).$$

Each one of the spaces $C^{CK}_{X,k}$ then is a projective finitely generated $\mathcal{C}^\infty(X \times X)$-module. Next we need a vector field $Z : X \times X \to T(S \times \mathbb{R}^n)$ which is defined like above by

$$Z(x,y) = \chi(x,y) \exp_y^{-1}(x) + i(1 - \chi(x,y))\, V(\pi(y)), \qquad x, y \in X.$$

Hereby exp is the exponential function of the Levi–Civita connection of a product metric formed by a Riemannian metric on S and the Euclidean scalar product on $\mathbb{R}^n$, and $\pi : S \times \mathbb{R}^n \to S$ is the canonical projection. One can insert the vector field Z into an element $\omega \in C^{CK}_{X,k}$ and obtains an element $i_Z \omega \in C^{CK}_{X,k-1}$.

6.4.6 Proposition *Let L, S and X be as above. Then the sequence*

$$C^{CK}_X : 0 \longrightarrow C^{CK}_{X,n} \xrightarrow{i_Z} \cdots$$
$$\xrightarrow{i_Z} C^{CK}_{X,1} \xrightarrow{i_Z} \mathcal{C}^\infty(X \times X) \xrightarrow{\Delta^*} \mathcal{C}^\infty(X) \longrightarrow 0$$

comprises a topologically projective resolution of the module $\mathcal{C}^\infty(X)$ over $\mathcal{C}^\infty(X \times X)$.

A continuous chain map over the identity from C^{CK}_X to $C^{Bar}_X := C^{Bar}_{\mathcal{C}^\infty(X)}$ is given by the morphism $F_k : C^{CK}_{X,k} \to C^{Bar}_{X,k}$ with

$$F_k \omega(x, y, x_1, \cdots, x_k) = \langle Z(x_1, x) \wedge \cdots \wedge Z(x_k, x), \omega(x,y) \rangle \quad x, y, x_i \in X, \ \omega \in C^{CK}_{X,k}. \tag{6.4.5}$$

PROOF: The proof can be performed almost exactly like for Proposition 6.4.2. One only has to regard that the there defined homotopy H is compatible with X, which by the convexity assumption on $CL \subset \mathbb{R}^n$ is satisfied indeed. □

Analogously like for manifolds we now obtain the following result.

6.4.7 Theorem *Let X be a stratified space of the form $S \times CL$, where $L \subset S^{n-1} \subset \mathbb{R}^n$ is an (A)-stratified space with the induced smooth structure and such that $CL \subset \mathbb{R}^n$ is convex. Then for every complete locally convex topological $\mathcal{C}^\infty(X)$-module one has the canonical isomorphy $H_\bullet(\mathcal{C}^\infty(X), \mathcal{M}) = \Omega^\bullet(X) \hat{\otimes}_{\mathcal{C}^\infty(X)} \mathcal{M}$. Moreover, if $\mathcal{M}$ is finitely generated projective, then $H^\bullet(\mathcal{C}^\infty(X), \mathcal{M}) = \Lambda^\bullet \operatorname{Der}(\mathcal{C}^\infty(X), \mathcal{M})$ holds. In particular one has $H_\bullet(\mathcal{C}^\infty(X), \mathcal{C}^\infty(X)) = \Omega^\bullet(X)$ and $H^\bullet(\mathcal{C}^\infty(X), \mathcal{C}^\infty(X)) = \Lambda^\bullet \operatorname{Der}(\mathcal{C}^\infty(X), \mathcal{C}^\infty(X))$.*

6.4.8 Remark The last result of this section is closely connected to the theorem proven by TELEMAN [165] that for the algebra of piecewise differentiable functions on a simplicial complex the Hochschild homology is isomorphic to the piecewise differentiable forms. By this and the result above we conjecture that a HOCHSCHILD–KOSTANT–ROSENBERG theorem holds for arbitrary cone spaces of class $\mathcal{C}^\infty$. Moreover, it remains to clarify the connection to the work BRASSELET–LEGRAND [23].

Appendix A

Supplements from linear algebra and functional analysis

A.1 The vector space distance

In this section we will give a natural distance function on the *Graßmannian* $\mathrm{Gr}_k(\mathbb{R}^n)$ of k-dimensional linear subspaces of $\mathbb{R}^n$. To keep notation reasonable we mention at this point that $\mathcal{V}, \mathcal{W}, \mathcal{V}_1, \cdots, \mathcal{V}_k, \cdots$ will always denote in this section vector subspaces of $\mathbb{R}^n$.

The *vector space distance* of $\mathcal{V}$ and $\mathcal{W}$ is defined by

$$d_{\mathrm{Gr}}(\mathcal{V}, \mathcal{W}) = \sup_{v \in \mathcal{V},\, \|v\|=1} \inf \{\|v - w\| \mid w \in \mathcal{W}\}.$$

In general d_{Gr} is not symmetric in $\mathcal{V}$ and $\mathcal{W}$, hence does not comprise a metric on the set of all subspaces of $\mathbb{R}^n$. But we will see that the restriction of d_{Gr} onto any Graßmannian comprises a metric.

Denote by $P_{\mathcal{V}}$ the orthogonal projection onto $\mathcal{V}$ with respect to the Euclidean scalar product on $\mathbb{R}^n$, $P_{\mathcal{W}}$ the orthogonal projection onto $\mathcal{W}$ and $Q_{\mathcal{W}} = \mathrm{id}_{\mathbb{R}^n} - P_{\mathcal{W}}$ the one onto the orthogonal space $\mathcal{W}^{\perp}$. Then one can write the vector space distance in the following form using the operator norm:

$$d_{\mathrm{Gr}}(\mathcal{V}, \mathcal{W}) = \|Q_{\mathcal{W}} P_{\mathcal{V}}\|. \qquad \text{(A.1.2)}$$

A.1.1 Proposition *The vector space distance has the following properties:*

(1) d_{Gr} *takes only values from* 0 *to* 1.

(2) *The relation* $d_{\mathrm{Gr}}(\mathcal{V}, \mathcal{W}) = 0$ *holds if and only if* $\mathcal{V} \subset \mathcal{W}$, *and* $d_{\mathrm{Gr}}(\mathcal{V}, \mathcal{W}) = 1$ *if and only if* $\mathcal{V} \cap \mathcal{W}^{\perp} \neq \{0\}$.

(3) *If* $\dim \mathcal{V} = \dim \mathcal{W}$, *then* $d_{\mathrm{Gr}}(\mathcal{V}, \mathcal{W}) = d_{\mathrm{Gr}}(\mathcal{W}, \mathcal{V})$.

(4) *For all* $\mathcal{V}_1, \mathcal{V}_2, \mathcal{V}_3$ *the triangle inequality holds:*

$$d_{\mathrm{Gr}}(\mathcal{V}_1, \mathcal{V}_3) \leq d_{\mathrm{Gr}}(\mathcal{V}_1, \mathcal{V}_2) + d_{\mathrm{Gr}}(\mathcal{V}_2, \mathcal{V}_3).$$

M.J. Pflaum: LNM 1768, pp.201-203, 2001

(5) *If* $\mathcal{V}_1, \cdots, \mathcal{V}_k$ *are pairwise orthogonal, then*

$$d_{Gr}(\mathcal{V}_1 \oplus \cdots \oplus \mathcal{V}_k, \mathcal{W}) \leq d_{Gr}(\mathcal{V}_1, \mathcal{W}) + \cdots + d_{Gr}(\mathcal{V}_k, \mathcal{W}).$$

PROOF: The properties (1), (2) and (5) follow immediately from (A.1.2). Under the assumption that $\dim \mathcal{V} = \dim \mathcal{W}$ the relation $d_{Gr}(\mathcal{V}, \mathcal{W}) = 0$ (resp. $d_{Gr}(\mathcal{V}, \mathcal{W}) = 1$) is by (2) fulfilled, if and only if $d_{Gr}(\mathcal{W}, \mathcal{V}) = 0$ (resp. $d_{Gr}(\mathcal{W}, \mathcal{V}) = 1$). That (3) holds for $0 < d_{Gr}(\mathcal{V}, \mathcal{W}) < 1$ as well, is a consequence of the following consideration. If $\dim \mathcal{V} = \dim \mathcal{W} = 1$, hence $\mathcal{V} = \operatorname{span} v$ and $\mathcal{W} = \operatorname{span} w$ with $\|v\| = \|w\| = 1$, then $d_{Gr}(\mathcal{V}, \mathcal{W}) = |\langle v, w\rangle| = d_{Gr}(\mathcal{W}, \mathcal{V})$ follows. If on the other hand $\dim \mathcal{V} = \dim \mathcal{W} > 1$, then choose a unit vector $v \in \mathcal{V}$ with $d_{Gr}(\mathcal{V}, \mathcal{W}) = \|v - P_{\mathcal{W}} v\|$. By $0 < d_{Gr}(\mathcal{V}, \mathcal{W}) < 1$ the relation $P_{\mathcal{W}} v \neq 0$ follows, hence $w = \frac{P_{\mathcal{W}} v}{\|P_{\mathcal{W}} v\|}$ is well-defined and $d_{Gr}(\mathcal{V}, \mathcal{W}) = |\langle w, v\rangle|$. By the same reason $P_{\mathcal{V}} w \neq 0$ follows. Hence altogether

$$\begin{aligned} d_{Gr}(\mathcal{W}, \mathcal{V}) &\geq \|w - P_{\mathcal{V}} w\| = \frac{|\langle P_{\mathcal{V}} w, w\rangle|}{\|P_{\mathcal{V}} w\|} = \|P_{\mathcal{V}} w\| = \frac{\|P_{\mathcal{V}} P_{\mathcal{W}} v\|}{\|P_{\mathcal{W}} v\|} \\ &\geq \frac{|\langle P_{\mathcal{V}} P_{\mathcal{W}} v, v\rangle|}{\|P_{\mathcal{W}} v\|} = |\langle w, v\rangle| = d_{Gr}(\mathcal{V}, \mathcal{W}). \end{aligned}$$

By reasons of symmetry $d_{Gr}(\mathcal{V}, \mathcal{W}) \geq d_{Gr}(\mathcal{W}, \mathcal{V})$ holds as well, hence we obtain (3).

For the proof of the triangle inequality (4) choose a normal $v \in \mathcal{V}_1$ with $d_{Gr}(\mathcal{V}_1, \mathcal{V}_3) = \|v - P_{\mathcal{V}_3} v\|$. Then

$$\begin{aligned} d_{Gr}(\mathcal{V}_1, \mathcal{V}_3) &= \|v - P_{\mathcal{V}_3} v\| \leq \|v - P_{\mathcal{V}_3} P_{\mathcal{V}_2} v\| \leq \|v - P_{\mathcal{V}_2} v\| + \|P_{\mathcal{V}_2} v - P_{\mathcal{V}_3} P_{\mathcal{V}_2} v\| \\ &\leq d_{Gr}(\mathcal{V}_1, \mathcal{V}_2) + d_{Gr}(\mathcal{V}_2, \mathcal{V}_3). \end{aligned}$$

This finishes the proof. □

A.2 Polar decomposition

The polar decomposition of a linear isomorphy is well-known in linear algebra and functional analysis. We will need this result several times in this work and will need in particular that the polar decomposition is differentiable. This special property is often not shown in the literature, hence we will prove it here.

A.2.1 Theorem *Let* $\mathcal{V}$ *be a finite dimensional (real or complex) vector space and* $\langle \cdot, \cdot \rangle$ *the (Euclidean or Hermitian) scalar product on* $\mathcal{V}$. *Then there exist smooth mappings* $u : GL(\mathcal{V}) \to GL(\mathcal{V})$ *and* $s : GL(\mathcal{V}) \to GL(\mathcal{V})$, *where* u *assumes only unitary maps as values and* s *only positive definite linear maps, such that for all* $g \in GL(\mathcal{V})$

$$g = u_g \, s_g.$$

Moreover, u *and* s *are uniquely determined by these properties.*

PROOF: Assign to every $N > 0$ the open set $U_N = \{g \in GL(\mathcal{V}) | \; \|g^* g - N \cdot \mathrm{id}_{\mathcal{V}}\| < N\}$, where $\| \cdot \|$ denotes the operator norm of $\langle \cdot, \cdot \rangle$ and g^* the adjoint operator of g. The U_N then provide an open covering of $GL(\mathcal{V})$. Note that $g^* g$ is a selfadjoint, positive definite operator on $\mathcal{V}$. We now define smooth functions $\mathfrak{u}_N : \mathfrak{U}_N \to \mathrm{GL}(\mathcal{V})$ and $\mathfrak{s}_N : \mathfrak{U}_N \to \mathrm{GL}(\mathcal{V})$ with the desired properties. To this end let us first determine the Taylor-coefficients of the analytic function $\mathfrak{h}_N : B_N(0) \to \mathbb{R}^{>0}$ $z \mapsto \sqrt{z+N}$ around the origin; they are given by $\mathfrak{h}_N^k = \left(\prod_{i=1}^k \left(\frac{3}{2} - i\right)\right) N^{1/2-k}$, $k \in \mathbb{N}$. By definition of $\mathfrak{U}_N$

$$\mathfrak{s}_{N,\mathfrak{g}} = \mathfrak{h}_N(\mathfrak{g}^* \mathfrak{g} - N \,\mathrm{id}_{\mathcal{V}}) := \sum_{k \in \mathbb{N}} \mathfrak{h}_N^k (\mathfrak{g}^* \mathfrak{g} - N \,\mathrm{id}_{\mathcal{V}})^k$$

is well-defined for all $\mathfrak{g} \in \mathfrak{U}_N$ and depends analytically, in particular smoothly on $\mathfrak{g}$. As $\mathfrak{h}_N$ assumes only values in the set of positive real numbers, $\mathfrak{s}_{N,\mathfrak{g}}$ must be positive definite. By definition of $\mathfrak{h}_N$ the relation $(\mathfrak{s}_{N,\mathfrak{g}})^2 = \mathfrak{g}^* \mathfrak{g}$ holds as well. Moreover, the operator $\mathfrak{u}_{N,\mathfrak{g}} := \mathfrak{g}\, \mathfrak{s}_{N,\mathfrak{g}}^{-1}$ depends analytically on $\mathfrak{g}$ and satisfies

$$\mathfrak{u}_{N,\mathfrak{g}}^* \mathfrak{u}_{N,\mathfrak{g}} = (\mathfrak{s}_{N,\mathfrak{g}}^*)^{-1} \mathfrak{g}^* \mathfrak{g}\, \mathfrak{s}_{N,\mathfrak{g}}^{-1} = \mathfrak{s}_{N,\mathfrak{g}}^{-1} (\mathfrak{s}_{N,\mathfrak{g}})^2 \mathfrak{s}_{N,\mathfrak{g}}^{-1} = \mathrm{id}_{\mathcal{V}},$$

that means $\mathfrak{u}_{N,\mathfrak{g}}$ is unitary. Thus we obtain the polar decomposition $\mathfrak{g} = \mathfrak{u}_{N,\mathfrak{g}}\, \mathfrak{s}_{N,\mathfrak{g}}$.

If we can yet show the uniqueness of the polar decomposition for every $\mathfrak{g} \in \mathrm{GL}(\mathcal{V})$, then the functions $\mathfrak{u}_N$ (resp. $\mathfrak{s}_N$) have to coincide on the intersections of their domains, thus define the desired functions $\mathfrak{u}$ and $\mathfrak{s}$. So let $\mathfrak{u}, \mathfrak{u}'$ be unitary and $\mathfrak{s}, \mathfrak{s}'$ be positive definite such that $\mathfrak{g} = \mathfrak{u}\,\mathfrak{s} = \mathfrak{u}'\,\mathfrak{s}'$. Then $\mathfrak{g}^* = (\mathfrak{u}\,\mathfrak{s})^* = \mathfrak{s}\,\mathfrak{u}^{-1}$, hence $\mathfrak{s}^2 = \mathfrak{s}\,\mathfrak{u}^{-1}\,\mathfrak{u}\,\mathfrak{s} = \mathfrak{g}^*\,\mathfrak{g} = \mathfrak{s}'^2$. As a positive definite operator has only one positive definite square root, $\mathfrak{s} = \mathfrak{s}'$ follows, hence $\mathfrak{u} = \mathfrak{u}'$. This proves the claim. □

A.3 Topological tensor products

One can consider many different locally convex topologies on the tensor product $\mathcal{V} \otimes \mathcal{W}$ of two locally convex $\mathbb{k}$-vector spaces $\mathcal{V}$ and $\mathcal{W}$ (with $\mathbb{k} = \mathbb{R}$ or $\mathbb{k} = \mathbb{C}$) such that these topologies are induced by the ones of $\mathcal{V}$ and $\mathcal{W}$. The most natural one is the π-topology that means the finest locally convex topology on $\mathcal{V} \otimes \mathcal{W}$ such that the natural mapping $\otimes : \mathcal{V} \times \mathcal{W} \to \mathcal{V} \otimes \mathcal{W}$ is continuous. $\mathcal{V} \otimes \mathcal{W}$ together with this topology will be denoted by $\mathcal{V} \otimes_\pi \mathcal{W}$, its completion by $\mathcal{V} \hat{\otimes}_\pi \mathcal{W}$. The π-topology has the following compatibility properties:

(TP1) $\otimes : \mathcal{V} \times \mathcal{W} \to \mathcal{V} \otimes \mathcal{W}$ is continuous.

(TP2) For every pair $(e, f) \in \mathcal{V}' \times \mathcal{W}'$ of continuous linear forms the mapping

$$e \otimes f : \mathcal{V} \otimes \mathcal{W} \to \mathbb{k}, \quad v \otimes w \mapsto e(v)\, f(w)$$

is continuous.

The π-topology then is the strongest and the ε-topology, which will not be explained here, is the weakest among the topologies on $\mathcal{V} \otimes \mathcal{W}$ compatible with $\otimes$ in this sense (see Grothendieck [70] or Trèves [171] for details).

At the end of this section let us note that the category of all complete locally convex topological vector spaces with $\hat{\otimes}_\pi$ as tensor product functor comprises a tensor category in the sense of Deligne [50].

Appendix B

Kähler differentials

B.1 The space of Kähler differentials

B.1.1 Let $\mathcal{R}$ be a commutative ring, $\mathcal{A}$ an $\mathcal{R}$-Algebra (with unit), and $\mathcal{M}$ an $\mathcal{A}$-module. A *derivation* on $\mathcal{A}$ over $\mathcal{R}$ with values in $\mathcal{M}$ then is an $\mathcal{R}$-linear mapping $\delta : \mathcal{A} \to \mathcal{M}$, such that $\delta(ab) = a\delta(b) + \delta(a)b$ for all $a, b \in \mathcal{A}$. The space of all such derivations will be denoted by $\mathrm{Der}_{\mathcal{R}}(\mathcal{A}, \mathcal{M})$.

By the space of *Kähler differentials* of $\mathcal{A}$ over $\mathcal{R}$ one understands an $\mathcal{A}$-module $\Omega_{\mathcal{A}/\mathcal{R}}$ together with a derivation $d : A \to \Omega_{\mathcal{A}/\mathcal{R}}$ called *Kähler derivative* such that the following universal property is satisfied:

(KÄ) For every $\mathcal{A}$-module $\mathcal{M}$ and every derivation $\delta : \mathcal{A} \to \mathcal{M}$ there exists a unique $\mathcal{R}$-linear mapping $i_\delta : \Omega_{\mathcal{A}/\mathcal{R}} \to M$ such that the diagram

$$\begin{array}{ccc} \mathcal{A} & \xrightarrow{\ \delta\ } & \mathcal{M} \\ {\scriptstyle d}\downarrow & \nearrow{\scriptstyle i_\delta} & \\ \Omega_{\mathcal{A}/\mathcal{R}} & & \end{array}$$

commutes.

The pair $(\Omega_{\mathcal{A}/\mathcal{R}}, d)$ is determined uniquely by this universal property; its existence is given by the following proposition. Thus

$$\mathrm{Der}_{\mathcal{R}}(\mathcal{A}, \mathcal{M}) = \mathrm{Hom}_{\mathcal{A}}(\Omega_{\mathcal{A}/\mathcal{R}}, \mathcal{M}). \tag{B.1.2}$$

B.1.2 Proposition *Let $\mathcal{A}$ be a commutative $\mathcal{R}$-algebra. Then the space $\Omega_{\mathcal{A}/\mathcal{R}}$ of Kähler differentials of $\mathcal{A}$ exists. It can be represented by either of the following spaces:*

(1) *Let Ω be the free $\mathcal{A}$-module over the symbols da with $a \in \mathcal{A}$, and $\mathfrak{J}$ the $\mathcal{A}$-submodule generated by the relations*

$$d(\lambda a + \mu b) - \lambda da - \mu db = 0, \quad \lambda, \mu \in \mathcal{R},\ a, b \in \mathcal{A},$$
$$d(ab) - adb - bda = 0, \quad a, b \in \mathcal{A}.$$

Then $\Omega^1_{\mathcal{A}/\mathcal{R}} = \Omega/\mathfrak{J}$ and $d : \mathcal{A} \to \Omega^1_{\mathcal{A}/\mathcal{R}}$, $a \mapsto da + \mathfrak{J}$.

M.J. Pflaum: LNM 1768, pp. 205 - 208, 2001

(2) *Let $\mathcal{B}$ be the ring $\mathcal{A}\otimes_{\mathcal{R}}\mathcal{A}$. Give $\mathcal{B}$ the structure of an $\mathcal{A}$-algebra by $a \to a\otimes 1$ and let $\varepsilon : \mathcal{B} \to \mathcal{A}$ be the homomorphism $(a, b) \to ab$. If $\mathcal{J}$ denotes the ideal $\mathcal{J} = \ker \varepsilon$, then the $\mathcal{A}$-module $\Omega^1_{\mathcal{A}/\mathcal{R}} = \mathcal{J}/\mathcal{J}^2$ forms a space of Kähler differentials, where the universal derivation is given by $d : \mathcal{A} \to \Omega^1_{\mathcal{A}/\mathcal{R}}$. The morphism $i_\delta : \Omega^1_{\mathcal{A}/\mathcal{R}} \to \mathcal{M}$ associated to a derivation $\delta : \mathcal{A} \to \mathcal{M}$ in the universal property has the form $\sum a_j \otimes b_j \mapsto \sum a_j\, \delta(b_j)$.*

(3) *Let $\mathcal{R} = \Bbbk$ be a field, $\mathcal{A}$ a local $\Bbbk$-algebra with maximal ideal $\mathfrak{m}$, and $j : \mathcal{A} \to \Bbbk$ a morphism such that the sequence*

$$0 \to \mathfrak{m} \to \mathcal{A} \overset{j}{\rightleftarrows} \Bbbk \to 0$$

is exact and splits. Then $\Omega^1_{\mathcal{A}/\Bbbk} = \mathfrak{m}/\mathfrak{m}^2$ is a module of Kähler differentials for $\mathcal{A}$ and $\mathcal{A} \to \mathfrak{m}/\mathfrak{m}^2$, $a \mapsto a - j(a) + \mathfrak{m}^2$ its universal derivation.

PROOF: MATSUMURA [125] □

B.1.3 Remark It follows by Proposition B.1.2 that every element of $\Omega_{\mathcal{A}/\mathcal{R}}$ can be written in the form $\sum_j a_j\, db_j$ with finitely many $a_j, b_j \in \mathcal{A}$.

B.1.4 First fundamental exact sequence *Let $\mathcal{B} \to \mathcal{A}$ be a homomorphism of commutative $\mathcal{R}$-algebras. Then there exists an exact sequence of $\mathcal{A}$-modules of the form*

$$\Omega_{\mathcal{B}/\mathcal{R}} \otimes_{\mathcal{B}} \mathcal{A} \xrightarrow{\alpha} \Omega_{\mathcal{A}/\mathcal{R}} \xrightarrow{\beta} \Omega_{\mathcal{A}/\mathcal{B}} \longrightarrow 0,$$

where $\alpha(db \otimes a) = a\,db$ and $\beta(da) = da$.

PROOF: WEIBEL [183, 9.2.6] □

B.1.5 Second fundamental exact sequence *Let $\mathcal{J} \subset \mathcal{A}$ be an ideal of the commutative $\mathcal{R}$-algebra $\mathcal{A}$. Then the sequence*

$$\mathcal{J}/\mathcal{J}^2 \xrightarrow{\delta} \Omega_{\mathcal{A}/\mathcal{R}} \otimes_{\mathcal{A}} \mathcal{A}/\mathcal{J} \xrightarrow{\alpha} \Omega_{(\mathcal{A}/\mathcal{J})/\mathcal{R}} \longrightarrow 0$$

is exact, where $\delta : \mathcal{J}/\mathcal{J}^2 \to \Omega_{\mathcal{A}/\mathcal{R}} \otimes_{\mathcal{A}} \mathcal{A}/\mathcal{J}$ is the $\mathcal{A}$-module map $a \mapsto da \otimes 1$.

PROOF: WEIBEL [183, 9.2.7] □

B.1.6 Starting from $\Omega_{\mathcal{A}/\mathcal{R}}$ one can build the k-th exterior product $\Omega^k_{\mathcal{A}/\mathcal{R}} = \Lambda^k \Omega_{\mathcal{A}/\mathcal{R}}$, where obviously $\Omega^0_{\mathcal{A}/\mathcal{R}} = \mathcal{A}$ and $\Omega^1_{\mathcal{A}/\mathcal{R}} = \Omega_{\mathcal{A}/\mathcal{R}}$. The direct sum $\Omega^\bullet_{\mathcal{A}/\mathcal{R}} = \bigoplus_{k\in\mathbb{N}} \Omega^k_{\mathcal{A}/\mathcal{R}}$ then becomes an $\mathcal{A}$-algebra which we call as usual the *exterior algebra* of $\mathcal{A}$. Moreover, the Kähler derivative $d : \mathcal{A} \to \Omega_{\mathcal{A}/\mathcal{R}}$ has a unique extension to an $\mathcal{R}$-linear mapping $d : \Omega^\bullet_{\mathcal{A}/\mathcal{R}} \to \Omega^\bullet_{\mathcal{A}/\mathcal{R}}$ such that

$$d(\alpha \wedge \beta) = d\alpha \wedge \beta + (-1)^k\, \alpha \wedge d\beta \qquad \alpha \in \Omega^k_{\mathcal{A}/\mathcal{R}},\ \beta \in \Omega^l_{\mathcal{A}/\mathcal{R}}$$

and $d \circ d = 0$. Thus we obtain the so-called *deRham complex* of $\mathcal{A}$:

$$(\Omega^\bullet_{\mathcal{A}/\mathcal{R}}, d) : \mathcal{A} \xrightarrow{d^0} \Omega^1_{\mathcal{A}/\mathcal{R}} \xrightarrow{d^1} \Omega^2_{\mathcal{A}/\mathcal{R}} \xrightarrow{d^2} \cdots \xrightarrow{d^{k-1}} \Omega^k_{\mathcal{A}/\mathcal{R}} \xrightarrow{d^k} \cdots .$$

The reader will find proofs and details about the exterior algebra in the above mentioned literature.

B.2 Topological version

We now consider the case that the $\Bbbk$-algebra $\mathcal{A}$ is defined over $\Bbbk = \mathbb{R}$ or $\mathbb{C}$, and that it carries a complete locally convex topology in the sense of Section 6.1. Then there exists for every complete locally convex $\mathcal{A}$-module $\mathcal{M}$ the space $\mathrm{Der_c}(\mathcal{A}, \mathcal{M})$ of continuous derivations from $\mathcal{A}$ to $\mathcal{M}$. The functor $\mathcal{M} \mapsto \mathrm{Der_c}(\mathcal{A}, \mathcal{M})$ is representable as well that means there exists a uniquely determined complete locally convex $\mathcal{A}$-module $\overline{\Omega}_{\mathcal{A}/\Bbbk}$ together with a continuous mapping $\mathrm{d} : \mathcal{A} \to \overline{\Omega}_{\mathcal{A}/\Bbbk}$ such that

$$\mathrm{Hom}_{\mathcal{A}}(\overline{\Omega}_{\mathcal{A}/\Bbbk}, \mathcal{M}) \to \mathrm{Der_c}(\mathcal{A}, \mathcal{M}), \quad \mathrm{i} \mapsto \mathrm{i} \circ \mathrm{d}$$

is an isomorphism. Note hereby that $\mathrm{Hom}_{\mathcal{A}}(-,-)$ means the space of continuous $\mathcal{A}$-linear morphisms. Applying Proposition B.1.2, point (2) we obtain a simple form, in which $\overline{\Omega}_{\mathcal{A}/\Bbbk}$ can be represented. First set $\overline{\Omega}_{\mathcal{A}/\Bbbk} = \overline{\mathcal{I}}/\overline{\mathcal{I}^2}$, where $\overline{\mathcal{I}}$ and $\overline{\mathcal{I}^2}$ are the topological closures of the ideals $\mathcal{I}$ and $\mathcal{I}^2$ in the completed tensor product $\mathcal{A}\hat{\otimes}_\pi\mathcal{A}$. Then the Kähler derivative $\mathrm{d} : \mathcal{A} \to \Omega_{\mathcal{A}/\Bbbk} \hookrightarrow \overline{\Omega}_{\mathcal{A}/\Bbbk}$ becomes a continuous mapping. Now assign to every continuous derivation $\delta : \mathcal{A} \to \mathcal{M}$ the continuous mapping

$$\mathrm{i}_\delta : \mathcal{A}\hat{\otimes}_\pi\mathcal{A} \to \mathcal{M}, \quad \sum_{j\in\mathbb{N}} a_j \otimes b_j \mapsto \sum a_j\, \delta(b_j),$$

which exists by the universal property of the π tensor product. Then $\delta = \mathrm{i}_\delta \circ \mathrm{d}$ holds and i_δ is uniquely determined by this property. In other words the thus defined $\overline{\Omega}_{\mathcal{A}/\Bbbk}$ represents the functor $\mathrm{Der_c}(\mathcal{A}, -)$ indeed. We call $\overline{\Omega}_{\mathcal{A}/\Bbbk}$ the space of *topological Kähler differentials.* The *topological Kähler differentials* of *order k* are given by the elements of the space $\overline{\Omega}^k_{\mathcal{A}/\Bbbk} = \overline{\Lambda}^k\overline{\Omega}_{\mathcal{A}/\Bbbk}$, where for any complete locally convex topological vector space $\mathcal{V}$ the symbol $\overline{\Lambda}^k\mathcal{V} \subset \mathcal{V}\hat{\otimes}_\pi \cdots \hat{\otimes}_\pi\mathcal{V}$ means the topological closure of $\Lambda^k\mathcal{V}$ in the completed k-th π-tensor product.

B.3 Application to locally ringed spaces

B.3.1 Next we consider a commutative locally ringed space $(X, \mathcal{O})$ over $\Bbbk = \mathbb{R}, \mathbb{C}$. Then we build for every point $x \in X$ the module $\Omega_x := \Omega_{\mathcal{O}_x/\Bbbk} = \mathfrak{m}_x/\mathfrak{m}_x^2$ of Kähler differentials of the stalk $\mathcal{O}_x$, where $\mathfrak{m}$ means the maximal ideal of $\mathcal{O}_x$. Give

$$\acute{\mathrm{E}}(\Omega_{X,\Bbbk}) := \bigcup_{x\in X} \Omega_{\mathcal{O}_x/\Bbbk}$$

the finest topology such that all mappings of the form

$$\overline{f} : V \to \acute{\mathrm{E}}(\Omega_{X,\Bbbk}), \quad x \mapsto \mathrm{d}[f]_x$$

are continuous. Hereby $V \subset X$ runs through all open sets in X, and f through all elements of the algebra $\mathcal{O}(V)$. Then $\acute{\mathrm{E}}(\Omega_{X,\Bbbk})$ becomes the espace étalé of a sheaf $\Omega_{X/\Bbbk}$ on X which is called the *sheaf of Kähler differentials* on $(X, \mathcal{O})$. (concerning details about the espace étalé see GODEMENT [60, Sec. II.1.2]). As the construction of the exterior product and the exterior derivative are functorial in $\mathcal{A}$ resp. $\Omega_{\mathcal{A}/\mathcal{R}}$, we obtain in a natural way the sheaf $\Omega^k_{X/\Bbbk}$ of k-forms on X and the sheaf $\Omega^\bullet_{X/\Bbbk}$. More precisely one constructs first presheaves $\widetilde{\Omega}^k_{X/\Bbbk}$ and $\widetilde{\Omega}^\bullet_{X/\Bbbk}$ on X by defining for all $U \subset X$:

$$\Omega^k_{X/\Bbbk}(U) := \Lambda^k(\Omega_{X/\Bbbk}(U)) \text{ and } \Omega^\bullet_{X/\Bbbk}(U) := \bigoplus_{k\in\mathbb{N}} \Omega^k_{X/\Bbbk}(U).$$

The associated sheaves then give the desired sheaves $\Omega^k_{X/\Bbbk}$ and $\Omega^\bullet_{X/\Bbbk}$. The Kähler derivative on the sectional spaces $d : \widetilde{\Omega}^\bullet_{X/\Bbbk} \to \widetilde{\Omega}^\bullet_{X/\Bbbk}$ provide a sheaf morphism $d : \Omega^\bullet_{X/\Bbbk} \to \Omega^\bullet_{X/\Bbbk}$. Altogether we thus obtain for every locally ringed space a complex of sheaves:

$$(\Omega^\bullet_{X/\Bbbk} d) : \mathcal{O} \xrightarrow{d^0} \Omega^1_{X/\Bbbk} \xrightarrow{d^1} \Omega^2_{X/\Bbbk} \xrightarrow{d^2} \cdots \xrightarrow{d^{k-1}} \Omega^k_{X/\Bbbk} \xrightarrow{d^k} \cdots .$$

B.3.2 Let us suppose that all stalks $\mathcal{O}_x$ of the locally ringed space $(X, \mathcal{O})$ are Noetherian or carry a complete locally convex topology. Then we denote by $T^Z_x X$ the *Zariski tangent space* of X over x that means the set of all linear resp. continuously linear mappings $\lambda : \Omega_x \to \Bbbk$. Hence, by the universal property (KÄ) the Zariski tangent space represents the set of all (continuous) derivations from $\mathcal{O}_x$ to $\Bbbk$.

Let $F = (f, \Phi) : (X, \mathcal{O}_X) \to (Y, \mathcal{O}_Y)$ be a morphism of locally ringed spaces such that in the topological case every canonical homomorphism $\Phi_x : \mathcal{O}_{Y,f(x)} \to \mathcal{O}_{X,x}$ is continuous. Then F induces a so-called *tangent map* $T^Z F : T^Z X \to T^Z Y$ by

$$T^Z_x X \ni \lambda \mapsto \lambda \circ \Phi_x \in T^Z_{f(x)} Y.$$

If $G = (g, \Psi) : (Y, \mathcal{O}_Y) \to (Z, \mathcal{O}_Z)$ is a further such morphism of locally ringed spaces then obviously $T^Z G \circ T^Z F = T^Z(G \circ F)$ holds.

Appendix C

Jets, Whitney functions and a few $\mathcal{C}^\infty$-mappings

C.1 Fréchet topologies for $\mathcal{C}^\infty$-functions

The algebra $\mathcal{C}^\infty(\mathsf{M})$ of smooth functions on a manifold M of dimension n possesses in a canonical way the structure of a Fréchet algebra. Let us indicate a sequence of seminorms defining the topology on $\mathcal{C}^\infty(\mathsf{M})$. To this end let $(\mathsf{U}_j)_{j\in\mathbb{N}}$ be a covering of M by chart domains $\mathsf{U}_j \subset \mathsf{M}$ such that there exist compact subsets $\mathsf{K}_j \subset \mathsf{U}_j$ with the property that the family of interior sets $\mathsf{K}_j^\circ \subset \mathsf{U}_j$ is a covering of M as well. Let further $(\mathsf{x}_j)_{j\in\mathbb{N}}$ be a family of differentiable charts $\mathsf{x}_j : \mathsf{U}_j \to \mathbb{R}^n$. Then

$$| \mathsf{f} |_m := \sum_{j,|\alpha|\leq k} \|\partial^\alpha_{\mathsf{x}_j} \mathsf{f}\|_{\mathsf{K}_j}, \qquad \mathsf{f} \in \mathcal{C}^\infty(\mathsf{M})$$

defines for every natural m a seminorm on $\mathcal{C}^\infty(\mathsf{M})$, where $\|\cdot\|_{\mathsf{K}_j}$ denotes the seminorm over the compact set K_j, and $\partial^\alpha_{\mathsf{x}_j}$ the higher partial derivatives in the coordinates given by x_j. With some patience and a few computations one now checks that the seminorms $|\cdot|_m$ turn the algebra $\mathcal{C}^\infty(\mathsf{M})$ into a Fréchet algebra, and that the Fréchet topology is independent of the special choice of the initial data. It is a well-known result from functional analysis that for two manifolds M and N the completed π-tensor product $\mathcal{C}^\infty(\mathsf{M})\hat{\otimes}_\pi\mathcal{C}^\infty(\mathsf{N})$ is canonically isomorphic to the Fréchet algebra $\mathcal{C}^\infty(\mathsf{M}\times\mathsf{N})$. A proof of this fact can be found for instance in TRÈVES [171].

Besides spaces of smooth functions one can also supply the space $\Omega(\mathsf{M})$ of smooth 1-forms on M or in other words the space of Kähler differentials of $\mathcal{C}^\infty(\mathsf{M})$ with a canonical topology such that $\Omega(\mathsf{M})$ becomes a finitely generated projective Fréchet $\mathcal{C}^\infty(\mathsf{M})$-module, hence in particular is topologically projective (see Section 6.3). In detail: via the above covering by charts $(\mathsf{x}_j)_{j\in\mathbb{N}}$ fix seminorms $|\cdot|_{m,1}$ inducing the Fréchet topology on $\Omega(\mathsf{M})$:

$$| \omega |_{m,1} := \sum_{j,|\alpha|\leq m} \sum_{l=1}^{n} \|\partial^\alpha_{\mathsf{x}_j} \omega_{j,l}\|_{\mathsf{K}_j}, \qquad \omega \in \Omega(\mathsf{M}),\ \omega_{|\mathsf{U}_j} = \sum_{l=1}^{n} \omega_{j,l} d\mathsf{x}_j^l.$$

Similarly the higher alternating products $\Omega^k(\mathsf{M})$ become finitely generated projective Fréchet $\mathcal{C}^\infty(\mathsf{M})$-modules.

M.J. Pflaum: LNM 1768, pp. 209 - 214, 2001

The Kähler derivative $d : \mathcal{C}^\infty(M) \to \Omega(M)$ is a continuous map with respect to the topologies defined above, hence induces by Section B.2 a uniquely determined continuous $\mathcal{C}^\infty(M)$-linear mapping $\Upsilon : \overline{\Omega}_{\mathcal{C}^\infty(M)/\mathbb{R}} \to \Omega(M)$. Note hereby that $\overline{\Omega}_{\mathcal{C}^\infty(M)/\mathbb{R}}$ has to be a Fréchet space, as $\mathcal{C}^\infty(M)$ is already one. The mapping Υ is surjective, as its image contains the generating system consisting of the forms df with $f \in \mathcal{C}^\infty(M)$, and injective, because by the universal property (KÄ) $\Omega(M)$ lies in $\overline{\Omega}_{\mathcal{C}^\infty(M)/\mathbb{R}}$. By the open mapping theorem $\Omega(M)$ and $\overline{\Omega}_{\mathcal{C}^\infty(M)/\mathbb{R}}$ then have to be topologically isomorphic.

C.2 Jets

C.2.1 In this monograph we will occasionally use Landau's notation, in particular in connection with jets and Whitney functions. Therefore we will briefly explain this notation. Let f, g be two real or complex valued functions on the topological space A, and $z \in A$. Then one writes

$$f(x) = O\left(g(x)\right), \qquad x \to z,$$

if there exists a neighborhood U of z such that $\left|\frac{f(x)}{g(x)}\right|$ is defined for all $x \in U \setminus \{z\}$ and bounded by $C > 0$. If additionally

$$\lim_{\substack{x \to z \\ x \in A\setminus\{z\}}} \frac{f(x)}{g(x)} = 0,$$

holds, one writes

$$f(x) = o\left(g(x)\right), \qquad x \to z.$$

C.2.2 Let A be a locally closed subset of $\mathbb{R}^n$, $m \in \mathbb{N} \cup \{\infty\}$ and $O \subset \mathbb{R}^n$ open such that $A \subset O$ closed. A family $F = \left(F^{(\alpha)}\right)_{|\alpha| \le k}$, $\alpha \in \mathbb{N}^n$ of continuous functions $F^{(\alpha)}$ on A is called *jet of order* m on A or briefly only m*-jet* on A. The space $J^m(A)$ of m-jets on A possesses in a natural way the structure of a real vector space. If $K \subset A$ is compact, then one obtains by

$$|F|_{K,m} = \sum_{|\alpha| \le m} \sup_{x \in K} \left|F^{(\alpha)}(x)\right|$$

a seminorm on $J^m(A)$. As there exists a compact exhaustion of A, i.e. a sequence of compact sets $K_j \subset A$ with $\bigcup_j K_j = A$ and $K_j \subset K_j^\circ$, the space $J^m(A)$ together with the locally convex topology defined by all seminorms $|\cdot|_{K,m}$ becomes a Fréchet space.

For all $x \in A$ and $F \in J^m(A)$ the value of F at x is defined by $F(x) := F^{(0)}(x)$. Moreover, one denotes for every $\beta \in \mathbb{N}^n$ with $|\beta| \le m$ the linear mapping $J^m(A) \to J^{m-|\beta|}(A)$, $\left(F^{(\alpha)}\right)_{|\alpha| \le k} \mapsto \left(F^{(\alpha+\beta)}\right)_{|\alpha| \le m-|\beta|}$ by D^β, and defines the so-called *jet mapping* $J^m : \mathcal{C}^m(O) \to J^m(A)$ by $g \mapsto \left(\left(\frac{\partial^{|\alpha|} g}{\partial x^\alpha}\right)_{|A}\right)_{|\alpha| \le k}$. Together with the well-known differential operator D^α, $|\alpha| \le m$ on $\mathcal{C}^m(O)$ we then have

$$D^\alpha \circ J^m = J^{m-|\alpha|} \circ D^\alpha.$$

The kernel of J^m has its own name; it is called the space of on A *flat functions* of *class* $\mathcal{C}^m$ and is denoted by $\mathcal{J}^m(A;O) := \ker J^m$ resp. $\mathcal{J}(A;O) := \mathcal{J}^\infty(A;O) := \ker J^\infty$.

Given an m-jet $F \in J^m(A)$ with $m < \infty$ and a point $z \in A$ one assigns to F and z a polynomial $T_z^m F : \mathbb{R}^n \to \mathbb{R}$ of order m:

$$T_z^m F(x) = \sum_{|\alpha| \le k} \frac{(x-z)^{(\alpha)}}{\alpha!} F^\alpha(z).$$

Furthermore one sets

$$R_z^m F = F - J^m(T_z^m F),$$

and interprets $R_z^m F$ as the "rest term" of F in the "Taylor expansion" $T_z^m F$ up to order m.

C.3 Whitney functions

C.3.1 Definition By a *Whitney function* on A of *class* $\mathcal{C}^m$, $m \in \mathbb{N}$, one understands a jet $F \in J^m(A)$ of order m, such that for every compact $K \subset A$

$$(R_x^m F)^{(\alpha)}(y) = o(|x-y|^{m-|\alpha|}), \qquad |x-y| \to 0,\ x, y \in K.$$

A jet $F \in J^\infty(A)$ is called a *Whitney function* of *class* $\mathcal{C}^\infty$, if the projection of F to $J^m(A)$ is for every $m \in \mathbb{N}$ a Whitney function. The space of all Whitney functions on A of class $\mathcal{C}^m$, $m \in \mathbb{N} \cup \{\infty\}$, is linear and will be denoted by $\mathcal{E}^m(A)$.

Then we have the following famous result:

C.3.2 Extension theorem of Whitney *For every locally closed set $A \subset \mathbb{R}^n$ and every open set $O \subset \mathbb{R}^n$ with $A \subset O$ closed the following sequence is exact:*

$$0 \longrightarrow \mathcal{J}^m(A;O) \longrightarrow \mathcal{C}^m(O) \longrightarrow \mathcal{E}^m(A) \longrightarrow 0.$$

In case $m < \infty$ this sequence splits topologically that means there exists a continuous section $\mathcal{E}^m(A) \to \mathcal{C}^m(O)$.

PROOF: See for instance MALGRANGE [118, Thm. 4.1]. □

In general $\mathcal{E}^m(A)$ need not be a closed subspace of $J^m(A)$. In other words $\mathcal{E}^m(A)$ is not always complete with respect to the seminorms $|\cdot|_{K,m}$ (but see Corollary 1.6.13 for criteria which entail that $\mathcal{E}^m(A)$ is closed in $J^m(A)$). Therefore one defines on the space of Whitney functions the seminorms $\|\cdot\|_{K,m}$ by

$$\|F\|_{K,m} := |F|_{K,m} + \sup_{\substack{x,y \in K,\ x \ne y \\ |\alpha| \le m}} \frac{(R_x^m F)^{(\alpha)}(y)}{|x-y|^{m-|\alpha|}}, \qquad F \in \mathcal{E}^m(A).$$

Now, if K runs through a compact exhaustion of A and m through all natural numbers, then the seminorms $\|\cdot\|_{K,m}$ provide a locally convex topology which turns $\mathcal{E}^m(A)$ into a Fréchet space.

C.3.3 Lemma *Under the prerequisites of Theorem C.3.2 the equality $\mathcal{J}^\infty(A;O) = \mathcal{J}^\infty(A;O)^2$ holds.*

PROOF: See TOUGERON [170, Lem. 2.4]. □

C.4 Smoothing of the angle

The goal of this section is to construct a function which intuitively smoothes the angle.

C.4.1 Lemma *There exists a smooth function $\varphi : [0,1] \times [0,1] \to [0,1] \times [0,1]$ with the following properties:*

(1) *φ is a homeomorphism onto its image.*

(2) *The restriction of φ to $([0,1] \times [0,1]) \setminus \{(\frac{1}{2},0)\}$ is smooth.*

(3) $D\varphi(s,t)$ *is bijective for all* $s,t \in [0,1]$ *with* $(s,t) \neq (\frac{1}{2},0)$.

(4) $\varphi\big(\frac{1}{2},0\big) = (0,0)$, $\varphi\big([0,\frac{1}{2}] \times \{0\}\big) \subset \{0\} \times [0,1]$ *and* $\varphi\big([\frac{1}{2},1] \times \{0\}\big) \subset [0,1] \times \{0\}$.

(5) $\varphi(s,t) = \big(\frac{t}{3}, 1-2s\big)$ *for* $0 \leq s \leq \frac{1}{6}$ *and* $\varphi(s,t) = \big(s-\frac{1}{2}, \frac{t}{3}\big)$ *for* $\frac{5}{6} \leq s \leq 1$.

Such a function φ can chosen to be either smooth on its domain or such that φ is tempered relative $\{(\frac{1}{2},0)\}$.

PROOF: Let $\chi : \mathbb{R} \to [0,1]$ be a smooth function such that $\chi(s) = 0$ for $s \leq 0$, $\chi'(s) > 0$ for $0 < s < 1$ and $\chi(s) = 1$ for $s \geq 1$. Besides that let $\overline{\chi} = 1 - \chi$. Then we define a smooth function $\tilde{\varphi} : [0,1] \times]0,1] \to [0,1] \times [0,1]$ by $\tilde{\varphi} = (\tilde{\varphi}_1, \tilde{\varphi}_2)$ with

$$\tilde{\varphi}_1(s,t) = \begin{cases} \frac{t}{3} & \text{for } s \in I_t^1, \\ \frac{t}{3}\,\overline{\chi}\big(\frac{6s+2t-3}{t}\big) + \big(\frac{2t}{3} - \frac{t}{3}\cos\big(\frac{\pi(6s+2t-3)}{8t}\big)\big)\,\chi\big(\frac{6s+2t-3}{t}\big) & \text{for } s \in I_t^2, \\ \big(\frac{2t}{3} - \frac{t}{3}\cos\big(\frac{\pi(6s+2t-3)}{8t}\big)\big) & \text{for } s \in I_t^3, \\ \big(s-\frac{1}{2}\big)\,\chi\big(\frac{6s-t-3}{t}\big) + \big(\frac{2t}{3} - \frac{t}{3}\cos\big(\frac{\pi(6s+2t-3)}{8t}\big)\big)\,\overline{\chi}\big(\frac{6s-t-3}{t}\big) & \text{for } s \in I_t^4, \\ s - \frac{1}{2} & \text{for } s \in I_t^5, \end{cases}$$

$$\tilde{\varphi}_2(s,t) = \begin{cases} (1-2s) & \text{for } s \in I_t^1, \\ (1-2s)\,\overline{\chi}\big(\frac{6s+2t-3}{t}\big) + \big(\frac{2t}{3} - \frac{t}{3}\sin\big(\frac{\pi(6s+2t-3)}{8t}\big)\big)\,\chi\big(\frac{6s+2t-3}{t}\big) & \text{for } s \in I_t^2, \\ \big(\frac{2t}{3} - \frac{t}{3}, \sin\big(\frac{\pi(6s+2t-3)}{8t}\big)\big) & \text{for } s \in I_t^3 \\ \frac{t}{3}\,\chi\big(\frac{6s-t-3}{t}\big) + \big(\frac{2t}{3} - \frac{t}{3}\sin\big(\frac{\pi(6s+2t-3)}{8t}\big)\big)\,\overline{\chi}\big(\frac{6s-t-3}{t}\big) & \text{for } s \in I_t^4, \\ \frac{t}{3} & \text{for } s \in I_t^5, \end{cases}$$

where $I_t^1 = [0, \frac{1}{2} - \frac{t}{3}]$, $I_t^2 = [\frac{1}{2} - \frac{t}{3}, \frac{1}{2} - \frac{t}{6}]$, $I_t^3 = [\frac{1}{2} - \frac{t}{6}, \frac{1}{2} + \frac{t}{6}]$, $I_t^4 = [\frac{1}{2} + \frac{t}{6}, \frac{1}{2} + \frac{t}{3}]$ and $I_t^5 := [\frac{1}{2} + \frac{t}{3}, 1]$. First check that $\tilde{\varphi}$ satisfies (5) by definition. By a lengthy, but easy computation one proves that $\tilde{\varphi}_1$ (resp. $\tilde{\varphi}_2$) is monotone increasing in t for fixed s and increasing (resp. decreasing) in s for fixed t. We show this for $\tilde{\varphi}_1$ and $s \in I_t^2$ in detail by differentiation; the other proofs of monotony are similar. Together with $\eta(t,s) = \frac{6s+2t-3}{t}$ we have

$$\begin{aligned}
\frac{\partial}{\partial t}\tilde{\varphi}_1(t,s) &= \\
&= \frac{1}{3}\overline{\chi}(\eta(t,s)) + \frac{1}{3}\left(2 - \cos\left(\frac{\pi}{8}\eta(t,s)\right) + \sin\left(\frac{\pi}{8}\eta(t,s)\right)\frac{\pi}{8}\frac{3-6s}{t}\right)\chi(\eta(t,s)) \\
&\quad + \frac{t}{3}\left(1 - \cos\left(\frac{\pi}{8}\eta(t,s)\right)\right)\frac{\partial}{\partial t}(\chi(\eta(t,s))) \\
&\geq \frac{1}{3} + \frac{\pi}{24}\sin\left(\frac{\pi}{8}\eta(t,s)\right)\chi(\eta(t,s)) > 0,
\end{aligned}$$

which implies $\tilde{\varphi}_1$ to be increasing with respect to t as long as $s \in I_t^2$. By the monotony of the component functions $\tilde{\varphi}_1$ and $\tilde{\varphi}_2$ one concludes that $\tilde{\varphi}$ has to be bijective. If one finally extends $\tilde{\varphi}$ by

$$\tilde{\varphi}(s,0) = \begin{cases} (0, 1-2s) & \text{for } s \leq \frac{1}{2}, \\ (s-\frac{1}{2}, 0) & \text{for } s \leq \frac{1}{2}, \end{cases}$$

to a continuous function on $[0,1] \times [0,1]$, then the thus extended $\tilde{\varphi}$ satisfies conditions (1) to (5) and is tempered relative $\{(\frac{1}{2}, 0)\}$.

Furthermore it is easy to see that the function

$$\varphi : [0,1] \times]0,1] \to [0,1] \times [0,1], \quad (s,t) \mapsto \chi\left(\left((s-\tfrac{1}{2})^2 + t^2\right) \cdot 100\right)\tilde{\varphi}(s,t)$$

can be extended to a smooth function φ on $[0,1] \times [0,1]$ and that φ is a homeomorphism onto its image. Another lengthy but obvious computation shows that the derivative $D\varphi(s,t)$ is bijective for all $(s,t) \neq (\frac{1}{2}, 0)$ and that $D\varphi(s,t)$ vanishes for $(s,t) = (\frac{1}{2}, 0)$. Altogether we thus obtain the claim. □

C.4.2 Proposition *Let* M *and* N *be two manifolds with nonempty boundaries* ∂M *and* ∂N. *Then there exists a (not unique) manifold structure on the topological product* $M \times N$ *such that* $M \times N$ *becomes a manifold-with-boundary, such that* $\partial(M \times N) = \partial M \times N \cup M \times \partial N$ *holds, and finally such that the differentiable structure on* $M^\circ \times N^\circ$ *coincides with the canonical product of the manifolds* M° *and* N°.

PROOF: Before we will come to our matter of concern let us first get a function

$$\varphi : [0,1] \times [0,1] \to [0,1] \times [0,1].$$

which smoothes the angle according to the preceding lemma. Then we supply the interior $M^\circ \times N^\circ$ of $M \times N$ with the natural differentiable structure of the product of the two manifolds M° and N°. Hence it only remains to provide a manifold structure for the boundary $\partial(M \times N) = \partial M \times N \cup M \times \partial N$, so we can assume loss of generality that $M = \partial M \times [0,1[$ and $N = \partial N \times [0,1[$. As both of the open sets $\partial M \times]0,1[\times \partial N \times [0,1[$ and $\partial M \times [0,1[\times \partial N \times]0,1[$ carry differentiable structures in a canonical way, one only has to find in the neighborhood of each of the points $(x, 0, y, 0)$ with $x \in \partial M$, $y \in \partial N$ a differentiable chart compatible with all other charts. More precisely we look for a homeomorphism $y : U \to \mathbb{R}^n \times \mathbb{R}^{\geq 0}$ of a neighborhood U of $(x, 0, y, 0)$ into Euclidean (half-) space such that y is a diffeomorphism around each of the points (x, s, y, t) with $(s, t) \neq (0, 0)$ and such that $y(\partial(M \times N)) \subset \mathbb{R}^{n-1} \times \{0\}$. Now the homeomorphism φ comes into the game; we denote the inverse function of φ by $\overline{\varphi} = (\overline{\varphi}_1, \overline{\varphi}_2)$. Then one defines a chart y on a sufficiently small U by

$$y(x', s, y', t) = \big(y_M(x'), y_N(y'), \overline{\varphi}_1(s,t), \overline{\varphi}_2(s,t)\big), \quad (x', s, y', t) \in U,$$

where y_M and y_N are differentiable charts of ∂M resp. ∂N around x resp. y. By the characteristic properties of φ the map y is a chart of the desired kind. But note that the thus defined differentiable structure on the product $M \times N$ depends near the critical points $(x, 0, y, 0)$ on the choice of the function φ. □

Bibliography

[1] R. Abraham and J.E. Marsden, *Foundations of Mechanics*, 2 ed., Addison-Wesley Publishing Company, 1978.

[2] H. Amann, *Gewöhnliche Differentialgleichungen*, Walter de Gruyter, 1983.

[3] N. Aronszajn, *Subcartesian and subriemannian spaces*, Notices Amer. Math. Soc. **14** (1967), 111.

[4] N. Aronszajn and P. Szeptycki, *Subcartesian spaces*, J. Differential Geom. **15** (1980), 393–416.

[5] M.F. Atiyah and R. Bott, *The Yang–Mills equations over Riemann surfaces*, Philos. Trans. Roy. Soc. London Ser. A **308** (1983), 524–615.

[6] L. Bates and E. Lerman, *Proper group actions and symplectic stratified spaces*, Pacific J. Math. **181** (1997), no. 2, 201–229.

[7] F. Bayen, M. Flato, C. Fronsdal, A. Lichnerowicz, and D. Sternheimer, *Deformation theory and quantization*, Ann. Physics **111** (1978), 61–151.

[8] K. Bekka, *Propriétés métriques et topologiques des espaces stratifiés*, Ph.D. thesis, Université de Paris-Sud, Centre d'Orsay, 1988.

[9] ———, *(c)-régularité et trivialité topologique*, Singularity Theory and its Applications (Warwick), LNM, vol. 1462, Springer-Verlag, 1989, pp. 42–62.

[10] ———, *Regular quasi-homogeneous stratifications*, Stratifications, Singularities and Differential Equations II, Stratifications and Topology of Singular Spaces (Honolulu) (D. Trotman & L.C. Wilson, ed.), Travaux en Cours, vol. 55, Hermann, 1997, pp. 1–14.

[11] K. Bekka and D. Trotman, *Propriétés métriques de familles ϕ-radiales des sous-variétés différentiables*, C. R. Acad. Sci. Paris Sér. I Math. **305** (1987), 389–392.

[12] K. Bekka and D. Trotman, *Weakly Whitney stratified sets*, Real and complex singularities. Proceedings of the 5th workshop, Sao Carlos, Brazil, July 27-31, 1998 (J. W. et al. Bruce, ed.), Res. Notes Math., vol. 412, Chapman Hall/CRC, 2000, pp. 1–15.

[13] E. Bierstone, *Lifting isotopies from orbit spaces*, Topology **14** (1975), 245–252.

[14] ———, *The Structure of Orbit Spaces and the Singularities of Equivariant Mappings*, Monografias de Matemática, vol. 35, Instituto de Matemática Pura e Aplicada, Rio de Janeiro, 1980.

[15] E. Bierstone and P.D. Milman, *Semianalytic and subanalytic sets*, Publ. Math. IHES **67** (1988), 5–42.

[16] R.L. Bishop and R.J. Crittenden, *Geometry of Manifolds*, Pure and Applied Mathematics, vol. XV, Academic Press, 1964.

[17] J. Block and E. Getzler, *Equivariant cyclic homology and equivariant differential forms*, Ann. Sci. École Norm. Sup. (4) **27** (1994), 493–527.

[18] T. Bloom and M. Herrera, *De Rham cohomology of an analytic space*, Invent. Math. **7** (1969), 275–296.

[19] A. Borel, *Fixed point theorems for elementary commutative groups I,II*, Seminar on Transformation Groups, Annals of Math. Studies, vol. 46, 1960.

[20] J.P. Brasselet, *De Rham theorems for singular varieties*, Differential Topology, Foliations, and Group Actions, Contemporary Mathematics, vol. 161, AMS, 1994, pp. 95–112.

[21] J.P. Brasselet, M. Goresky, and R. MacPherson, *Simplicial differential forms with poles*, Amer. J. Math. **113** (1991), 1019–1052.

[22] J.P. Brasselet, G. Hector, and M. Saralegi, *Théorème de De Rham pour les variétés stratifiées*, Ann. Global Anal. Geom. **9** (1991), no. 3, 211–243.

[23] J.P. Brasselet and A. Legrand, *Differential forms on singular varieties and cyclic homology*, Proceedings of the European singularities conference, Liverpool, UK (Bill Bruce et al., ed.), Lond. Math. Soc. Lect. Note Ser., vol. 263, Cambridge University Press, 1999, pp. 175–187.

[24] J.P. Brasselet and M.H. Schwartz, *Sur les classes de Chern d'un ensemble analytique complexe*, Astérisque **82-83** (1981), 93–147.

[25] G.E. Bredon, *Introduction to Compact Transformation Groups*, Academic Press, New York, 1972.

[26] J. Brüning and E. Heintze, *Representations of compact Lie groups and elliptic operators*, Invent. Math. **50** (1979), 169–203.

[27] ———, *The asymptotic expansion of Minakshisundaram-Pleijel in the equivariant case*, Duke Math. J. **51** (1984), 959–980.

[28] J. Brüning and M. Lesch, *Hilbert complexes*, J. Funct. Anal. **108** (1992), no. 1, 88–132.

[29] ———, *Kähler–Hodge theory for conformal complex cones*, Geom. Funct. Anal. **3** (1993), 439–473.

[30] ———, *On the spectral geometry of algebraic curves*, J. Reine Angew. Math. **474** (1996), 25–66.

[31] J. Brüning and R. Seeley, *Regular singular asymptotics*, Adv. in Math. **58** (1985), 133–148.

[32] ———, *The resolvent expansion for second order regular singular operators*, J. Funct. Anal. **73** (1987), 369–429.

[33] ———, *An index theorem for first order regular singular operators*, Amer. J. Math. **110** (1988), 659–714.

[34] ———, *The expansion of the resolvent near a singular stratum of conical type*, J. Funct. Anal. **95** (1991), 255–290.

[35] J.L. Brylinski, *Cyclic homolgy and equivariant theories*, Ann. Inst. Fourier (Grenoble) **37** (1987), no. 4, 15–28.

[36] H. Busemann, *The Geometry of Geodesics*, Pure and Applied Mathematics, vol. VI, Academic Press, 1955.

[37] ———, *Recent Synthetic Differential Geometry*, Ergebnisse der Mathematik und ihrer Grenzgebiete, vol. 54, Springer Verlag, 1970.

[38] J. Cheeger, *On the spectral geometry of spaces cone-like singularities*, Proc. Nat. Acad. Sci. U.S.A. **76** (1979), 2103–2106.

[39] ———, *On the Hodge theory of Riemannian pseudomanifolds*, Proc. Sympos. Pure Math. **36** (1980), 21–45.

[40] ———, *Hodge theory of complex cones*, Analyse et topologie sur les espaces singuliers, Astérisque, vol. 102, Soc. Math. France, 1983, pp. 118–134.

[41] ———, *Spectral geometry of singular Riemannian spaces*, J. Differential Geom. **18** (1983), 575–657.

[42] J. Cheeger, M. Goresky, and R. MacPherson, $L_{(2)}$-*Cohomology and intersection homology for singular algebraic varieties*, Semin. differential geometry, Ann. of Math. Studies, vol. 102, Princeton University Press, 1982, pp. 303–340.

[43] J. Cheeger, W. Müller, and R. Schrader, *On the curvature of piecewise flat spaces*, Comm. Math. Phys. **92** (1984), 405–454.

[44] S. Cohn-Vossen, *Existenz kürzester Wege*, Dokl. Akad. Nauk SSSR **3** (1935), 339–342.

[45] A. Connes, *Noncommutative differential geometry*, Inst. Hautes Études Sci. Publ. Math. **62** (1985), 257–360.

[46] S. C. Coutinho, *A Primer of Algebraic* D*-modules*, Students Text, vol. 33, London Mathematical Society, 1995.

[47] R.H. Cushman and L.M. Bates, *Global Aspects of Classical Integrable Systems*, Birkhäuser, 1997.

[48] M. De Wilde and P. Lecomte, *Existence of star-products and of formal deformations of the Poisson Lie algebra of arbitrary symplectic manifolds*, Lett. Math. Phys. **7** (1983), 487–496.

[49] ______, *Formal deformations of the Poisson Lie algebra of symplectic manifold and star-products. Existence, equivalence, derivations*, Deformation Theory of Algebras and Structures and Applications (Dordrecht) (M. Hazewinkel and M. Gerstenhaber, eds.), Kluwer Acad. Pub., 1988, pp. 897–960.

[50] P. Deligne, *Categories tannakiennes*, The Grothendieck Festschrift, Collect. Artic. in Honor of the 60th Birthday of A. Grothendieck. Vol. II, Progress in Mathematics, vol. 87, Birkhäuser, 1990, pp. 111–195.

[51] Z. Denkowska and K. Wachta, *Une construction de la stratification sous-analytique avec la condition (w)*, Bull. Polish Acad. Sci. Math. **35** (1987), 401–405.

[52] K. Dovermann and R. Schultz, *Equivariant surgery theories and their periodicity properties*, Lecture Notes in Mathematics, vol. 1443, Springer–Verlag, 1990.

[53] B. Fedosov, *A simple geometrical construction of deformation quantization*, J. Differential Geom. **40** (1994), 213–238.

[54] ______, *Deformation Quantization and Index Theory*, Akademie Verlag, 1996.

[55] M. Ferrarotti, *Volume on stratified sets*, Ann. Mat. Pura Appl. (4) **144** (1986), 183–201.

[56] ______, G*-manifolds and stratification*, Rend. Ist. Mat. Univ. Trieste **26** (1994), no. 1–2, 211–232.

[57] ______, *A complex of stratified forms satisfying De Rhams theorem*, Stratifications, Singularities and Differential Equations II, Stratifications and Topology of Singular Spaces (Honolulu) (D. Trotman & L.C. Wilson, ed.), Travaux en Cours, vol. 55, Hermann, 1997, pp. 25–38.

[58] ______, *Some results about volume of stratified sets*, Stratifications, Singularities and Differential Equations II, Stratifications and Topology of Singular Spaces (Honolulu) (D. Trotman & L.C. Wilson, ed.), Travaux en Cours, vol. 55, Hermann, 1997, pp. 39–43.

[59] M. Ferrarotti and L.C. Wilson, *Generalized Hesténès' Lemma and extension of functions*, Trans. Amer. Math. Soc. **350** (1998), no. 5, 1957–1975.

[60] R. Godement, *Topologie Algébrique et Theorie des Faisceaux*, Hermann, Paris, 1958.

[61] W.M. Goldman, *The symplectic nature of the fundamental groups of surfaces*, Adv. in Math. **54** (1984), 200–225.

[62] M. Goresky, *Triangulation of stratified objects*, Proc. Amer. Math. Soc. **72** (1978), 193–200.

[63] M. Goresky and R. MacPherson, *Intersection homology theory*, Topology **19** (1980), 135–162.

[64] ———, *Intersection homology II*, Invent. Math. **72** (1983), 77-129.

[65] ———, *Stratified Morse Theory*, Springer-Verlag, New York, 1988.

[66] A. Gray, *Tubes*, Addison-Wesley Publishing Company, 1990.

[67] D. Grieser, *Local geometry of singular real analytic surfaces*, preprint ESI 659 (1999), Erwin-Schrödinger-Institut, Wien.

[68] D. Grieser and M. Lesch, *On the L^2-Stokes theorem and Hodge theory for singular algebraic varieties*, preprint SFB 288 (1999).

[69] M. Gromov, *Metric Structures for Riemannian and Non-Riemannian Spaces*, Progress in Mathematics, vol. 152, Birkhäuser, 1998.

[70] A. Grothendieck, *Produits tensoriels topologiques et espace nucléaires*, Memoirs of the AMS, vol. 16, Amer. Math. Soc., 1955.

[71] ———, *On the De Rham cohomology of algebraic varieties*, Publ. Math. IHES **29** (1966), 351–359.

[72] ———, *Elements de géométrie algébrique, EGA IV, Etude locale des schémas et des morphismes de schémas*, Publ. Math. IHES **32** (1967).

[73] ———, *Esquisse d'un programme*, Geometric Galois Theory 1. Around Grothendieck's Esquisse d'un Programme (L. Schneps & P. Lochak, ed.), Lond. Math. Soc. Lect. Notes, vol. 242, Cambridge University Press, 1997, pp. 5–48.

[74] K. Guruprasad, J. Huebschmann, L. Jeffrey, and A. Weinstein, *Group systems, groupoids, and moduli spaces of parabolic bundles*, Duke Math. J. **89** (1997), no. 2, 377–412.

[75] M. Gutfleisch, *Stratifizierte Räume und Quotienten in der Analysis*, Diplomarbeit, Ludwig-Maximilians-Universität, München, November 1995.

[76] Z. Hajto, *On the equivalence of Whitney (b)-regularity and (b_s)-regularity*, Wyż. Szkoła Ped. Krakow. Rocznik Nauk.-Dydakt. Prace Mat. **10** (1982), 77–86.

[77] R. Hardt, *Stratifications of real analytic mappings and images*, Invent. Math. **28** (1975), 193–208.

[78] ———, *Topological properties of subanalytic sets*, Trans. Amer. Math. Soc. **211** (1975), 57–70.

[79] ———, *Triangulation of subanalytic sets and proper light subanalytic maps*, Invent. Math. **38** (1977), 207–217.

[80] R. Hartshorne, *On the De Rham cohomology of algebraic varieties*, Inst. Hautes Études Sci. Publ. Math. **45** (1975), 5–99.

[81] A. Hassell, R. Mazzeo, and R.B. Melrose, *A signature formula for manifolds with corners of codimension two*, Topology **36** (1997), no. 5, 1055–1075.

[82] M. Herrera, *De Rham theorems on semianalytic sets*, Bull. Amer. Math. Soc. **73** (1967), 414–418.

[83] M. Herrera and D. Lieberman, *Duality and the DeRham cohomology of infinitesimal neighborhoods*, Invent. Math. **13** (1971), 97–124.

[84] M. Hesténès, *Extension of the range of a differentiable function*, Duke Math. J. **8** (1941), 183–192.

[85] P.J. Hilton and U. Stammbach, *A Course in Homological Algebra*, Graduate Texts in Mathematics, Springer-Verlag, New York, 1971.

[86] H. Hironaka, *Subanalytic sets*, Number Theory, Algebraic Geometry and Commutative Algebra (Kinokuniya, Tokyo), 1973, volume in honor of A. Akizuki, pp. 453–493.

[87] ———, *Triangulation of algebraic sets*, Proc. Sympos. Pure Math. **29** (1975), 165–185.

[88] G. Hochschild, B. Kostant, and A. Rosenberg, *Differential forms on regular affine algebras*, Trans. Amer. Math. Soc. **102** (1962), 383–408.

[89] H. Holmann and H. Rummler, *Alternierende Differentialformen*, Graduate Texts in Mathematics, B.I.-Wissenschaftsverlag, Mannheim, Wien, Zürich, 1981.

[90] H. Hopf and W. Rinow, *Über den Begriff der vollständigen differentialgeometrischen Fläche*, Comment. Math. Helv. **3** (1932), 209–225.

[91] J. Huebschmann, *Symplectic and Poisson structures of certain moduli spaces*, Duke Math. J. **80** (1995), 737–756.

[92] ———, *Poisson geometry of flat connections for* SU(2)*-bundles on surfaces*, Math. Z. **221** (1996), 243–259.

[93] ———, *Smooth structures on certain moduli spaces for bundles on a surface*, J. Pure Appl. Algebra **126** (1998), 183–221.

[94] B. Hughes, L.R. Taylor, S. Weinberger, and B. Williams, *Neighborhoods in stratified spaces with two strata*, Topology **39** (2000), no. 5, 873–919.

[95] K. Jänich, *Differenzierbare G-Mannigfaltigkeiten*, Lecture Notes in Mathematics, vol. 59, Springer-Verlag, Berlin, Heidelberg, New York, 1968.

[96] F. Johnson, *On the triangulation of stratified sets and singular varieties*, Trans. Amer. Math. Soc. **275** (1983), 333–343.

[97] Y. Karshon, *An algebraic proof for the symplectic structure of moduli spaces*, Proc. AMS **116** (1992), 591–605.

[98] M. Kashiwara and P. Schapira, *Sheaves on Manifolds*, Grundlehren der mathematischen Wissenschaften, vol. 292, Springer, Berlin, Heidelberg, 1990.

[99] T. Kawasaki, *The signature theorem for V-manifolds*, Topology **17** (1978), 75–83.

[100] ———, *The index of elliptic operators over V-manifolds*, Nagoya Math. J. **84** (1981), 135–157.

[101] W. Klingenberg, *Riemannian Geometry*, Studies in Mathematics, vol. 1, de Gruyter, Berlin, New York, 1982.

[102] M. Kontsevich, *Deformation quantization of Poisson manifolds, I*, q-alg/9709040 (1997).

[103] ———, *Formality conjecture*, Deformation Theory and Symplectic Geometry (D. Sternheimer et al., ed.), Kluwer Academic Publishers, 1997, pp. 139–156.

[104] J.L. Koszul, *Sur certains groupes de transformation de Lie*, Colloque de Géométrie Différentielle, Collogues du CNRS, vol. 71, 1953, pp. 137–141.

[105] T.C. Kuo, *The ratio test for analytic Whitney stratifications*, Liverpool Singularity Symposium I, LNM, vol. 192, Springer-Verlag, 1971, pp. 141–149.

[106] M. Kuppe, *Integralgeometrie Whitney-stratifizierter Mengen*, Ph.D. thesis, Universität Münster, 1999.

[107] K. Kurdyka and P. Orro, *Distance géodésique sur un sous-analytique*, Revista Mat. Univ. Complutense de Madrid **10** (1997), no. Suplementario, 173–182.

[108] S. Lang, *Differential Manifolds*, Springer-Verlag, Berlin, Heidelberg, New York, 1972.

[109] E. Lerman, R. Montgomery, and R. Sjamaar, *Examples of singular reduction*, Symplectic Geometry (D. Salamon, ed.), Lond. Math. Soc. Lect. Note Ser., vol. 192, Cambridge University Press, 1993, pp. 127–155.

[110] M. Lesch, *Die Struktur von Orbiträumen*, Vortragsausarbeitungen eines Vortrags bei der Max-Planck-Arbeitsgruppe Potsdam, September 1992.

[111] ———, *Operators of Fuchs type, conical singularities, and asymptotic methods*, Teubner Texte zur Mathematik, vol. 136, B. G. Teubner, Leipzig, 1997.

[112] J.-L. Loday, *Cyclic Homology*, Grundlehren der mathematischen Wissenschaften, vol. 301, Springer-Verlag, Berlin, Heidelberg, 1992.

[113] S. Łojasiewicz, *Sur le problème de la division*, Studia Math. **18** (1959), 87–136.

[114] ———, *Triangulation of semi-analytic sets*, Ann. Scuola Norm. Sup. Pisa **18** (1964), 449–474.

[115] ———, *Ensemble semi-analytique*, Mimeographié, Institute des Hautes Études Scientifique, Bures-sur-Yvette, France, 1965.

[116] S. Łojasiewicz, J. Stasica, and K. Wachta, *Stratification sous-analytiques, Condition de Verdier*, Bull. Polish Acad. Sci. Math. **34** (1986), 531–539.

[117] J. Lott, *Signatures and higher signatures of S^1-quotients*, math. DG/9804105 (1998).

[118] B. Malgrange, *Ideals of Differentiable Functions*, Tata Institute of Fundamental Research Studies in Mathematics, Oxford University Press, 1966.

[119] J.E. Marsden and A. Weinstein, *Reduction of symplectic manifolds with symmetry*, Rep. Math. Phys. **5** (1974), 121–130.

[120] C. Marshall, *Calculus on subcartesian spaces*, J. Differential Geom. **10** (1975), 551–573.

[121] D. Massey, *Critical points of functions on singular spaces*, Topology **103** (2000), 55–93.

[122] J.N. Mather, *Notes on topological stability*, Mimeographed Lecture Notes, Harvard, 1970.

[123] ———, *Stratifications and mappings*, Dynamical Systems (M. M. Peixoto, ed.), Academic Press, 1973, pp. 195–232.

[124] ———, *Differentiable invariants*, Topology **16** (1977), 145–155.

[125] H. Matsumura, *Commutative Algebra*, Benjamin/Cummings, London, etc., 1980.

[126] R.B. Melrose, *The Atiyah-Patodi-Singer Index Theorem*, Research Notes in Mathematics, vol. 4, A K Peters, Wellesley, Massachusetts, 1993.

[127] ———, *Differential Analysis on Manifolds with Corners*, in preparation, 1999 (preliminary version).

[128] R.B. Melrose and V. Nistor, *Homology of pseudodifferential operators I. Manifolds with boundary*, funct-an/9606005 (1996).

[129] K. Meyer, *Symmetries and integrals in mathematics*, Dynamical Systems (M.M. Peixoto, ed.), Academic Press, New York, 1973, pp. 259–273.

[130] T. Mostowski, *Lipschitz Equisingularity*, Dissertationes Math. (Rozprawy Mat.) CCXLIII (1985).

[131] M. Nagase, *$\mathcal{L}^2$-cohomology and intersection cohomology of stratified spaces*, Duke Math. J. **50** (1983), 329–368.

[132] ———, *Sheaf theoretic $\mathcal{L}^2$-cohomology*, Adv. Stud. Pure Math. **8** (1987), 273–279.

[133] R. Nest and B. Tsygan, *The algebraic index theorem*, Comm. Math. Phys. **172** (1995), 223–262.

[134] L. Noirel and D. Trotman, *Subanalytic and semialgebraic realisations of abstract stratified sets*, Contemporary Mathematics **253** (2000), 203–207.

[135] F. Norguet, *Dérivées partielles et résidus de formes différentielles*, Séminaire P. Lelong 1958-1959, no. Exp.no.10, Secrétariat mathématique, Paris, 1959.

[136] T. Ohsawa, *Cheeger–Goresky–MacPherson conjecture for varieties with isolated singularities*, Math. Z. **206** (1991), 219–224.

[137] J.-P. Ortega and T.S. Ratiu, *Singular reduction of Poisson manifolds*, Lett. Math. Phys. **46** (1998), no. 4, 359–372.

[138] R.S. Palais, *On the existence of slices for actions of non-compact Lie groups*, Anal. Math. **73** (1961), 295–323.

[139] W. Pardon and M. Stern, *Pure Hodge structure on the L^2-cohomology of varieties with isolated singularities*, alg-geom/9711003 (1997).

[140] A. Parusiński, *Unveröffentlichte Arbeiten zur Integration von Krümmungsformen über subanalytischen Mengen.*

[141] ———, *Lipschitz stratification of subanalytic sets*, Ann. Sci. École Norm. Sup. (4) t. 27 (1994), 661–696.

[142] N. Perkal, *On proving the geometric versions of Whitney regularity*, J. London Math. Soc. (2) **29** (1984), 343–351.

[143] M.J. Pflaum, *On continuous Hochschild homology and cohomology groups*, Lett. Math. Phys. **44** (1998), 43–51.

[144] ———, *Smooth structures on stratified spaces*, Quantization of singular symplectic spaces (N. P. Landsman, M. J. Pflaum, and M. Schlichenmaier, eds.), Birkhäuser-Verlag, to appear 2001.

[145] H.-J. Reiffen, *Das Lemma von Poincaré für holomorphe Differentialformen auf komplexen Räumen*, Math. Z. **101** (1967), 269–284.

[146] I. Satake, *On a generaliztion of the notion of manifold*, Proc. Nat. Acad. Sci. U.S.A. **42** (1956), 359–363.

[147] M. Schlessinger and J. Stasheff, *The Lie algebra structure of tangent cohomology and deformation theory*, J. Pure Appl. Algebra **38** (1985), no. 2–3, 313–322.

[148] B.-W. Schulze, *Pseudo-differential Operators on Manifolds with Singularities*, North-Holland, Amsterdam, 1991.

[149] ______, *Pseudo-differential Boundary Value Problems, Conical Singularities, and Asymptotics*, Akademie-Verlag, Berlin, 1994.

[150] ______, *Operator algebras with symbol hierarchies on manifolds with singularities*, preprint 99/30 (1999), Arbeitsgruppe "Partielle Differentialgleichungen und Komplexe Analysis", Universität Potsdam.

[151] M.-H. Schwartz, *Espaces pseudo-fibrés et systèmes obstructeurs*, Bull. Soc. Math. France **88** (1960), 1–55.

[152] ______, *Lectures on stratification of complex analytic sets*, Tata Institute of Fundamental Research Lectures on Mathematics and Physics, vol. 38, Bombay: Tata Institute of Fundamental Research, 1966.

[153] ______, *Local theory of analytic functions of several complex variables. (notes by J.F. Price)*, Semin. Schwartz, Notes Pure Math, vol. 7 Part II, 1973.

[154] ______, *Stratification et conditions de Whitney*, Acta Math. Vietnam. (1976).

[155] ______, *Champs radiaux sur une stratification analytique*, Travaux en cours, Hermann, Paris, 1991.

[156] G.W. Schwarz, *Smooth functions invariant under the action of a compact Lie group*, Topology **14** (1975), 63–68.

[157] R. Seeley, *Extension of $\mathcal{C}^\infty$ functions defined in a half space*, Proc. Amer. Math. Soc. **15** (1964), 625–626.

[158] M. Shiota, *Geometry of Subananlytic and Semialgebraic Sets*, Progress in Mathematics, vol. 150, Birkhäuser, 1997.

[159] R. Sikorski, *Abstract covariant derivative*, Colloq. Math. **18** (1967), 251–272.

[160] ______, *Differential modules*, Colloq. Math. **24** (1971), 45–79.

[161] R. Sjamaar, *Singular Orbit Spaces in Riemannian and Symplectic Geometry*, Ph.D. thesis, Rijksuniversiteit te Utrecht, 1990.

[162] R. Sjamaar and E. Lerman, *Stratified symplectic spaces and reduction*, Ann. of Math. (2) **134** (1991), 375–422.

[163] K. Spallek, *Differenzierbare Räume*, Math. Ann. **180** (1969), 269–296.

[164] J.L. Taylor, *Homology and cohomology of topological algebras*, Advances in Math. **9** (1972), 137–182.

[165] N. Teleman, *Microlocalisation de l'homologie de Hochschild*, C. R. Acad. Sci. Paris Sér. I Math. **326** (1998), 1261–1264.

[166] R. Thom, *Les singularités des applications différentiables*, Ann. Inst. Fourier (Grenoble) **6** (1955–56), 43–87.

[167] ———, *La stabilité topologique des applications polynomiales*, Enseign. Math. (2) **8** (1962), 24–33.

[168] ———, *Local topological properties of differentiable mappings*, Colloquium on Differential Analysis, Tata Inst. Bombay, Oxford Univ. Press, 1964, pp. 191–202.

[169] ———, *Ensembles et morphismes stratifiés*, Bull. Amer. Math. Soc. (N.S.) **75** (1969), 240–284.

[170] J.-C. Tougeron, *Idéaux de Fonctions Différentiables*, Ergebnisse der Mathematik und ihrer Grenzgebiete, vol. 71, Springer-Verlag, Berlin-Heidelberg-New York, 1972.

[171] F. Trèves, *Topological Vector Spaces, Distributions and Kernels*, Academic Press, Inc., New York, 1967.

[172] D. Trotman, *Geometric versions of Whitney regularity for smooth stratifications*, Ann. Sci. École Norm. Sup. (4) **12** (1979), 453–463.

[173] ———, *On the canonical Whitney stratification of algebraic hypersurfaces*, Séminaire sur la géométrie réelle (J.-J. Risler, ed.), vol. 24, Publ. Math. Univ. Paris VII, no. 1, 1987, pp. 123–152.

[174] ———, *Espaces stratifiés réels*, Stratifications, Singularities and Differential Equations II, Stratifications and Topology of Singular Spaces (Honolulu) (D. Trotman & L.C. Wilson, ed.), Travaux en Cours, vol. 55, Hermann, 1997, pp. 93–107.

[175] J.-L. Verdier, *Stratifications de Whitney et théorème de Bertini-Sard*, Invent. Math. **36** (1976), 295–312.

[176] A. Verona, *Le théorème de De Rham pour les prestratifications abstraites*, C. R. Acad. Sci. Paris Sér. I Math. **273** (1971), 886–889.

[177] ———, *Homological properties of abstract préstratifications*, Rev. Roumaine Math. Pures Appl. Tome XVII, N° **7** (1972), 1109–1121.

[178] ———, *Integration on Whitney prestratifications*, Rev. Roumaine Math. Pures Appl. Tome XVII, N° **9** (1972), 1473–1480.

[179] ———, *Sur la cohomologie de De Rham des préstratifications de Whitney*, C. R. Acad. Sci. Paris Sér. I Math. **274** (1972), 1340–1343.

[180] ———, *Stratified Mappings - Structures and Triangulability*, Lecture Notes in Mathematics, vol. 1102, Springer Verlag, 1984.

[181] A.A. Voronov, *Quantizing Poisson manifolds*, Contemporary Mathematics **214** (1991), 189–195.

[182] A. Wassermann, *Cyclic cohomology of algebras of smooth functions on orbifolds*, Operator algebras and application. Vol. 1: Structure theory; K-theory, geometry and topology, Lond. Math. Soc. Lect. Note, vol. 135, Cambridge University Press, 1988, pp. 229–244.

[183] C.A. Weibel, *An Introduction to Homological Algebra*, Cambridge University Press, 1994.

[184] S. Weinberger, *The Topological Classification of Stratified Spaces*, Chicago Lectures in Mathematics, The University of Chicago Press, 1994.

[185] A. Weinstein, *Lectures on Symplectic Manifolds*, CBMS Regional Conference Series, vol. 29, American Mathematical Society, 1976.

[186] H. Weyl, *The Classical Groups*, Princeton University Press, Princeton, 1946.

[187] H. Whitney, *Analytic extensions of differentiable functions defined in closed sets*, Trans. Amer. Math. Soc. **36** (1934), 63–89.

[188] ———, *Complexes of manifolds*, Proc. Nat. Acad. Sci. U.S.A. **33** (1946), 10–11.

[189] ———, *On singularities of mappings of Euclidean spaces. I. Mappings of the plane into the plane*, Ann. of Math. **62** (1955), no. 3, 374–410.

[190] ———, *Elementary structure of real algebraic varieties*, Ann. of Math. **66** (1957), no. 3, 545–556.

[191] ———, *Tangents to an analytic variety*, Ann. of Math. **81** (1964), 496–549.

[192] ———, *Local properties of analytic varieties*, Differentiable and Combinatorial Topology (Princeton), Princeton University Press, 1965.

[193] S. Zucker, *Hodge theory with degenerating coefficients: L_2-cohomology in the Poincaré metric*, Ann. of Math. **109** (1979), 415–476.

[194] ———, *L^2-cohomology of warped products and arithmetic groups*, Invent. Math. **70** (1982), 169–218.

Index

Vol. 1726: V. Marić, Regular Variation and Differential Equations. X, 127 pages. 2000.

Vol. 1727: P. Kravanja, M. Van Barel, Computing the Zeros of Analytic Functions. VII, 111 pages. 2000.

Vol. 1728: K. Gatermann, Computer Algebra Methods for Equivariant Dynamical Systems. XV, 153 pages. 2000.

Vol. 1729: J. Azéma, M. Émery, M. Ledoux, M. Yor, Séminaire de Probabilités XXXIV. VI, 431 pages. 2000.

Vol. 1730: S. Graf, H. Luschgy, Foundations of Quantization for Probability Distributions. X, 230 pages. 2000.

Vol. 1731: T. Hsu, Quilts: Central Extensions, Braid Actions, and Finite Groups,. XII, 185 pages. 2000.

Vol. 1732: K. Keller, Invariant Factors, Julia Equivalences and the (Abstract) Mandelbrot Set. X, 206 pages. 2000.

Vol. 1733: K. Ritter, Average-Case Analysis of Numerical Problems. IX, 254 pages. 2000.

Vol. 1734: M. Espedal, A. Fasano, A. Mikelić, Filtration in Porous Media and Industrial Applications. Cetraro 1998. Editor: A. Fasano. 2000.

Vol. 1735: D. Yafaev, Scattering Theory: Some Old and New Problems. XVI, 169 pages. 2000.

Vol. 1736: B. O. Turesson, Nonlinear Potential Theory and Weighted Sobolev Spaces. XIV, 173 pages. 2000.

Vol. 1737: S. Wakabayashi, Classical Microlocal Analysis in the Space of Hyperfunctions. VIII, 367 pages. 2000.

Vol. 1738: M. Émery, A. Nemirovski, D. Voiculescu, Lectures on Probability Theory and Statistics. XI, 356 pages. 2000.

Vol. 1739: R. Burkard, P. Deuflhard, A. Jameson, J.-L. Lions, G. Strang, Computational Mathematics Driven by Industrial Problems. Martina Franca, 1999. Editors: V. Capasso, H. Engl, J. Periaux. VII, 418 pages. 2000.

Vol. 1740: B. Kawohl, O. Pironneau, L. Tartar, J.-P. Zolesio, Optimal Shape Design. Tróia, Portugal 1999. Editors: A. Cellina, A. Ornelas. IX, 388 pages. 2000.

Vol. 1741: E. Lombardi, Oscillatory Integrals and Phenomena Beyond all Algebraic Orders. XV, 413 pages. 2000.

Vol. 1742: A. Unterberger, Quantization and Non-holomorphic Modular Forms. VIII, 253 pages. 2000.

Vol. 1743: L. Habermann, Riemannian Metrics of Constant Mass and Moduli Spaces of Conformal Structures. XII, 116 pages. 2000.

Vol. 1744: M. Kunze, Non-Smooth Dynamical Systems. X, 228 pages. 2000.

Vol. 1745: V. D. Milman, G. Schechtman, Geometric Aspects of Functional Analysis. VIII, 289 pages. 2000.

Vol. 1746: A. Degtyarev, I. Itenberg, V. Kharlamov, Real Enriques Surfaces. XVI, 259 pages. 2000.

Vol. 1747: L. W. Christensen, Gorenstein Dimensions. VIII, 204 pages. 2000.

Vol. 1748: M. Růžička, Electrorheological Fluids: Modeling and Mathematical Theory. XV, 176 pages. 2001.

Vol. 1749: M. Fuchs, G. Seregin, Variational Methods for Problems from Plasticity Theory and for Generalized Newtonian Fluids. VI, 269 pages. 2001.

Vol. 1750: B. Conrad, Grothendieck Duality and Base Change. X, 296 pages. 2001.

Vol. 1751: N. J. Cutland, Loeb Measures in Practice: Recent Advances. XI, 111 pages. 2001.

Vol. 1752: Y. V. Nesterenko, P. Philippon, Introduction to Algebraic Independence Theory. XIII, 256 pages. 2001.

Vol. 1753: A. I. Bobenko, U. Eitner, Painlevé Equations in the Differential Geometry of Surfaces. VI, 120 pages. 2001.

Vol. 1754: W. Bertram, The Geometry of Jordan and Lie Structures. XVI, 269 pages. 2001.

Vol. 1755: J. Azéma, M. Émery, M. Ledoux, M. Yor, Séminaire de Probabilités XXXV. VI, 427 pages. 2001.

Vol. 1756: P. E. Zhidkov, Korteweg de Vries and Nonlinear Schrödinger Equations: Qualitative Theory. VII, 147 pages. 2001.

Vol. 1757: R. R. Phelps, Lectures on Choquet's Theorem. VII, 124 pages. 2001.

Vol. 1758: N. Monod, Continuous Bounded Cohomology of Locally Compact Groups. X, 214 pages. 2001.

Vol. 1759: Y. Abe, K. Kopfermann, Toroidal Groups. VIII, 133 pages. 2001.

Vol. 1760: D. Filipović, Consistency Problems for Heath-Jarrow-Morton Interest Rate Models. VIII, 134 pages. 2001.

Vol. 1761: C. Adelmann, The Decomposition of Primes in Torsion Point Fields. VI, 142 pages. 2001.

Vol. 1762: S. Cerrai, Second Order PDE's in Finite and Infinite Dimension. IX, 330 pages. 2001.

Vol. 1763: J.-L. Loday, A. Frabetti, F. Chapoton, F. Goichot, Dialgebras and Related Operads. IV, 132 pages. 2001.

Vol. 1764: A. Cannas da Silva, Lectures on Symplectic Geometry. XII, 217 pages. 2001.

Vol. 1765: T. Kerler, V. V. Lyubashenko, Non-Semisimple Topological Quantum Field Theories for 3-Manifolds with Corners. VI, 379 pages. 2001.

Vol. 1766: H. Hennion, L. Hervé, Limit Theorems for Markov Chains and Stochastic Properties of Dynamical Systems by Quasi-Compactness. VIII, 145 pages. 2001.

Vol. 1767: J. Xiao, Holomorphic Q Classes. VIII, 112 pages. 2001.

Vol. 1768: M. J. Pflaum, Analytic and Geometric Study of Stratified Spaces. VIII, 230 pages. 2001.

Recent Reprints and New Editions

Vol. 1200: V. D. Milman, G. Schechtman, Asymptotic Theory of Finite Dimensional Normed Spaces - Corrected Second Printing 2001. X, 156 pages. 1986.

Vol. 1618: G. Pisier, Similarity Problems and Completely Bounded Maps - Second, Expanded Edition VII, 198 pages. 2001.

Vol. 1629: J. D. Moore, Lectures on Seiberg-Witten Invariants - Second Edition. VIII, 121 pages. 2001.

Vol. 1638: P. Vanhaecke, Integrable Systems in the realm of Algebraic Geometry - Second Edition. X, 256 pages. 2001.

Vol. 1702: J. Ma, J. Yong, Forward-Backward Stochastic Differential Equations and Their Applications - Corrected Second Printing 2000. XIII, 270 pages. 1999.

Vol. 1676: P. Cembranos, J. Mendoza, Banach Spaces of Vector-Valued Functions. VIII, 118 pages. 1997.

Vol. 1677: N. Proskurin, Cubic Metaplectic Forms and Theta Functions. VIII, 196 pages. 1998.

Vol. 1678: O. Krupková, The Geometry of Ordinary Variational Equations. X, 251 pages. 1997.

Vol. 1679: K.-G. Grosse-Erdmann, The Blocking Technique. Weighted Mean Operators and Hardy's Inequality. IX, 114 pages. 1998.

Vol. 1680: K.-Z. Li, F. Oort, Moduli of Supersingular Abelian Varieties. V, 116 pages. 1998.

Vol. 1681: G. J. Wirsching, The Dynamical System Generated by the 3n+1 Function. VII, 158 pages. 1998.

Vol. 1682: H.-D. Alber, Materials with Memory. X, 166 pages. 1998.

Vol. 1683: A. Pomp, The Boundary-Domain Integral Method for Elliptic Systems. XVI, 163 pages. 1998.

Vol. 1684: C. A. Berenstein, P. F. Ebenfelt, S. G. Gindikin, S. Helgason, A. E. Tumanov, Integral Geometry, Radon Transforms and Complex Analysis. Firenze, 1996. Editors: E. Casadio Tarabusi, M. A. Picardello, G. Zampieri. VII, 160 pages. 1998

Vol. 1685: S. König, A. Zimmermann, Derived Equivalences for Group Rings. X, 146 pages. 1998.

Vol. 1686: J. Azéma, M. Émery, M. Ledoux, M. Yor (Eds.), Séminaire de Probabilités XXXII. VI, 440 pages. 1998.

Vol. 1687: F. Bornemann, Homogenization in Time of Singularly Perturbed Mechanical Systems. XII, 156 pages. 1998.

Vol. 1688: S. Assing, W. Schmidt, Continuous Strong Markov Processes in Dimension One. XII, 137 page. 1998.

Vol. 1689: W. Fulton, P. Pragacz, Schubert Varieties and Degeneracy Loci. XI, 148 pages. 1998.

Vol. 1690: M. T. Barlow, D. Nualart, Lectures on Probability Theory and Statistics. Editor: P. Bernard. VIII, 237 pages. 1998.

Vol. 1691: R. Bezrukavnikov, M. Finkelberg, V. Schechtman, Factorizable Sheaves and Quantum Groups. X, 282 pages. 1998.

Vol. 1692: T. M. W. Eyre, Quantum Stochastic Calculus and Representations of Lie Superalgebras. IX, 138 pages. 1998.

Vol. 1694: A. Braides, Approximation of Free-Discontinuity Problems. XI, 149 pages. 1998.

Vol. 1695: D. J. Hartfiel, Markov Set-Chains. VIII, 131 pages. 1998.

Vol. 1696: E. Bouscaren (Ed.): Model Theory and Algebraic Geometry. XV, 211 pages. 1998.

Vol. 1697: B. Cockburn, C. Johnson, C.-W. Shu, E. Tadmor, Advanced Numerical Approximation of Nonlinear Hyperbolic Equations. Cetraro, Italy, 1997. Editor: A. Quarteroni. VII, 390 pages. 1998.

Vol. 1698: M. Bhattacharjee, D. Macpherson, R. G. Möller, P. Neumann, Notes on Infinite Permutation Groups. XI, 202 pages. 1998.

Vol. 1699: A. Inoue, Tomita-Takesaki Theory in Algebras of Unbounded Operators. VIII, 241 pages. 1998.

Vol. 1700: W. A. Woyczyński, Burgers-KPZ Turbulence, XI, 318 pages. 1998.

Vol. 1701: Ti-Jun Xiao, J. Liang, The Cauchy Problem of Higher Order Abstract Differential Equations, XII, 302 pages. 1998.

Vol. 1702: J. Ma, J. Yong, Forward-Backward Stochastic Differential Equations and Their Applications. XIII, 270 pages. 1999.

Vol. 1703: R. M. Dudley, R. Norvaiša, Differentiability of Six Operators on Nonsmooth Functions and p-Variation. VIII, 272 pages. 1999.

Vol. 1704: H. Tamanoi, Elliptic Genera and Vertex Operator Super-Algebras. VI, 390 pages. 1999.

Vol. 1705: I. Nikolaev, E. Zhuzhoma, Flows in 2-dimensional Manifolds. XIX, 294 pages. 1999.

Vol. 1706: S. Yu. Pilyugin, Shadowing in Dynamical Systems. XVII, 271 pages. 1999.

Vol. 1707: R. Pytlak, Numerical Methods for Optimal Control Problems with State Constraints. XV, 215 pages. 1999.

Vol. 1708: K. Zuo, Representations of Fundamental Groups of Algebraic Varieties. VII, 139 pages. 1999.

Vol. 1709: J. Azéma, M. Émery, M. Ledoux, M. Yor (Eds), Séminaire de Probabilités XXXIII. VIII, 418 pages. 1999.

Vol. 1710: M. Koecher, The Minnesota Notes on Jordan Algebras and Their Applications. IX, 173 pages. 1999.

Vol. 1711: W. Ricker, Operator Algebras Generated by Commuting Projećtions: A Vector Measure Approach. XVII, 159 pages. 1999.

Vol. 1712: N. Schwartz, J. J. Madden, Semi-algebraic Function Rings and Reflectors of Partially Ordered Rings. XI, 279 pages. 1999.

Vol. 1713: F. Bethuel, G. Huisken, S. Müller, K. Steffen, Calculus of Variations and Geometric Evolution Problems. Cetraro, 1996. Editors: S. Hildebrandt, M. Struwe. VII, 293 pages. 1999.

Vol. 1714: O. Diekmann, R. Durrett, K. P. Hadeler, P. K. Maini, H. L. Smith, Mathematics Inspired by Biology. Martina Franca, 1997. Editors: V. Capasso, O. Diekmann. VII, 268 pages. 1999.

Vol. 1715: N. V. Krylov, M. Röckner, J. Zabczyk, Stochastic PDE's and Kolmogorov Equations in Infinite Dimensions. Cetraro, 1998. Editor: G. Da Prato. VIII, 239 pages. 1999.

Vol. 1716: J. Coates, R. Greenberg, K. A. Ribet, K. Rubin, Arithmetic Theory of Elliptic Curves. Cetraro, 1997. Editor: C. Viola. VIII, 260 pages. 1999.

Vol. 1717: J. Bertoin, F. Martinelli, Y. Peres, Lectures on Probability Theory and Statistics. Saint-Flour, 1997. Editor: P. Bernard. IX, 291 pages. 1999.

Vol. 1718: A. Eberle, Uniqueness and Non-Uniqueness of Semigroups Generated by Singular Diffusion Operators. VIII, 262 pages. 1999.

Vol. 1719: K. R. Meyer, Periodic Solutions of the N-Body Problem. IX, 144 pages. 1999.

Vol. 1720: D. Elworthy, Y. Le Jan, X-M. Li, On the Geometry of Diffusion Operators and Stochastic Flows. IV, 118 pages. 1999.

Vol. 1721: A. Iarrobino, V. Kanev, Power Sums, Gorenstein Algebras, and Determinantal Loci. XXVII, 345 pages. 1999.

Vol. 1722: R. McCutcheon, Elemental Methods in Ergodic Ramsey Theory. VI, 160 pages. 1999.

Vol. 1723: J. P. Croisille, C. Lebeau, Diffraction by an Immersed Elastic Wedge. VI, 134 pages. 1999.

Vol. 1724: V. N. Kolokoltsov, Semiclassical Analysis for Diffusions and Stochastic Processes. VIII, 347 pages. 2000.

Vol. 1725: D. A. Wolf-Gladrow, Lattice-Gas Cellular Automata and Lattice Boltzmann Models. IX, 308 pages. 2000.

4. Lecture Notes are printed by photo-offset from the master-copy delivered in camera-ready form by the authors. Springer-Verlag provides technical instructions for the preparation of manuscripts. Macro packages in TEX, LATEX2e, LATEX2.09 are available from Springer's web-pages at

http://www.springer.de/math/authors/b-tex.html.

Careful preparation of the manuscripts will help keep production time short and ensure satisfactory appearance of the finished book.

The actual production of a Lecture Notes volume takes approximately 12 weeks.

5. Authors receive a total of 50 free copies of their volume, but no royalties. They are entitled to a discount of 33.3 % on the price of Springer books purchase for their personal use, if ordering directly from Springer-Verlag.

Commitment to publish is made by letter of intent rather than by signing a formal contract. Springer-Verlag secures the copyright for each volume. Authors are free to reuse material contained in their LNM volumes in later publications: A brief written (or e-mail) request for formal permission is sufficient.

Addresses:

Professor J.-M. Morel
CMLA, Ecole Normale Supérieure de Cachan
61 Avenue du Président Wilson
94235 Cachan Cedex France
E-mail: Jean-Michel.Morel@cmla.ens-cachan.fr

Professor B. Teissier
Université Paris 7
UFR de Mathématiques
Equipe Géométrie et Dynamique
Case 7012
2 place Jussieu
75251 Paris Cedex 05
E-mail: Teissier@ens.fr

Professor F. Takens, Mathematisch Instituut,
Rijksuniversiteit Groningen, Postbus 800,
9700 AV Groningen, The Netherlands
E-mail: F.Takens@math.rug.nl

Springer-Verlag, Mathematics Editorial, Tiergartenstr. 17
D-69121 Heidelberg, Germany
Tel.: *49 (6221) 487-701
Fax: *49 (6221) 487-355
E-mail: lnm@Springer.de